PHARMACOKINETICS FOR THE PHARMACEUTICAL SCIENTIST

HOW TO ORDER THIS BOOK

BY PHONE: 800-233-9936 or 717-291-5609, 8AM–5PM Eastern Time

BY FAX: 717-295-4538

BY MAIL: Order Department
Technomic Publishing Company, Inc.
851 New Holland Avenue, Box 3535
Lancaster, PA 17604, U.S.A.

BY CREDIT CARD: American Express, VISA, MasterCard

PHARMACOKINETICS FOR THE PHARMACEUTICAL SCIENTIST

John G. Wagner, Ph.D.

Professor Emeritus of Pharmacology and
John G. Searle Professor Emeritus of Pharmaceutics
The University of Michigan

CRC Press is an imprint of the
Taylor & Francis Group, an **informa** business

CRC Press
Taylor & Francis Group
6000 Broken Sound Parkway NW, Suite 300
Boca Raton, FL 33487-2742

First issued in paperback 2019

ISBN-13: 978-1-56676-032-4 (hbk)
ISBN-13: 978-0-367-40240-2 (pbk)

Main entry under title:
Pharmacokinetics for the Pharmaceutical Scientist

A Technomic Publishing Company book
Bibliography: p.
Includes index p. 313

Library of Congress Catalog Card No. 93-60366

Visit the Taylor & Francis Web site at
http://www.taylorandfrancis.com

and the CRC Press Web site at
http://www.crcpress.com

CONTENTS

INTRODUCTION

One purpose of pharmacokinetics is to optimize the design of human and animal drug studies. The study of drug delivery has become a critical discipline, and at the doctor-patient level this new knowledge has improved patient benefits from drugs and has allowed professionals to maximize drug utilization. The pharmacokineticist in industry and academia requires similar knowledge, but at a more sophisticated level. Drug delivery scientists should be able to estimate *in vivo* absorption rates following the administration of different dosage forms of the same drug. These scientists must have a firm basis in pharmacokinetic theory in order to perform effective research.

Almost all textbooks in pharmacokinetics cover only a few compartment models, and some virtually none. But compartment models are still the backbone of pharmacokinetics, and many problems in the field cannot be solved without them. This book covers many more compartment models than have ever been covered before, and Chapter 12 integrates physiologically based flow models with the usual compartment models.

The author has worked fifteen years in the pharmaceutical industry and twenty-five years in academia, and has performed such research for forty years. He realizes how important it is to know not only what pharmacokinetic theory is, but how to apply it. This book attempts to show researchers how to do this.

CORRELATION OF DRUG BLOOD CONCENTRATIONS AND PHARMACODYNAMIC EFFECT OR EFFICACY

This section will show some examples of where drug blood (plasma or serum) concentrations have been correlated with pharmacodynamic effect or efficacy–a process that provides one reason for measuring drug blood concentrations. Another reason for measuring such concentrations is the Food and Drug Administration's requirements to provide pharmacokinetic data, and its evaluation which is needed in order to obtain an approved New Drug Application for a specific drug. Chapter 14, entitled "Pharmacokinetic-Pharmacodynamic Modeling" has more examples where drug blood concentrations have been correlated indirectly with pharmacodynamic effect.

Brodie, in a review ("Physiochemical and Biochemical Aspects of Pharmacology," *J. Am. Med. Assoc.*, 202:600–609, 1967), stated:

> I first became aware of the importance of relating drug effects to plasma levels, rather than dosage, when I participated in the clinical screening of antimalarial drugs under the direction of James A. Shannon. These studies showed that the antimalarial effects of the cinchona alkaloids, quinacrine, and many other agents, are highly correlated with plasma levels but not with dosage. By relating effects to plasma levels, only a few patients were needed to gain a definitive assay of activity, compared to the large numbers required when effects are related to dosage.

Sometimes one can plot a pharmacodynamic effect on the ordinate *versus* the drug blood or plasma concentration on the abscissa and draw a straight or curved line through the data points. Two of the following four figures illustrate such simple direct correlations.

Figure I.1 is an example where ethyl alcohol is the drug used.

Figure I.2 is based on a single male patient, age sixty-three, with coronary heart disease. The drug quinidine prevented attacks of paroxysmal ventricular tachycardia in this patient, providing the serum quinidine concentration was above 4 mg/L. When the serum quinidine concentration fell below this level the patient experienced ventricular tachycardia. Additional doses of drug were administered at these times; these doses raised the serum quinidine concentration above the critical level of 4 mg/L and the tachycardia disappeared. The critical serum quinidine concentration needed to convert to sinus rhythm varied with the patient from 1 to 16 mg/L.

Hirulog-1 (BG8967) is a direct thrombin inhibitor built by rational design using the protein hirudin as a model. Its pharmacodynamic effects include prolongation of activated partial thromboplastin time (APTT) and thrombin time. In a clinical study, one group of six subjects, Group 9, received a single bolus intravenous injection of 0.3 mg/kg. Eight other groups received different treatments. Figure I.3 is a plot of mean (APTT-baseline) *versus* mean Hirulog-1 (BG8967) plasma concentrations with a smooth curved line drawn through the points.

Schentag has done work on an extremely important job, correlating pharmacokinetic and microbiological parameters to the efficacy of antibiotics.

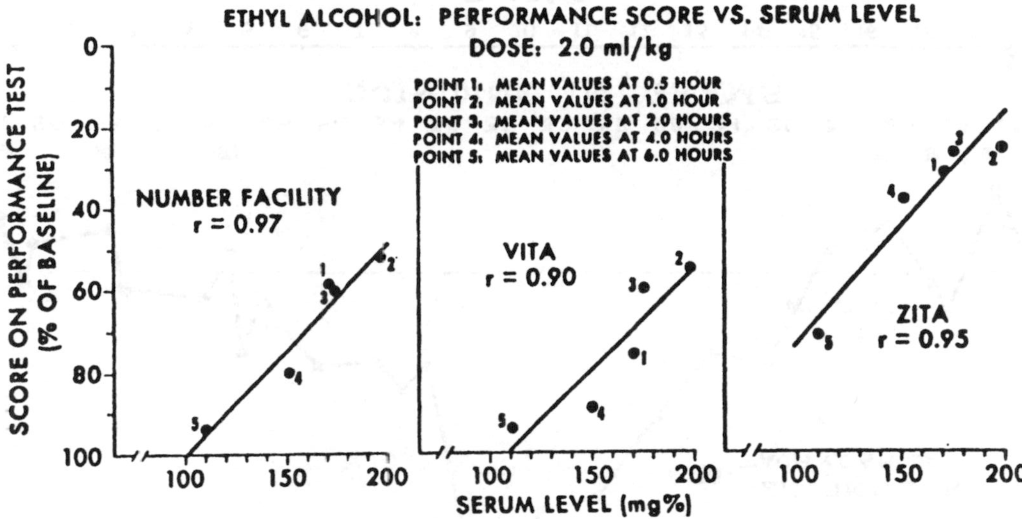

FIGURE I.1. Correlation of three different performance scores with serum alcohol concentration (F. R. Sidell and J. E. Pless. "Ethyl Alcohol: Blood Levels and Performance Decrements after Oral Administration to Man," *Psychopharmacologia*, 19:246–261, 1971; with permission from Springer-Verlag).

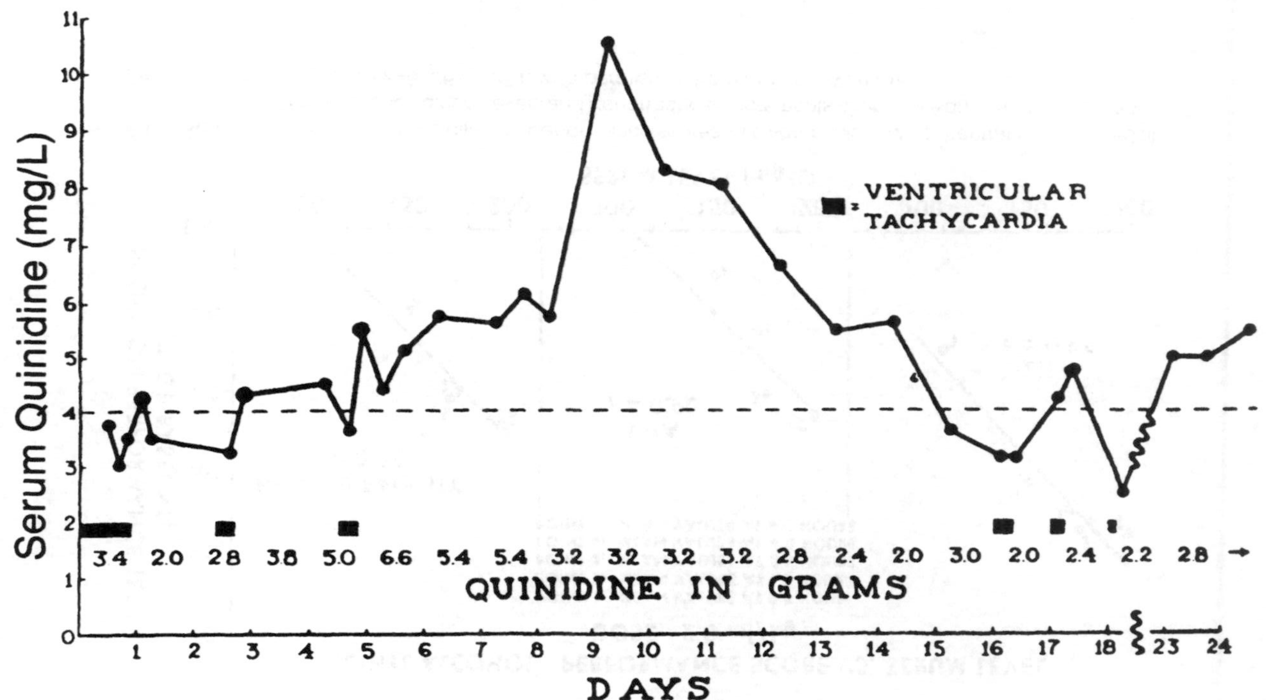

FIGURE I.2. Illustrating the importance of determining the critical serum concentration of quinidine to prevent attacks of paroxysmal ventricular tachycardia (M. Sokolow and A. L. Edgar. "Blood Quinidine Concentration as a Guide in the Treatment of Cardiac Arrhythmias," *Circulation*, 1:576–592, 1950; with permission from the American Heart Association).

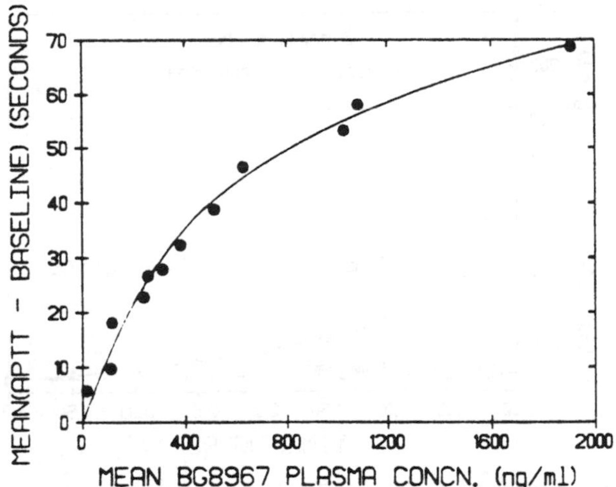

FIGURE I.3. Correlation of mean (APTT-baseline) with mean plasma concentration of Hirulog-1 (BG8967) following bolus I.V. injection of 0.3 mg/kg (I. Fox, A. Dawson, P. Loynds, J. Eisner, K. Findlen, E. Levin, D. Hanson, T. Mant, J. Wagner and J. Maraganore. "Hirulog-1: A Direct Thrombin Inhibitor with Potent Anticoagulant Properties in Humans," *Thrombosis and Haemostasis,* 69:157–163, 1993 [© F. K. Schattauer Verlagsgesellschaft mbH (Stuttgart)].

For β-lactams and quinolones he has shown that the amount of time that the antiobiotic serum concentration was greater than the MIC (minimum inhibitory concentration) was predictive of bacterial eradication times. He has also shown that the DRC_{50} concentration was highly correlated with the MIC. The DRC_{50} is an *in vitro* antibiotic concentration/effect parameter measured using the Abbott MS-2 Research System. The MIC was defined as the lowest concentration preventing visually detectable growth after eighteen hours incubation at 37°C. The correlation was:

$$\log DRC_{50} = 1.058 \log MIC - 0.254 \quad (r = 0.90)$$

The area under the antibiotic serum concentration-time (AUC) above the DRC_{50} was shown to be predictive of antibiotic efficacy. Figure I.4 illustrates this area.

NUMBERING OF POLYEXPONENTIAL EXPONENTS

Bolus intravenous blood concentration-time curves are often fitted to sums of exponential terms with the general expression:

$$C = \sum_{i=1}^{n} C_i e^{-\lambda_i t} \quad (i = 1,2,\ldots,n)$$

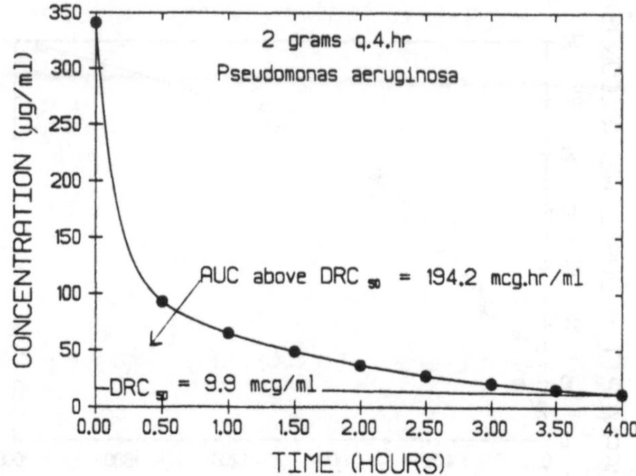

FIGURE I.4. Illustrating the AUC above DRC_{50} when cefmenoxime was given at a 2 gram dose every four hours in the treatment of pneumonia caused by *Pseudomonas aeruginosa* (adapted from data in Figure 5 of J. J. Schentag, "Correlation of Pharmacokinetic Parameters to Efficacy of Antibiotics: Relationships between Serum Concentrations, MIC Values, and Bacterial Eradication Times in Patients with Gram-Negative Pneumonia," *Scand. J. Infect. Dis. Suppl.*, 74:218–234, 1991; with permission from Almquist & Wiksell Periodical Co.).

There are various methods of numbering or naming the λ_i in relation to the magnitudes of the numerical values. The smallest λ_i value is associated with the data at the very tail end of the curve (largest times); the next λ_i is larger and is associated with data closer to the front end of the curve, etc. Table I.1 summarizes some of the systems.

The symbolism used in this book makes most sense to the author since First = 1, Second = 2, etc. The Rowland method makes First = z, Second = 1, Third = 2, etc. The classical method is problematic since beta is the second letter of the Greek alphabet but is assigned to First below

TABLE I.1. Order Starting at the Tail End of the C,t Curve.

	Fourth*	Third	Second	First**
This book	λ_4	λ_3	λ_2	λ_1
Rowland	λ_3	λ_2	λ_1	λ_z
Classical	δ	γ	α	β

*Largest numerical value.
**Smallest numerical value.

or the smallest λ_i and alpha is the first letter of the Greek alphabet but is assigned to the second largest λ_i.

SHAPES OF CURVES

Examples of real or simulated data are given throughout this book. Both concentration-time and absorption curves are illustrated for many models. The reader should relate the shapes of the curves to the model as he/she reads.

CHAPTER 1

Simple Linear Models

MODEL I. ONE-COMPARTMENT OPEN MODEL WITH BOLUS INTRAVENOUS ADMINISTRATION

This is the simplest of all pharmacokinetic models and there are few data sets that obey the model. One such set is shown in Figure 1.1 published by Smith et al. [1]. The equation to which these data were fitted is:

$$C = C_o e^{-Kt} \tag{1}$$

The diagram of this model is shown in Scheme 1.1.

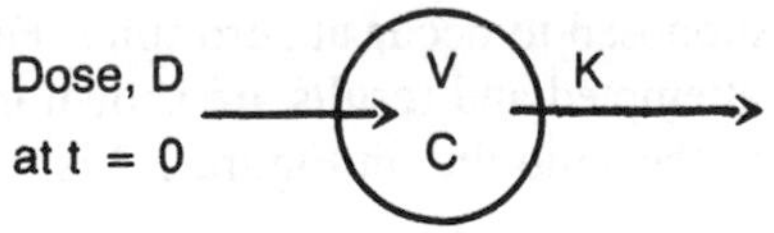

Dose, D
at t = 0
V
C
K

SCHEME 1.1.

The parameters estimated by nonlinear least squares method with the program MINSQ [2] were $C_o = 2.44$ (0.121) μg/ml and $K = 0.399$ (0.0273) hr^{-1} where the numbers without parentheses are the estimates and the numbers with parentheses are the standard deviations of the estimates. During the fitting of these data the concentrations were weighted reciprocally, i.e., $1/C_i$. The data were obtained by administering 10 ml of a 3.75 mg/ml sterile aqueous solution of the drug (specific activity 1.33 μCi/mg) by bolus intravenous injection to six adult male volunteers.

There are problems with the fit shown in Figure 1.1—specifically, there are systematic deviations. Note that the second to the fourth points lie below

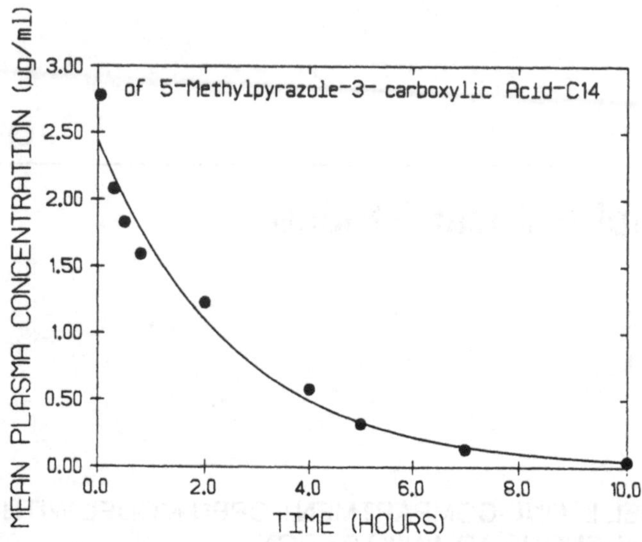

FIGURE 1.1. Mean plasma concentrations of 5-methylpyrazole-3-carboxylic acid-^{14}C computer-fitted with a monoexponential equation and the parameters C_o = 2.4 μg/ml and K = 0.399 hr^{-1}.

the fitted line while the next two points lie above the fitted line. Also, the intercept, $C_o = 2.44$, is quite a bit below the first point, $C = 2.78$. The theoretical model shown in Scheme 1.1 assumes that the dose administered at zero time distributes instantaneously in the volume V and hence the highest concentration is supposed to occur at zero time. For these reasons a biexponential fit was attempted and results are shown in Figure 1.2.

The equation for the solid line in Figure 1.2 is:

$$C = 2.28e^{-0.381t} + 0.952e^{-11.3t} \quad (2)$$

This gave a better fit to the data and solved the problems presented by Figure 1.1. Hence, although this data set appeared monoexponential at first glance, it really is biexponential data.

A better example of the model is shown in Figure 1.3. This figure shows plasma concentrations of tritium after bolus intravenous administration of 3 mg/kg of tritiated Hirulog-1 (BG8967), which is a synthetic thrombin inhibitor [3]. The data were fitted with the program MINSQ [2] to Equation (1). The estimated parameters were C_o = 3325 (37.8) cpm/ml and K = 2.73 (0.0635) hr^{-1}.

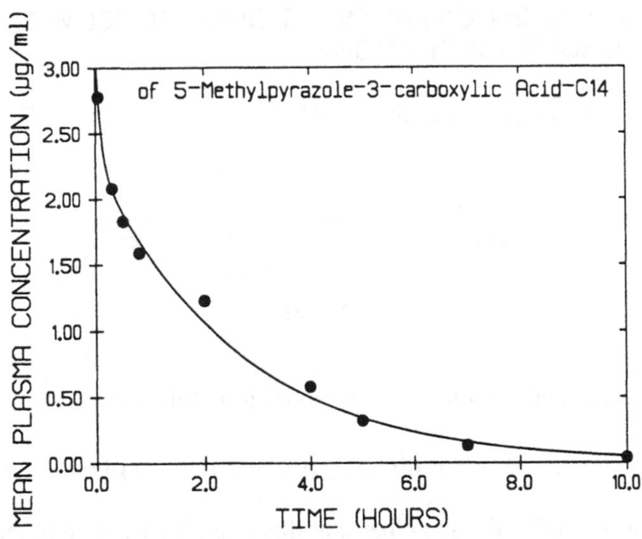

FIGURE 1.2. Mean plasma concentrations of 5-methylpyrazole-3-carboxylic acid-^{14}C computer-fitted with the biexponential Equation (2).

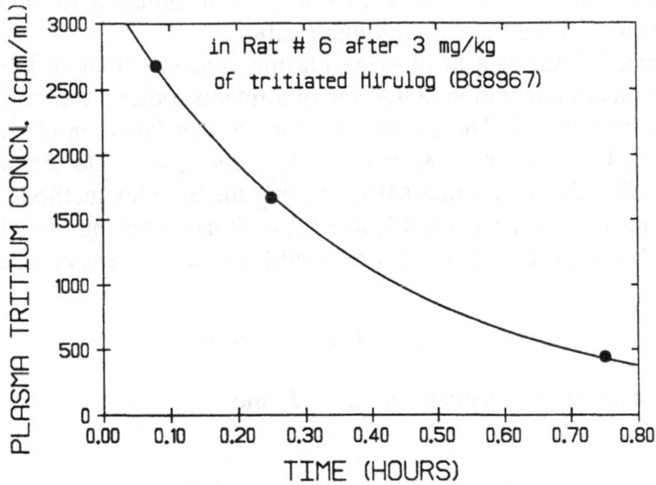

FIGURE 1.3. Plasma tritium concentration in rat #6 after 3 mg/kg of tritiated Hirulog after bolus intravenous injection fitted with Equation (1). Estimated parameters were C_o = 3325 cpm/ml and K = 2.73 hr^{-1}.

MODEL II. ONE-COMPARTMENT OPEN MODEL WITH FIRST-ORDER ABSORPTION

The model is shown as Scheme 1.2.

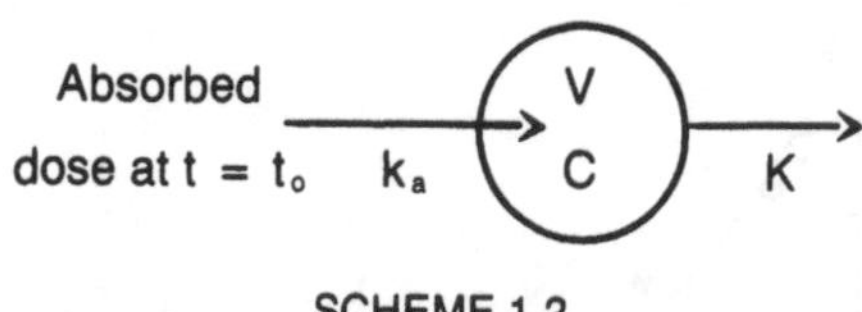

SCHEME 1.2.

The integrated equation for the model is as follows:

$$C = (FD/V)\{k_a/(k_a - K)\}[\exp\{-K(t - t_o)\} - \exp\{-k_a(t - t_o)\}] \quad (3)$$

In Equation (3), F is the bioavailability (i.e., the fraction of the dose that reaches the circulation), D is the administered dose, k_a is the first-order input or absorption rate constant, K is the first-order elimination rate constant, V is the volume of distribution, and t_o is the lag time. Hence FD is the amount of drug that reaches the circulation. In Equation (3) and all future equations, t is the time after administration.

Figure 1.4 shows a fit of mean plasma concentrations of tolmetin acid after oral administration of 400 mg of tolmetin sodium to twenty-four normal volunteers [4]. The estimated parameters of Equation (3) are: V/F = 10.48 L, k_a = 13.5 hr^{-1}, K = 0.584 hr^{-1}, and t_o = 0.0981 hr, which were obtained with the program MINSQ using the Simplex method.

If one used the program RSTRIP [2] to fit data obeying Model II instead of MINSQ and Equation (3), one would have an expression like Equation (4):

$$C = B_1 e^{-Kt} + B_2 e^{-k_a t} \quad (4)$$

Here one usually assumes that $k_a > K$ and

$$B_1 = C_o\left\{\frac{k_a}{(k_a - K)}\right\} e^{+Kt_o} \quad (5)$$

$$B_2 = C_o\left\{\frac{k_a}{(k_a - K)}\right\} e^{+k_a t_o} \quad (6)$$

$$C_o = \frac{FD}{V} \quad (7)$$

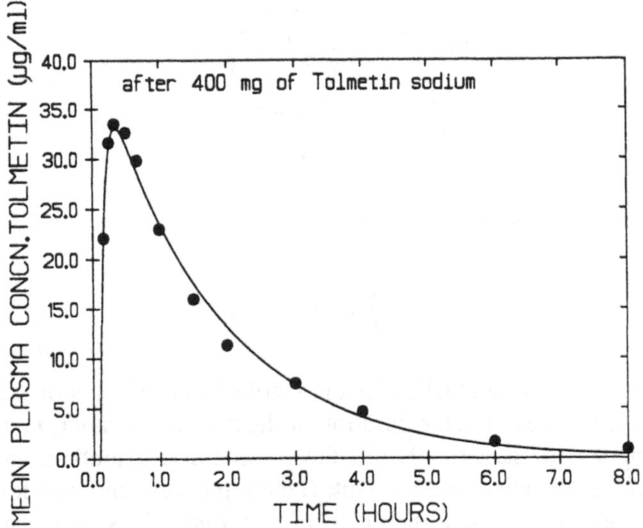

FIGURE 1.4. Mean plasma concentrations of tolmetin after oral administration of 400 mg of tolmetin sodium to twenty-four normal volunteers computer-fitted to Equation (3). Estimated parameters were V/F = 10.48 L, k_a = 13.5 hr^{-1}, K = 0.584 hr^{-1}, and t_o = 0.0981 hr.

Hence:

$$t_o = \frac{(\ln B_2 - \ln B_1)}{(k_a - K)} \tag{8}$$

And,

$$\frac{V}{F} = \frac{D}{C_o} = \frac{Dk_a \exp(+Kt_o)}{(k_a - K)B_1} \tag{9}$$

Hence, the parameters of the model may be estimated either by computer fitting with Equation (3) or by fitting data with a difference of two exponential terms and then using Equations (8) and (9) to estimate t_o and V/F and assigning the exponents of the two exponential terms as k_a and K. If $B_1 = B_2$ then $t_o = 0$.

MODEL III. ONE-COMPARTMENT OPEN MODEL WITH ZERO-ORDER INPUT

During input the schematic model is as shown in Scheme 1.3, and Equation (10) applies. When $t > T$, Scheme 1.1 applies.

$$C = \frac{k_o}{VK}\{1 - e^{-Kt}\} \tag{10}$$

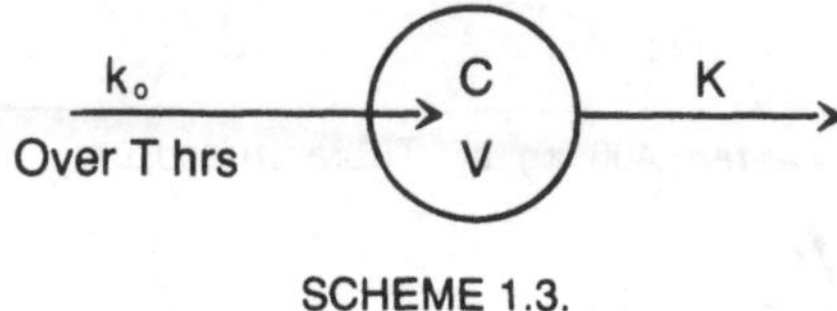

SCHEME 1.3.

Post-input, Equation (11) applies.

$$C = \frac{k_o}{VK}\{1 - e^{-KT}\}e^{-K(t-T)} \tag{11}$$

In Equations (10) and (11), C is the whole blood (plasma or serum) concentration at time t, T is the duration of the zero-order input, k_o is the zero-order input rate (mass/time), V is the volume of distribution, and K is the first-order elimination rate constant. If the input was continued long enough (i.e., T was large enough) then C would approach the steady-state concentration equal to k_o/VK, where VK is the clearance.

Wagner and Alway [5] reported serum concentrations of the antibiotic lincomycin when the drug was infused for two hours at a rate of 150,000 μg/hr and blood was sampled at two, six, and twelve hours. From these three concentrations, the nonlinear least squares method with Equation (1) gave $K = 0.221$ hr^{-1}. Substitution of this value and $C = 9.7$ μg/ml when $t = 2$ hours into Equation (10) gave $k_o/V = 5.77$ μg/ml/hr, hence $V = 150{,}000/5.77 = 25{,}997$ ml. Using both Equations (10) and (11), results are shown in Figure 1.5.

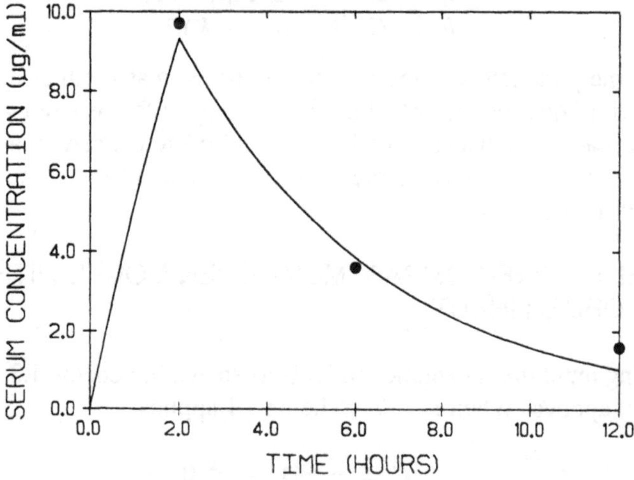

FIGURE 1.5. Serum concentrations of lincomycin measured at 2, 6, and 12 hr when the antibiotic was infused at a constant rate of 150,000 μg/hr. See text regarding fitted lines.

To produce Figure 1.5, Equations (10) and (11) were introduced into the computer program shown below and used with MINSQ [2]:

```
A: =KO*(1-EXP((-K)*T))/K
B: =KO/K*(1-EXP((-K)*TO))*EXP((-K)*(T-TO))
FLAG: =UNIT(T-TO)
C: =A*(1-FLAG)+B*FLAG
```

In the above program, KO is equivalent to k_o/V, T is equivalent to t, and TO is equivalent to the infusion time T.

MODEL IV. ONE-COMPARTMENT OPEN MODEL WITH BOLUS INTRAVENOUS INJECTION AND INTRAVENOUS INFUSION

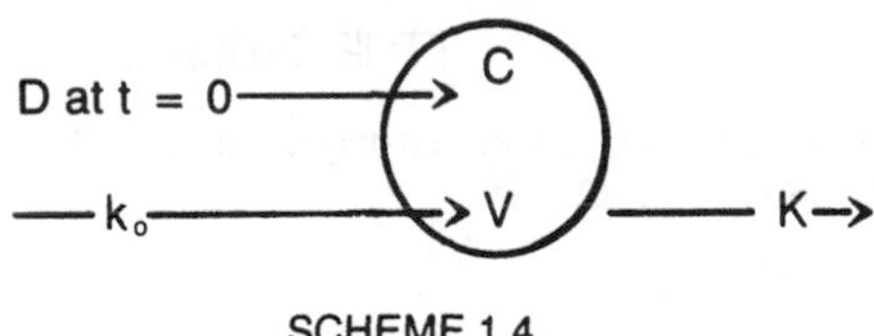

SCHEME 1.4.

In Scheme 1.4, D is the loading dose administered by bolus intravenous injection and k_o is the infusion rate (mass/time) for the infusion started at the same time as the bolus is given. Equation (12) is based on Scheme 1.4.

$$C = \left(\frac{D}{V}\right)e^{-Kt} + \frac{k_o}{VK}\{1 - e^{-Kt}\} = \frac{k_o}{VK} - \left(\frac{k_o}{VK} - \frac{D}{V}\right)e^{-Kt} \quad (12)$$

To produce the example shown in Figure 1.6, Equation (12) was programmed on a Hewlett Packard 27S scientific calculator and the parameter values $D = 300$ mg, $k_o = 150$ mg/hr, $V = 26$ L, and $K = 0.221$ hr^{-1} were employed with $t = 0, 0.25, 0.5, 0.75, 1, 1.25, 1.5, 1.75$, and 2 hours.

In Figure 1.6 the serum concentration continues to increase during the two-hour infusion time. However, this model allows one to maintain a constant concentration. The right-hand side of Equation (12) indicates that if $D = k_o/K$, then $C = C_{ss} = k_o/VK$, where C_{ss} is the steady-state concentration, which is established as soon as the loading dose, D, is given and is maintained as long as the infusion is maintained. Figure 1.7 shows results of a simulation with lincomycin parameters, namely $k_o = 41.67$ mg/hr, $K = 0.221$ hr^{-1}, $V = 26$ L, $D = k_o/K = 189$ mg, and $C_{ss} = k_o/VK = 189/26 = 7.27$ μg/ml. There appears to be little, if any, real data where this model applies. Future Models XV and XVI based on the two-compartment

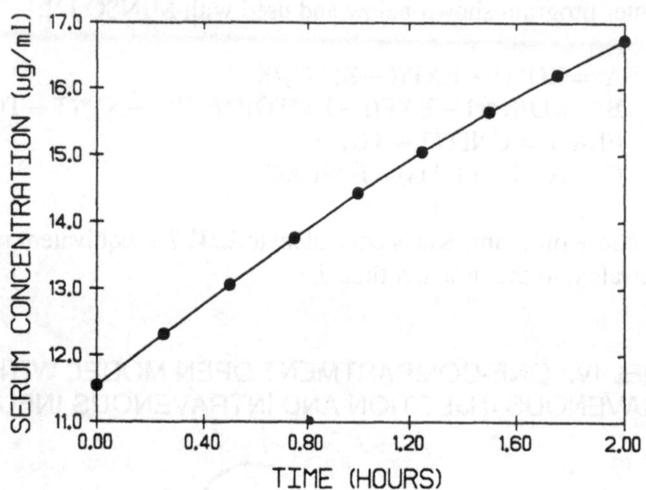

FIGURE 1.6. Simulated lincomycin data using Equation (12) with $D = 300$ mg, $k_o = 150$ mg/hr, $V = 26$ L, and $K = 0.221$ hr^{-1}.

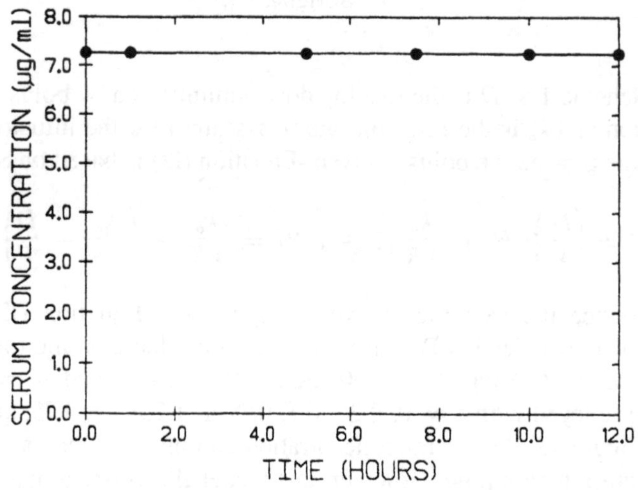

FIGURE 1.7. Simulated lincomycin data using Equation (12) with $D = 189$ mg, $k_o = 150$ mg/hr, $V = 26$ L, and $K = 0.221$ hr^{-1}.

open disposition model are more realistic and have had several applications in the real world.

MODEL V. ONE-COMPARTMENT OPEN MODEL WITH TWO PARALLEL FIRST-ORDER INPUT SITES

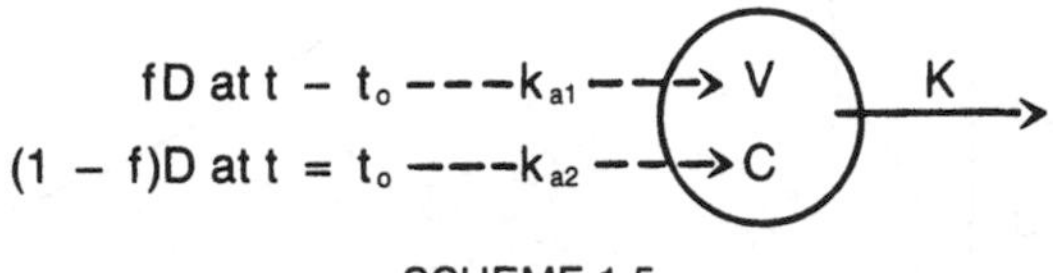

fD at t – t_o ---k_{a1} --> V —K—>
(1 – f)D at t = t_o ---k_{a2} --> C

SCHEME 1.5.

In Scheme 1.5, f is the fraction of the *absorbed* dose, *FD,* which is absorbed starting from zero time from a site with first-order rate constant k_{a1}, and $1 - f$ is the fraction of the *absorbed* dose, *FD,* which is absorbed starting at time t_o from a site with first-order rate constant k_{a2}. Here F is the bioavailability, so that the volume of distribution estimated is V/F. Hence, the amount of drug absorbed from site #1 is fFD and the amount of drug absorbed from site #2 is $(1 - f)FD$. The equation for the model is shown as Equation (13):

$$C = \left(\frac{fFD}{V}\right)\left\{\frac{k_{a1}}{(k_{a1} - K)}\right\}(e^{-Kt} - e^{-k_{a1}t})$$

$$+ \left\{\frac{(1 - f)FD}{V}\right\}\left\{\frac{k_{a2}}{(k_{a2} - K)}\right\}\{e^{-K(t-t_o)} - e^{-k_{a2}(t-t_o)}\} \qquad (13)$$

For this model to apply to real data, the two absorption sites need not be at physically different places in the gastrointestinal tract. Absorption with rate constant k_{a1} could occur, for example, from the small intestine while the major amount of the unabsorbed drug is in the stomach and small amounts are emptying from the stomach; absorption at rate constant k_{a2} could occur from the small intestine after most of the drug has emptied from the stomach. Hence, t_o is the same as the major stomach emptying time.

A simulation was performed using Equation (13) and the parameter values $f = 0.5$, $D = 100$ mg, $V = 10$ L, $k_{a1} = 0.5$ hr^{-1}, $k_{a2} = 2$ hr^{-1}, $K = 0.25$ hr^{-1}, $F = 1$, and $t_o = 1.5$ hr. Results are shown in Figure 1.8.

If real data were obtained, like the curve in Figure 1.8, it would be very useful to prepare a Wagner-Nelson [6] plot since this would aid in elu-

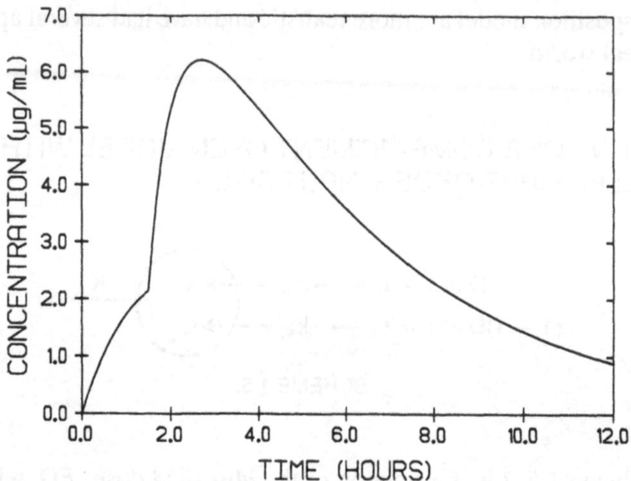

FIGURE 1.8. Simulated data for Model V using Equation (13) with $f = 0.5$, $D = 100$ mg, $V = 10$ L, $k_{a_1} = 0.5\ \text{hr}^{-1}$, $k_{a_2} = 2\ \text{hr}^{-1}$, $K = 0.25\ \text{hr}^{-1}$, $F = 1$, and $t_o = 1.5$ hr.

cidating the kinetics involved. Equation (14) gives the fraction absorbed (FA) as a function of time for this Model V.

$$\text{FA} = f(1 - e^{-k_{a_1}t}) + (1 - f)\{1 - e^{-k_{a_2}(t-t_o)}\} \tag{14}$$

The first term on the right is used when t is less than or equal to t_o, and both terms are used when $t > t_o$. Figure 1.9 shows the Wagner-Nelson FA *versus* t plot corresponding to the C *versus* t plot shown in Figure 1.8.

MODEL VI. ONE-COMPARTMENT OPEN MODEL WITH TWO CONSECUTIVE FIRST-ORDER INPUT RATE CONSTANTS

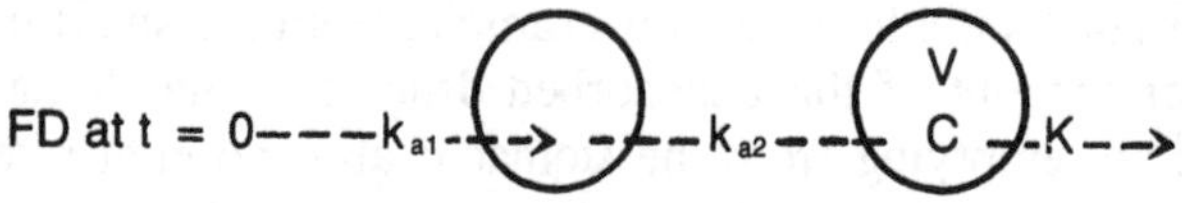

FD at t = 0---k_{a1}--> --k_{a2}--> (V, C) --K-->

SCHEME 1.6.

In relation to the gastrointestinal tract, k_{a1} could be the hybrid first-order rate constant representing dissolution and stomach emptying, and k_{a2} could be the first-order rate constant for absorption across the gastrointestinal barrier. F is the fraction of the dose, D, which is absorbed, V is the volume of distribution, K is the first-order elimination rate constant, and C is the concentration at time t. The estimated volume is V/F.

The concentration at time t is given by Equation (15).

$$C = k_{a1}k_{a2}\left(\frac{FD}{V}\right)\left[\frac{\exp(-Kt)}{(k_{a1} - K)(k_{a2} - K)} + \frac{\exp(-k_{a1}t)}{(k_{a2} - k_{a1})(K - k_{a1})} + \frac{\exp(-k_{a2}t)}{(k_{a1} - k_{a2})(K - k_{a2})}\right] \quad (15)$$

The Wagner-Nelson equation corresponding to Equation (15) is:

$$\text{FA} = 1 - [k_{a1}e^{-k_{a2}t} - k_{a2}e^{-k_{a1}t}]/(k_{a1} - k_{a2}) \quad (16)$$

If we substitute into Equation (15) $K = 0.1$ and $FD/V = 100$ and in one case $k_{a1} = 2$ and $k_{a2} = 0.5$ and in the other case $k_{a1} = 0.5$ and $k_{a2} = 2$ we get the same polyexponential equation that is:

$$C = 35.09e^{-2t} - 166.67e^{-0.5t} + 131.58e^{-0.1t} \quad (17)$$

Hence one cannot determine which is the largest rate constant—k_{a1} or k_{a2}—and in all cases the negative coefficient is associated with the intermediate-sized rate constant. The latter is different than the usual case ($k_a > \lambda_2 > \lambda_1$) with Model X below where the negative sign is associated with the largest rate constant k_a. Also, with Model VI there is no "nose" on

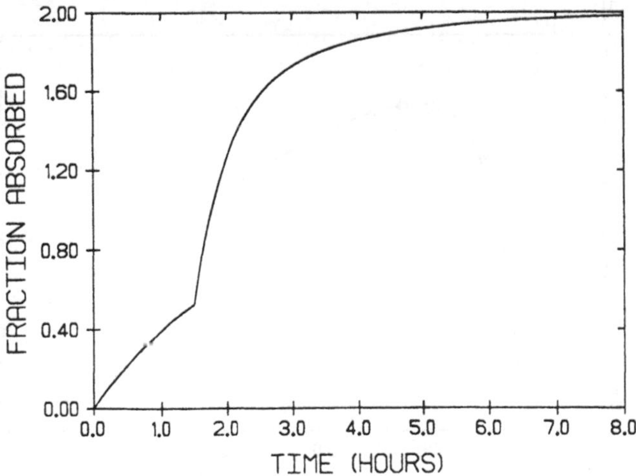

FIGURE 1.9. Simulated Wagner-Nelson plot using Equation (14) with the same parameter values as listed for Figure 1.8.

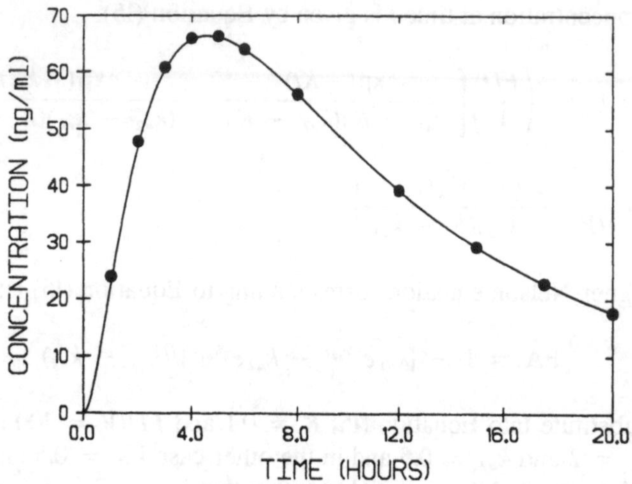

FIGURE 1.10. Rectilinear plot of *C versus t* based on Equation (17).

a semilogarithmic plot of *C versus t*, i.e., the early points all lie below the terminal log-linear line extrapolated back to zero time; whereas, with Model X and the above relative size of the rate constants there *is* a "nose" and the early points lie above the terminal log-linear line extrapolated back to zero time. As an example, Figures 1.10 and 1.11 show plots of Equation (17) on rectilinear and semilogarithmic graph paper, respectively.

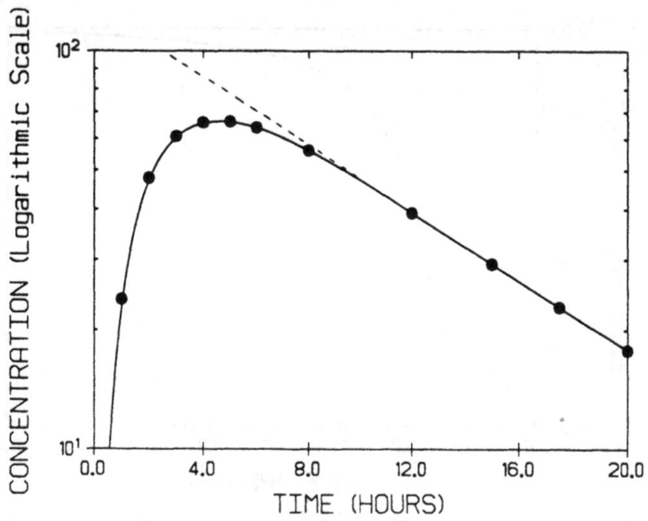

FIGURE 1.11. Semilogarithmic plot based on Equation (17).

MODEL VII. ONE-COMPARTMENT OPEN MODEL WITH PARALLEL FIRST- AND ZERO-ORDER INPUTS

D_I at t = 0---k_a ---> V
D_s at t = 0---k_o ---> C ----K-->

SCHEME 1.7.

Scheme 1.7 may be used to simulate C,t data for a sustained-release dosage form where D_I is the immediate-release dose, which is absorbed with a fast first-order rate constant k_a, and D_s is the dose released zero-order with rate constant k_o over T hours. K, V, and C have the same meaning as before. A_B is the amount of drug in the body.

Equation (18) applies to this model:

$$A_B = \left\{\frac{D_I k_a}{k_a - K}\right\}(e^{-Kt} - e^{-k_a t}) + \frac{k_o}{K}\{1 - e^{-Kt}\} \tag{18}$$

After input ends, Equation (19) applies to the example below, where $T = 12$ hours.

$$A_B = (A_B)_{12\,\text{hr}} e^{-K(t-T)} \tag{19}$$

Suppose we had a drug with an elimination half-life of three hours and a

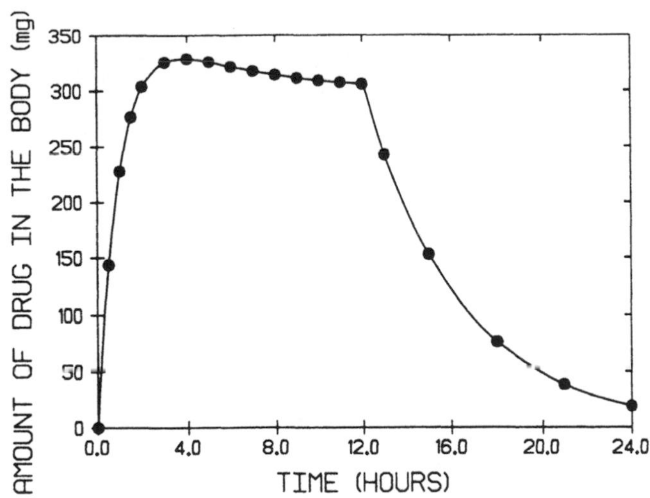

FIGURE 1.12. Simulated data for Model VII based on Equations (18) and (19) with D_I = 300 mg, K = 0.231 hr^{-1}, k_o = 69.33 mg/hr, and $(A_B)_{12\,\text{hr}}$ = 306 mg.

therapeutic dose of 300 mg. Also suppose we wished to make a sustained-release dosage form that would produce a therapeutic effect for about twelve hours. What would the requirements be, based on this model?

Since the half-life is three hours, the elimination rate constant K = 0.693/3 = 0.231 hr^{-1}, make D_I = 300 mg. Then $D_s = D_I TK$ = (300)(12)(0.231) = 832 mg and $k_o = D_s/T$ = 832/12 = 69.33 mg/hr. Substitution of these values into Equation (18) with t = 12 hours gives $(A_B)_{12\,hr}$ = 306 mg. Substitution of all numerical values into Equations (18) and (19) gives the plot shown in Figure 1.12 on the previous page.

The results with this trial simulation are quite good. The amount of drug in the body is 304 to 306 mg from two to twelve hours. The initial rise in the first hour or two would be faster if k_a was increased. The results tell one something about the properties of a dosage form needed to provide certain desirable characteristics.

CHAPTER 2

More Complicated Linear Models

MODEL VIII. THE CLASSICAL TWO-COMPARTMENT OPEN MODEL WITH CENTRAL COMPARTMENT ELIMINATION AND BOLUS INTRAVENOUS ADMINISTRATION

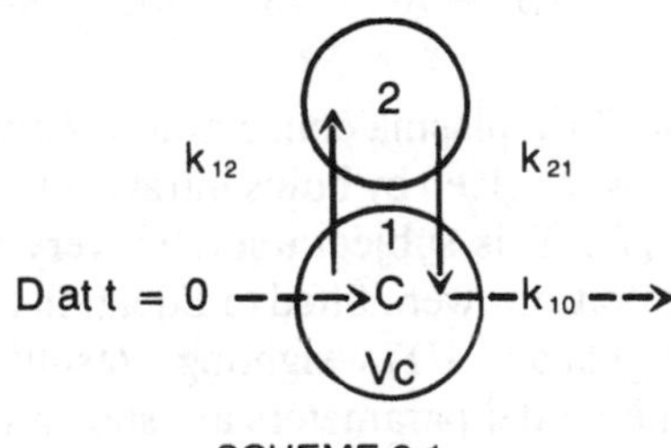

SCHEME 2.1.

Equation (20) gives the concentration in the central compartment #1 as a function of time.

$$C = D[(k_{21} - \lambda_1)e^{-\lambda_1 t} + (k_{21} - \lambda_2)e^{-\lambda_2 t}]/Vc(\lambda_2 - \lambda_1) \quad (20)$$

Equation (20) may also be written as Equation (21):

$$C = C_1 e^{-\lambda_1 t} + C_2 e^{-\lambda_2 t} \quad (21)$$

where

$$C_1 = D(k_{21} - \lambda_1)/Vc(\lambda_2 - \lambda_1) \quad (22)$$

$$C_2 = D(k_{21} - \lambda_2)/Vc(\lambda_1 - \lambda_2) \quad (23)$$

and Vc is the volume of the central compartment #1, D is the single dose, and λ_1 and λ_2 are given by Equations (24) and (25):

$$\lambda_1 = 0.5[k_{12} + k_{21} + k_{10} - \{(k_{12} + k_{21} + k_{10})^2 - 4k_{21}k_{10}\}^{1/2}] \quad (24)$$

$$\lambda_2 = 0.5[k_{12} + k_{21} + k_{10} + \{(k_{12} + k_{21} + k_{10})^2 - 4k_{21}k_{10}\}^{1/2}] \quad (25)$$

One may fit data to Equation (20) and estimate Vc, k_{12}, k_{21}, and k_{10}, or, alternatively, one may fit data to Equation (21) and then estimate the parameters of the model with Equations (26) through (29):

$$Vc = D/(C_1 + C_2) \quad (26)$$

$$k_{21} = (C_1\lambda_2 + C_2\lambda_1)/(C_1 + C_2) \quad (27)$$

$$k_{10} = \lambda_1\lambda_2/k_{21} \quad (28)$$

$$k_{12} = \lambda_1 + \lambda_2 - k_{21} - k_{10} \quad (29)$$

Examples are warfarin plasma concentrations after doses of 50, 100, and 200 mg of warfarin were given by bolus intravenous injection to subject N-2 by O'Reilly et al. [7]. This subject tolerated very high doses of warfarin. The plasma concentrations were fitted to Equation (21) with RSTRIP [2] using both equal weights and $1/Y_i$ weighting. Results are shown in Table 2.1 and Figure 2.1. The model parameters are shown in Table 2.2.

A semilogarithmic plot of the plasma concentrations after the 200 mg dose with the fitted line based on $1/Y_i$ weights is shown in Figure 2.2. The fact that the three sets of data were fitted with biexponential equations without systematic deviations indicated that the data obeyed linear kinetics.

TABLE 2.1. Parameters Estimated by Nonlinear Least Squares with RSTRIP [2] from Warfarin Plasma Concentrations (μg/ml) of Subject N-2 [7].

Dose (mg)	Weights	C_1	λ_1	C_2	λ_2
50	Equal	6.253	0.0170	5.654	1.720
100	Equal	10.806	0.0170	6.939	0.4413
200	Equal	26.951	0.0151	28.997	2.830
50	$1/Y_i$	6.024	0.0176	5.574	1.479
100	$1/Y_i$	10.872	0.0174	6.907	0.4545
200	$1/Y_i$	27.003	0.0152	29.166	2.862

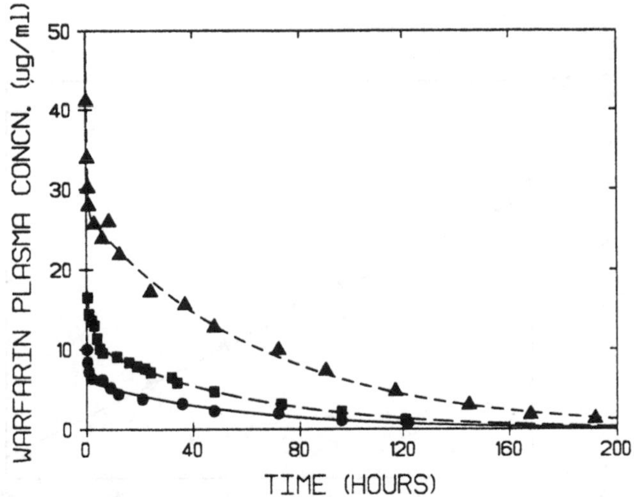

FIGURE 2.1. Examples of Model VIII. These are warfarin plasma concentrations after bolus intravenous doses of 50, 100, and 200 mg of warfarin to subject N-2 [7] and fitting to Equation (21) with RSTRIP [2]. See estimated parameters in Table 2.1.

However, another test of a linear pharmacokinetic system is the linearity of an area-dose plot. Figure 2.3 shows such a plot based on $1/Y_i$ weighting of the warfarin plasma concentration data of subject N-2. The two lowest doses give a straight line through the origin with slope equal to 6.52, which is the reciprocal of the clearance. Hence CL = 1/6.52 = 0.153 L/hr. The clearance from the high dose of 200 mg is *D*/AUC = 200/1,781.3 = 0.112 L/hr, which is appreciably lower. Hence, these results indicate that the system is nonlinear. One would have to study the system further to make a definitive conclusion.

Another example for Model VIII is the spectinomycin data of Wagner et

TABLE 2.2. Parameters of Model VIII Estimated from Data in Table 2.1 and with Equations (26) to (29).

Dose (mg)	Weights	*Vc* (L)	k_{12} (hr^{-1})	k_{21} (hr^{-1})	k_{10} (hr^{-1})
50	Equal	4.20	0.791	0.912	0.0359
100	Equal	5.64	0.156	0.275	0.0273
200	Equal	3.57	1.443	1.371	0.0311
50	$1/Y_i$	4.31	0.686	0.776	0.0335
100	$1/Y_i$	5.62	0.159	0.284	0.0277
200	$1/Y_i$	3.56	1.462	1.383	0.0315

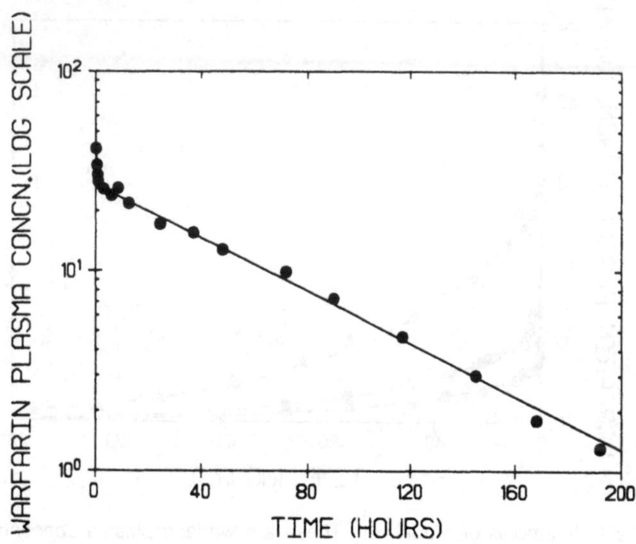

FIGURE 2.2. Semilogarithmic plot of warfarin plasma concentrations following the 200 mg bolus I.V. dose of warfarin to subject N-2 and RSTRIP fitting to Equation (21) with $1/Y_i$ weighting.

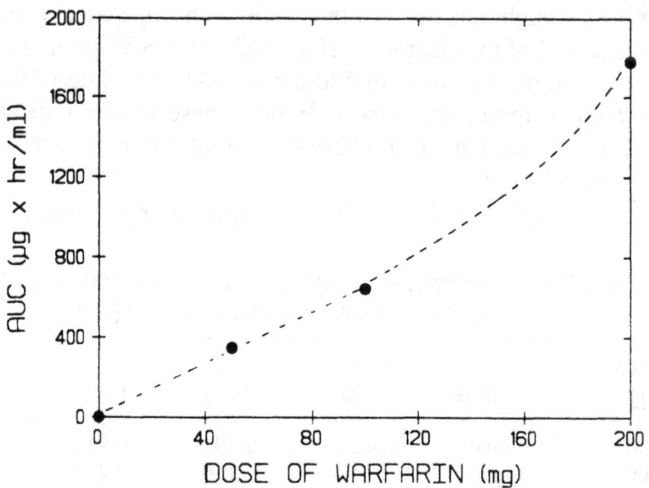

FIGURE 2.3. Warfarin area-dose plot for subject N-2 given three doses of warfarin bolus I.V. The curvature indicates nonlinearity.

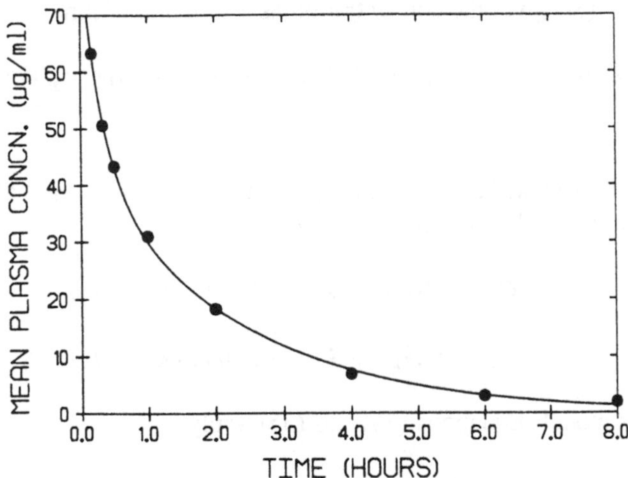

FIGURE 2.4. Mean plasma concentrations of spectinomycin fitted with Equation (30) with $1/Y_i$ weighting. This is an example of Model VIII.

al. [8]. Six subjects received the dihydrochloride pentahydrate salt of the antibiotic spectinomycin by bolus intravenous injection and intramuscularly in crossover fashion. The mean plasma concentrations are used as an example of Model VIII here while the I.M. data are used as an example of Model IX later. Five milliliters of a sterile aqueous solution containing 100 mg/ml were administered intravenously. Figure 2.4 shows the biexponential fit to Equation (21) with $1/Y_i$ weighting. The particular biexponential equation is shown as Equation (30).

$$C = 37.8e^{-3.21t} + 43.9e^{-0.438t} \tag{30}$$

Use of Equations (26) to (30) with the dose of 500 mg of the drug gave the model parameters $Vc = 6.12$ L, $k_{12} = 0.991$ hr^{-1}, $k_{21} = 1.93$ hr^{-1}, and $k_{10} = 0.729$ hr^{-1}.

MODEL IX. THE ROWLAND TWO-COMPARTMENT OPEN MODEL WITH PERIPHERAL ELIMINATION AND BOLUS INTRAVENOUS INJECTION

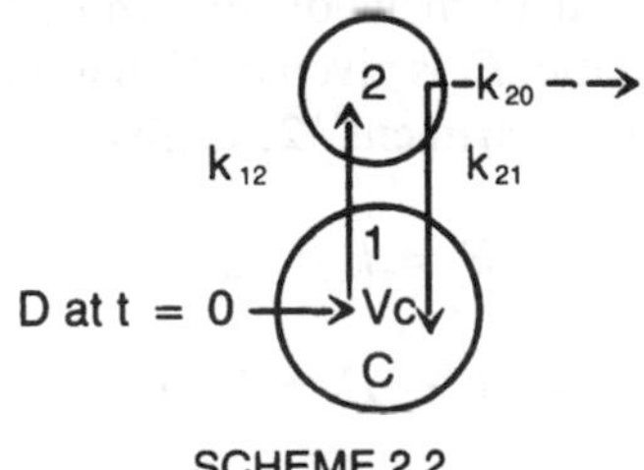

SCHEME 2.2.

Equation (31) gives the concentration as a function of time.

$$C = D[(k_{21} + k_{20} - \lambda_1)e^{-\lambda_1 t} + (k_{21} + k_{20} - \lambda_2)e^{-\lambda_2 t}]/Vc(\lambda_2 - \lambda_1) \tag{31}$$

Equation (31) may be written as Equation (20), where C_1 and C_2 are given by Equations (32) and (33).

$$C_1 = D[k_{21} + k_{20} - \lambda_1]/Vc(\lambda_2 - \lambda_1) \tag{32}$$

$$C_2 = D[k_{21} + k_{21} - \lambda_2]/Vc(\lambda_1 - \lambda_2) \tag{33}$$

For this model, Equations (34) and (35) apply.

$$\lambda_1 + \lambda_2 = k_{12} + k_{21} + k_{20} \tag{34}$$

$$\lambda_1\lambda_2 = k_{12}k_{20} \tag{35}$$

It should be noted that for Model VIII, k_{21} replaces k_{12} in Equation (35) and k_{10} replaces k_{20} in Equations (34) and (35). If data are fitted to Equation (20), then Model IX parameters may be obtained with Equations (26) and (36) to (39).

$$A = (C_1\lambda_2 + C_2\lambda_1)/(C_1 + C_2) = k_{21} + k_{20} \tag{36}$$

$$k_{12} = \lambda_1 + \lambda_2 - A \tag{37}$$

$$k_{20} = \lambda_1\lambda_2/k_{12} \tag{38}$$

$$k_{21} = A - k_{20} \tag{39}$$

For Model IX the dose is placed in the central compartment #1 at time zero for bolus intravenous administration, and the drug must distribute to compartment #2 to get out of the body. The corresponding oral model would have drug input into the peripheral compartment #2 and all of the drug would be exposed to metabolism. Thus, this is a first-pass model where the extraction ratio, E, is given by Equation (40) and the bioavailability, F, for input into compartment #2, is given by Equation (41).

$$E = k_{20}/(k_{20} + k_{21}) \tag{40}$$

$$F = 1 - E = k_{21}/(k_{20} + k_{21}) \tag{41}$$

TABLE 2.3. Biexponential Parameters from Rowland and Riegelman [9].

Subject	Dose (mg)	C_1	λ_1	C_2	λ_2
A	325	15.5	0.048	42	0.26
A	650	29	0.050	90	0.24
B	650	37	0.050	56	0.31
C	650	40	0.048	60	0.29
D	650	33	0.050	67	0.23
E	650	22	0.037	64	0.19

It should be noted that the classical Model VIII has no first-pass effect since both intravenous and oral administrations are into the central compartment #1 from which elimination occurs.

The aspirin data of Rowland and Riegelman [9] will serve as examples of Model IX. They administered acetylsalicylic acid by rapid intravenous injection to five subjects and measured the drug by a specific assay method. They fitted each set of plasma aspirin concentrations to biexponential equations [Equation (21)] and results are listed in Table 2.3.

Using Equations (26) and (36) to (39) and the data in Table 2.3, the Model IX parameters were calculated and are shown in Table 2.4.

Another way to interpret Model IX is that the intrinsic metabolic clearance, CL_m, is given by Equation (42) and the blood flow rate, Q, is given by Equation (43).

$$CL_m = V_2 k_{20} \tag{42}$$

$$Q = Vck_{12} = V_2 k_{21} \tag{43}$$

*TABLE 2.4. Model IX Parameters for Acetylsalicylic Acid.**

Subject	Dose	*Vc* (L)	k_{12}	k_{21}	k_{20}	*E*
A	325	5.65	0.203	0.0436	0.0615	0.585
A	650	5.46	0.194	0.0344	0.0620	0.643
B	650	6.99	0.207	0.0784	0.0750	0.489
C	650	6.50	0.193	0.0728	0.0720	0.497
D	650	6.50	0.171	0.0420	0.0674	0.616
E	650	7.56	0.151	0.0295	0.0466	0.616

*k_{12}, k_{21}, and k_{20} are in min^{-1} and E is dimensionless.

The total body clearance, CL_{TB}, is given by Equation (44).

$$CL_{TB} = D/AUC = QE = 1/(1/Q + 1/CL_m) = QCL_m/(Q + CL_m) \tag{44}$$

Hence, total body clearance is a function of both blood flow rate to the site of metabolism and metabolic (enzyme) activity.

MODEL X. THE CLASSICAL TWO-COMPARTMENT OPEN MODEL WITH FIRST-ORDER ABSORPTION

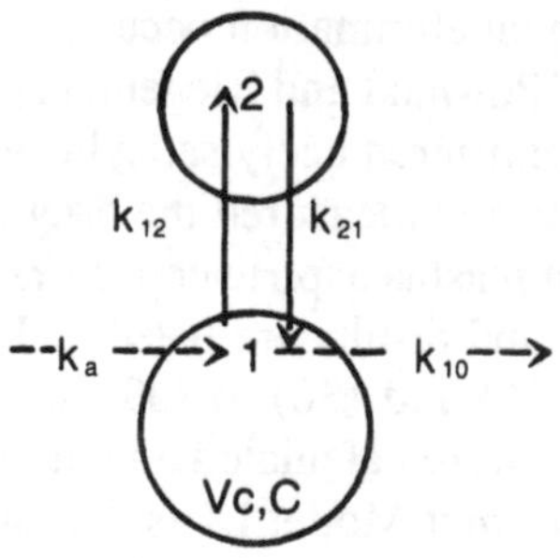

SCHEME 2.3.

In Scheme 2.3, k_{12} and k_{21} are first-order distribution rate constants, k_a is the first-order absorption rate constant, k_{10} is the first-order elimination rate constant, Vc is the volume of the central compartment and FD is the amount of drug absorbed, where F is the fraction absorbed, and D is the dose.

The concentration, C, in the central compartment at time t is given by Equation (45).

$$C = \{k_a FD/Vc\}\left[\frac{(k_{21} - \lambda_1)e^{-\lambda_1 t}}{(\lambda_2 - \lambda_1)(k_a - \lambda_1)} + \frac{(k_{21} - \lambda_2)e^{-\lambda_2 t}}{(k_a - \lambda_2)(\lambda_1 - \lambda_2)} + \frac{(k_{21} - k_a)e^{-k_a t}}{(\lambda_1 - k_a)(\lambda_2 - k_a)}\right] \tag{45}$$

Equation (45) may be written as Equation (46).

$$C = B_1 \exp(-\lambda_1 t) + B_2 \exp(-\lambda_2 t) + B_3 \exp(-k_a t) \tag{46}$$

where

$$B_1 = k_aFD(k_{21} - \lambda_1)/(\lambda_2 - \lambda_1)(k_a - \lambda_1)Vc \tag{47}$$

$$B_2 = k_aFD(k_{21} - \lambda_2)/(k_a - \lambda_2)(\lambda_1 - \lambda_2)Vc \tag{48}$$

$$B_3 = k_aFD(k_{21} - k_a)/(\lambda_1 - k_a)(\lambda_2 - k_a)Vc \tag{49}$$

If data are fitted to Equation (46) the model parameters may be obtained from the coefficients and exponents with Equations (50) through (53).

$$Vc/F = k_aD/\{B_1(k_a - \lambda_1) + B_2(k_a - \lambda_2)\} \tag{50}$$

$$k_{21} = \{B_2\lambda_1k_a + B_1\lambda_2k_a + B_3\lambda_1\lambda_2\}/\{B_2(k_a - \lambda_2) + B_1(k_a - \lambda_1)\} \tag{51}$$

$$k_{10} = \lambda_1\lambda_2/k_{21} \tag{52}$$

$$k_{12} = \lambda_1 + \lambda_2 - k_{21} - k_{10} \tag{53}$$

An example is the administration of 1.25 ml of an aqueous solution-suspension of spectinomycin containing 400 mg/ml (D = 500 mg) to six volunteers intramuscularly [8]. The mean serum concentrations were fitted with the triexponential Equation (54) and the fit is shown in Figure 2.5.

$$C = 50.2e^{-0.400t} + 34.2e^{-1.48t} - 84.4e^{-2.17t} \tag{54}$$

Using Equations (50) through (53) and Equation (54) the parameters obtained were: Vc/F = 6.57 L, k_{12} = 0.624 hr^{-1}, k_{21} = 1.25 hr^{-1}, and k_{10} = 0.694 hr^{-1}.

In Equation (20) for Model VIII the exponents are λ_1 and λ_2, and in Equation (45) for Model X the exponents are λ_1, λ_2, and k_a. In the spectinomycin case, the corresponding equations with numerical values are Equation (30) with λ_1 = 0.438 hr^{-1} and λ_2 = 3.21 hr^{-1} and Equation (54) with exponents 0.400, 1.48, and 2.17 hr^{-1}. One can assume that λ_1 = 0.400 hr^{-1} in the I.M. equation but it is difficult to assign 1.48 and 2.17 as λ_2 or k_a. Preparing an Exact Loo-Riegelman absorption plot [10] aids in these decisions. For the classical two-compartment open model the Exact Loo-Reigelman equation [10] is Equation (55).

$$A_T/Vc = C_T + k_{10}\int_0^T Cdt + k_{12}e^{-k_{21}T}\int_0^T C\exp(k_{21}t)dt \tag{55}$$

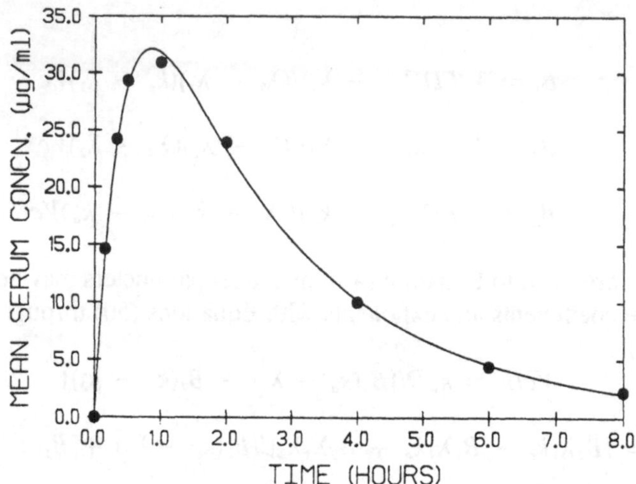

FIGURE 2.5. Example of Model X. Mean serum concentrations of spectinomycin following the administration of 500 mg of spectinomycin intramuscularly to six volunteers [8] and after fitting data to Equation (54).

Equation (55) with the Proost modification [11] was applied to the intramuscular spectinomycin data (see Figure 2.5) and the fraction absorbed (FA) at various times obtained with Equation (56).

$$\text{FA} = (A_T/Vc)/(A_{\text{max}}/Vc) \tag{56}$$

where A_{max} was the mean value of A_T/Vc at four, six, and eight hours and the asymptotic value of A_T/Vc. The FA, t values were fitted to the first-order function [Equation (57)] using MINSQ [2]. The fit is shown in Figure 2.6.

$$\text{FA} = 1 - e^{-k_a t} \tag{57}$$

The estimated k_a was 1.52 hr^{-1}, which agrees closely with the value of 1.48 in Equation (54); thus, the exponent 2.17 in Equation (54) may be assigned to λ_2. Intrasubject variation can explain the differences in λ_1 and λ_2 after I.V. and I.M. administrations; the fact that mean concentrations were used also may have had an effect on the values.

Substituting $B_1 = 50.2$, $\lambda_1 = 0.400$, $B_2 = -84.4$, $\lambda_2 = 2.178$, $B_3 = 34.2$, and $k_a = 1.48$ into Equations (50) through (53) gave $Vc/F = 6.58$ L, $k_{12} = 0.624$ hr^{-1}, $k_{21} = 1.25$ hr^{-1}, and $k_{10} = 0.693$ hr^{-1}. Thus, this example was an unusual case where $\lambda_2 > k_a > \lambda_1$.

Another example for Model X is one of the sets of flurbiprofen plasma concentrations published by Gonzalez [12] following the oral administra-

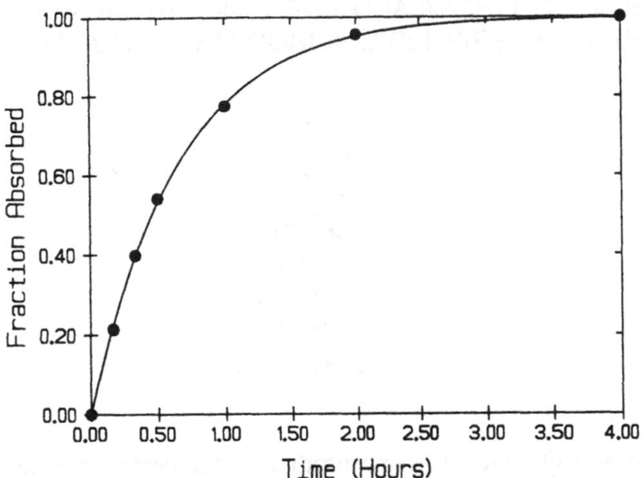

FIGURE 2.6. Exact Loo-Riegelman plot [10] based on the I.M. spectinomycin data and Equations (55) and (56). Points were fitted with Equation (57) with $k_a = 1.52$ hr^{-1}.

tion of 100 mg of flurbiprofen (as the sodium salt) in aqueous solution. The data of subject #5 were fitted to the triexponential Equation (58).

$$C = 3.236e^{-0.1142t} + 9.110e^{-0.4895t} - 12.346e^{-12.46t} \tag{58}$$

The fit is shown in Figure 2.7.

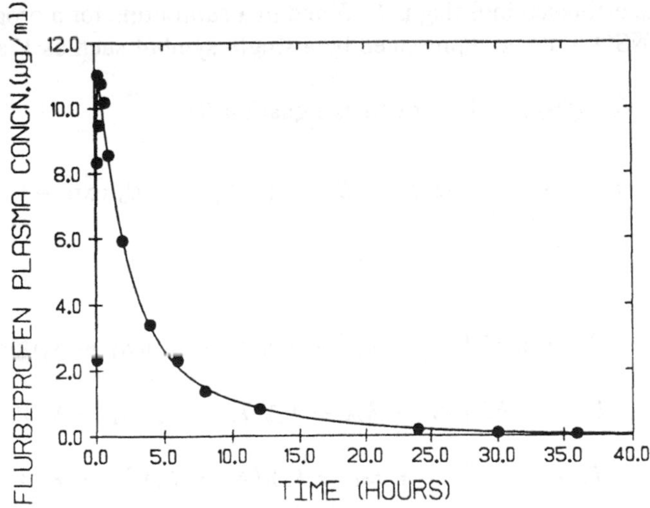

FIGURE 2.7. Fit of flurbiprofen plasma concentrations to Equation (58) as another example of Model X. One hundred milligrams of flurbiprofen (as the sodium salt) was administered orally to subject #5 [12].

MODEL XI. THE ROWLAND TWO-COMPARTMENT OPEN MODEL WITH PERIPHERAL ABSORPTION AND ELIMINATION

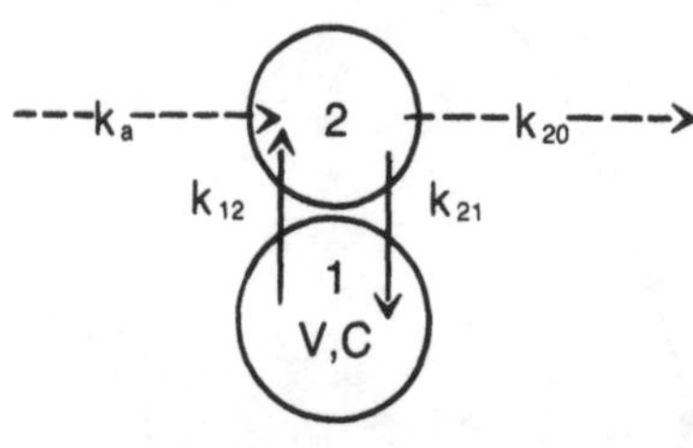

SCHEME 2.4.

Equation (59) gives the concentration in the central compartment as a function of time.

$$C = (k_a FD/Vc)[(k_{21} + k_{20} - \lambda_1)\exp(-\lambda_1 t)/(k_a - \lambda_1)(\lambda_2 - \lambda_1)$$
$$+ (k_{21} + k_{20} - \lambda_2)\exp(-\lambda_2 t)/(k_a - \lambda_2)(\lambda_1 - \lambda_2)$$
$$+ (k_{21} + k_{20} - k_a)\exp(-k_a t)/(\lambda_1 - k_a)(\lambda_2 - k_a)] \quad (59)$$

The amount of drug originally at the absorption site at zero time is *FD*. The volume estimated in fitting is *Vc*/*F* and in a subroutine for a program such as MINSQ must be represented by a single symbol such as *V* and not as *Vc*/*F*.

Equation (59) may be written as Equation (60).

$$C = B_1 \exp(-\lambda_1 t) + B_2 \exp(-\lambda_2 t) + B_3 \exp(-k_a t) \quad (60)$$

where

$$B_1 = k_a FD[(k_{21} + k_{20} - \lambda_1)/(k_a - \lambda_1)(\lambda_2 - \lambda_1)Vc \quad (61)$$

$$B_2 = k_a FD[(k_{21} + k_{20} - \lambda_2)/(k_a - \lambda_2)(\lambda_1 - \lambda_2)Vc \quad (62)$$

$$B_3 = k_a FD[(k_{21} + k_{20} - k_a)/(\lambda_1 - k_a)(\lambda_2 - k_a)Vc \quad (63)$$

If data are fitted to the triexponential Equation (60) then the model parameters may be obtained with Equations (64) to (68).

$$A = \{B_2\lambda_1 k_a + B_1\lambda_2 k_a + B_3\lambda_1\lambda_2\}/\{B_1(k_a - \lambda_1) + B_2(k_a - \lambda_2)\}$$

$$= k_{20} + k_{21} \quad (64)$$

$$k_{12} = \lambda_1 + \lambda_2 - A \quad (65)$$

$$k_{20} = \lambda_1\lambda_2/k_{12} \quad (66)$$

$$k_{21} = A - k_{20} \quad (67)$$

$$Vc/F = (k_a D/B_1)(k_{21} + k_{20} - \lambda_1)/(k_a - \lambda_1)(\lambda_2 - \lambda_1) \quad (68)$$

For a simulation example, the following assignments were used: $D = 100$, $Vc/F = 50$, $k_a = 4$, $k_{12} = 2$, $k_{21} = 1$, and $k_{20} = 0.1$. Thus:

$$\lambda_1 + \lambda_2 = k_{12} + k_{20} + k_{21} = 3.1 \quad (69)$$

and

$$\lambda_1\lambda_2 = k_{12}k_{20} = 0.2 \quad (70)$$

Thus:

$$\lambda_1 = 0.5[\lambda_1 + \lambda_2 - \{(\lambda_1 + \lambda_2)^2 - 4\lambda_1\lambda_2\}^{1/2}] = 0.0659 \quad (71)$$

$$\lambda_2 = 0.5[\lambda_1 + \lambda_2 + \{(\lambda_1 + \lambda_2)^2 - 4\lambda_1\lambda_2\}^{1/2}] = 3.0341 \quad (72)$$

Substituting the above values into Equation (60) gives Equation (73).

$$C = 0.7085e^{-0.0659t} + 5.3969e^{-3.0341t} - 6.1054e^{-4t} \quad (73)$$

Figure 2.8 shows the plot of concentration *versus* time based on Equation (73).

MODEL XII. THE CLASSICAL TWO-COMPARTMENT OPEN MODEL WITH CENTRAL COMPARTMENT ELIMINATION AND ZERO-ORDER INPUT

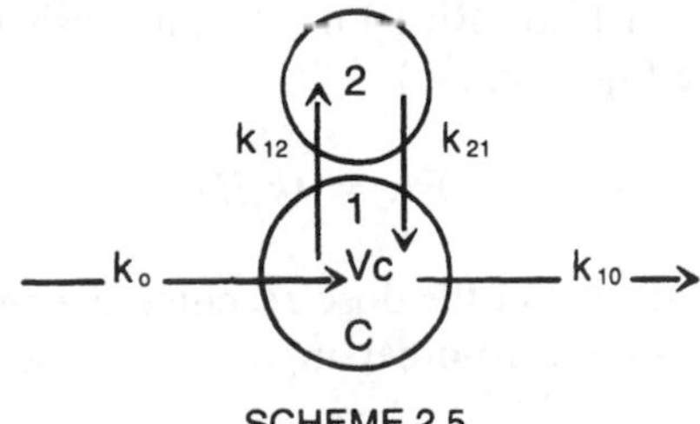

SCHEME 2.5.

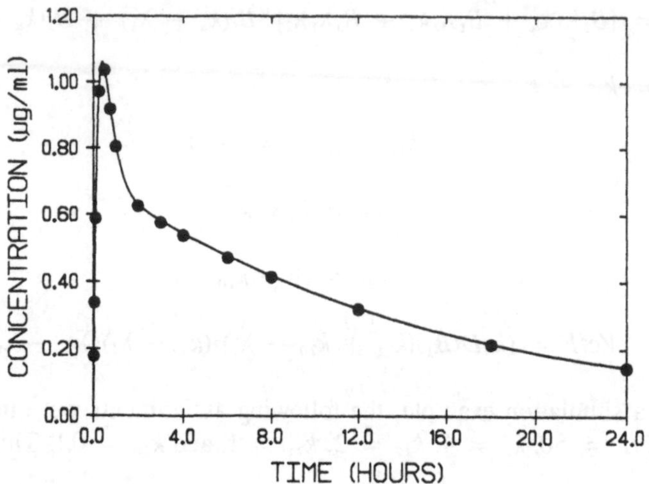

FIGURE 2.8. Simulated data for Model XI based on Equation (73) obtained by substituting $D = 100$, $V_c/F = 50$, $k_a = 4$, $k_{12} = 2$, $k_{21} = 1$, and $k_{20} = 0.1$ into Equation (59).

During the zero-order input, such as an intravenous infusion, the concentration, C, in the central compartment at time t is given by Equation (74). This is during the interval $0 < t < T$ where T is the duration of the zero-order input.

$$C = k_o(k_{21} - \lambda_1)\{1 - \exp(-\lambda_1 t)\}/\lambda_1(\lambda_2 - \lambda_1)Vc$$
$$+ k_o(k_{21} - \lambda_2)\{1 - \exp(-\lambda_2 t)\}/\lambda_2(\lambda_1 - \lambda_2)Vc \quad (74)$$

After the zero-order input ceases, the concentration is given by Equation (75), i.e., when $t > T$.

$$C = k_o(k_{21} - \lambda_1)(e^{+\lambda_1 T} - 1)e^{-\lambda_1 t}/\lambda_1(\lambda_2 - \lambda_1)Vc$$
$$+ k_o(k_{21} - \lambda_2)(e^{+\lambda_2 T} - 1)e^{-\lambda_1 t}/\lambda_2(\lambda_1 - \lambda_2)Vc \quad (75)$$

The results of Exact Loo-Riegelman [10] analysis with data obeying this model would give Equation (76):

$$\text{FA} = (k_o/D)t \quad (76)$$

where FA is the fraction of the dose reaching the central compartment in time t and k_o/D is the zero-order input rate constant with dimension of

1/time. Application of Equation (55) (see Model X) to data obeying this Model XII would give Equation (77), where k_o/Vc has dimensions mass/(volume × time).

$$A_T/Vc = (k_o/Vc)t \tag{77}$$

Method for Intestinal Infusion

In the case of an intestinal infusion or oral sustained-release dosage form which provides zero-order input, the Vc in Equations (74), (75), and (77) should be replaced by Vc/F and the D in Equation (76) replaced by FD. In such cases one can determine all parameters without intravenous data.

The mean residence time (MRT) for Model VIII (i.e., the disposition portion of Model XII) is given by Equation (78).

$$\mathrm{MRT} = \mathrm{AUMC}/\mathrm{AUC} = \int_0^{\mathrm{Inf}} tCdt \Big/ \int_0^{\mathrm{Inf}} Cdt \tag{78}$$

The mean input time (MIT) for zero-order input is $T/2$ where T is the duration of the zero-order input.

For Model XII, when administration is oral or by intestinal infusion, the $\mathrm{MRT_{po}}$ is given by Equation (79).

$$\begin{aligned}\mathrm{MRT_{po}} &= \mathrm{MRT} + \mathrm{MIT}\\ &= \left\{\int_0^{T'} tCdt + C_{T'}T/\lambda_1 + C_{T'}/\lambda_1^2\right\} \Big/ \int_0^{\mathrm{Inf}} Cdt\end{aligned} \tag{79}$$

where $C_{T'}$ is the estimated concentration at time T' in the terminal log-linear phase. The integrals in Equation (79) may be estimated with the linear and logarithmic trapezoidal rules where the decision to use each rule is made with the criterion of Proost [11]. For a given set of data $\mathrm{MRT_{po}}$ may be obtained with Equation (79) and MIT subtracted to give MRT as shown in Equation (80).

$$\mathrm{MRT} = \mathrm{MRT_{po}} - \mathrm{MIT} \tag{80}$$

Wagner [13] showed that the MRT for Model VIII is also given by Equation (81).

$$\mathrm{MRT} = (\lambda_1 + \lambda_2 - k_{10})/\lambda_1\lambda_2 \tag{81}$$

Rearrangement of Equation (81) gives Equation (82).

$$k_{10} = \lambda_1 + \lambda_2 - \lambda_1\lambda_2(\text{MRT}) \tag{82}$$

Also,

$$k_{21} = \lambda_1\lambda_2/k_{10} \tag{83}$$

$$k_{12} = \lambda_1 + \lambda_2 - k_{10} - k_{21} \tag{84}$$

$$\frac{FD}{Vc} = k_{10}\int_0^{\text{Inf}} C dt \tag{85}$$

$$\frac{Vc}{F} = k_o T/(FD/Vc) \tag{86}$$

since $k_o T$ is equal to the dose, D, administered.

Example of Intestinal Infusion

An example of intestinal infusion is the labetalol data of Wagner et al. [14]. Subject #6 (as well as eight other subjects) was administered 200 mg of labetalol in 40 ml of solution by intestinal infusion over a four-hour period via a small bowel intubation tube. Plasma concentrations of labetalol were measured at 0, 0.5, 1, 1.5, 2, 2.5, 3, 4, 4.5, 5, 6, 8, 10, 12, 16, and 24 hours.

Plasma concentrations of subject #6 from four to twenty-four hours (i.e., post-input) were fitted with a biexponential equation with $1/Y_i$ weighting using the program RSTRIP [2]. The resulting equation is shown as Equation (87) and the fit is shown in Figure 2.9.

$$C = 32.851e^{-0.076542t} + 10{,}824e^{-1.5458t} \tag{87}$$

Hence, $\lambda_1 = 0.076542\ \text{hr}^{-1}$ and $\lambda_2 = 1.5458\ \text{hr}^{-1}$. Equation (79) gave $\text{MRT}_{\text{po}} = 12.46$ hr and since MIT = 2 hr then Equation (80) gave MRT = 12.46 − 2 = 10.46 hr. Substituting into Equations (82) through (86) gave: $k_{10} = 0.3847\ \text{hr}^{-1}$, $k_{21} = 0.3076\ \text{hr}^{-1}$, $k_{12} = 0.9300\ \text{hr}^{-1}$, $FD/Vc = 178.1$ ng/ml since AUC = 462.9 ng × hr/ml. Also, $k_o = (200 \times 103)/4 = 50{,}000\ \mu\text{g/hr}$ and $Vc/F = 1123$ L. Using the above values of k_{12}, k_{21}, and k_{10} the Exact Loo-Riegelman method [10] [see Equation (55) under Model X] gave the points shown in Figure 2.10. The least squares line

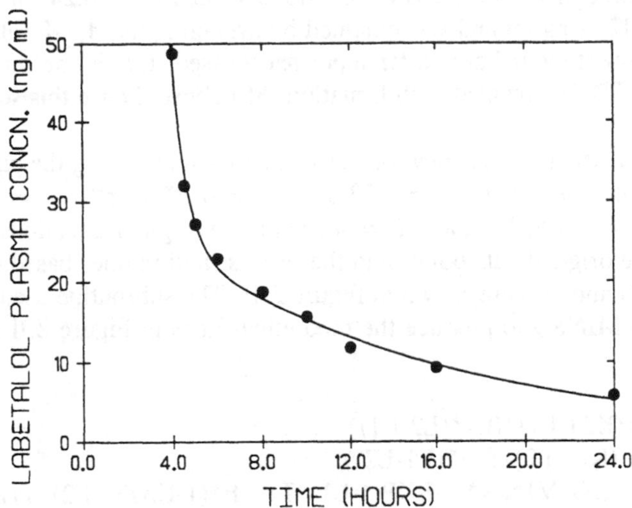

FIGURE 2.9. Post-input plasma concentrations of labetalol following 200 mg of labetalol in 40 ml of solution given by intestinal infusion over a four-hour period. Data were fitted with Equation (87) with $1/Y_i$ weighting.

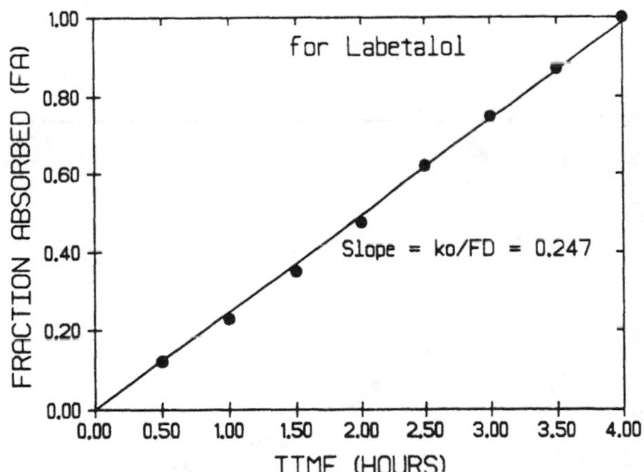

FIGURE 2.10. The Exact Loo-Riegelman plot [10] based on the labetalol data is fitted with a least squares line forced through the origin with slope equal to 0.247. This shows that the labetalol was infused at a constant rate as known experimentally.

forced through the origin gave the slope = k_o/FD = 0.247 hr^{-1}. Also, A_{max} = 178.2 ng/ml and was obtained by averaging the A_T/Vc values from four to twenty-four hours after input had ceased; this is the same as the value of FD/Vc obtained with Equation (85) above, hence this serves as a check.

A reconstruction may now be performed by substituting the parameters k_o = 50,000 μg/hr, Vc/F = 1123 L, k_{21} = 0.3076 hr^{-1}, λ_1 = 0.076542 hr^{-1}, λ_2 = 1.5458 hr^{-1}, and T = 4 hr into the original Equations (74) and (75). The original data points and the reconstruction lines based on Equations (74) and (75) are shown in Figure 2.11. The subroutine used with the program MINSQ to produce the theoretical lines in Figure 2.11 is shown below:

```
A:=(K21-L1)/(L1*(L2-L1))
B:= (K21-L2)/(L2*(L1-L2))
C1:=KO/V1*(A*(1-EXP((-L1)*T)+B*(1-EXP((-L2)*T)))
CT1:=KO/V1*(A*(1-EXP((-L1)*TAU))
CT2:=KO/V1*(B*(1-EXP((-L2)*TAU))
C2:=CT1*EXP((-L1)*(T-TAU))+CT2*EXP((-L2)*(T-TAU))
FLAG:=UNIT(T-TAU) C:=C1*(1-FLAG)+C2*FLAG
```

Since T = 4 hr in the above example, the theoretical value of k_o/FD is

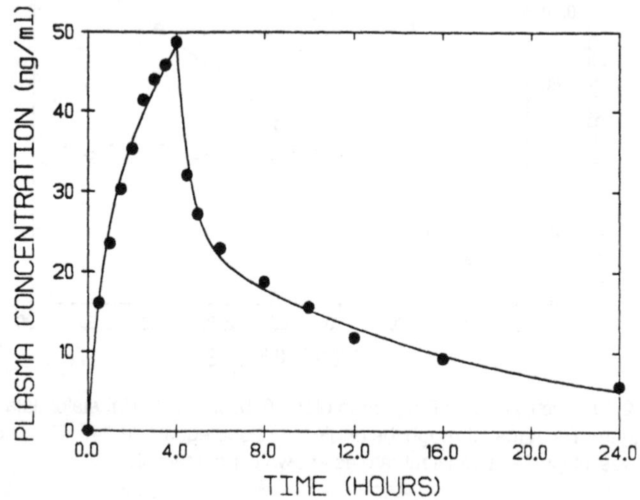

FIGURE 2.11. A reconstruction of the labetalol data based on Equations (74) and (75) and the parameters for subject #6 derived from the plots shown in Figures 2.9 and 2.10.

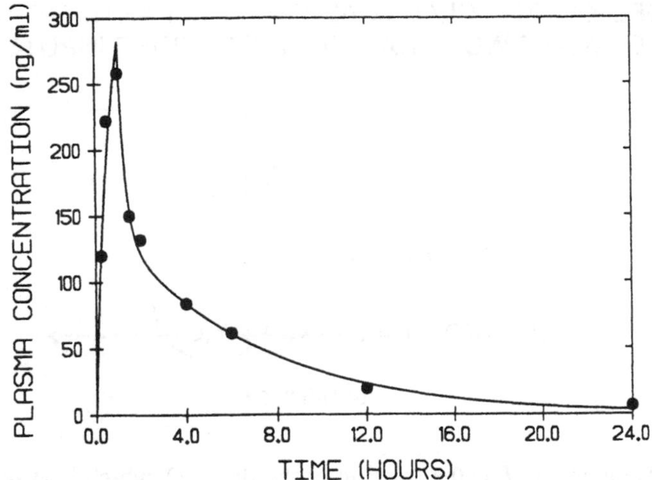

FIGURE 2.12. Intravenous infusion example for Model XII. Plasma concentrations of diphenhydramine in subject #1 administered 50 mg of diphenhydramine by I.V. infusion over a one-hour period. The *C,t* data were fitted to Equations (74) and (75) using the MINSQ [2] subroutine shown above.

$1/4 = 0.250\ hr^{-1}$. The observed value for subject #6 was $0.247\ hr^{-1}$, which supports the fact that the protocol was followed and that the drug was delivered at a constant rate over four hours.

In the development of a sustained-release dosage form for oral administration, and prior to formulation studies, it would be wise to perform an intestinal infusion study, such as was performed with labetalol and described above. This would, if results were like those with labetalol, ensure that there was no pharmacokinetic reason for not making such a dosage form.

Intravenous Infusion Example for Model XII

An example of intravenous infusion for this model is the plasma diphenhydramine concentrations published by Albert et al. [15]. Subject #1 was administered 50 mg of diphenhydramine in 240 ml of sterile solution, which was infused intravenously at a rate of 4 ml/min. Diphenhydramine plasma concentrations were measured by a specific GLC method.

The plasma concentrations were fitted to Equations (74) and (75) using the subroutine shown above and the program MINSQ with $1/Y_i$ weighting. Estimated parameters were: $\lambda_1 = 0.161\ hr^{-1}$, $\lambda_2 = 3.32\ hr^{-1}$, $k_{21} = 0.874\ hr^{-1}$, and $Vc = 87.8$ L. The fit is shown in Figure 2.12.

MODEL XIII. THE CLASSICAL TWO-COMPARTMENT OPEN MODEL WITH TWO PARALLEL FIRST-ORDER INPUTS

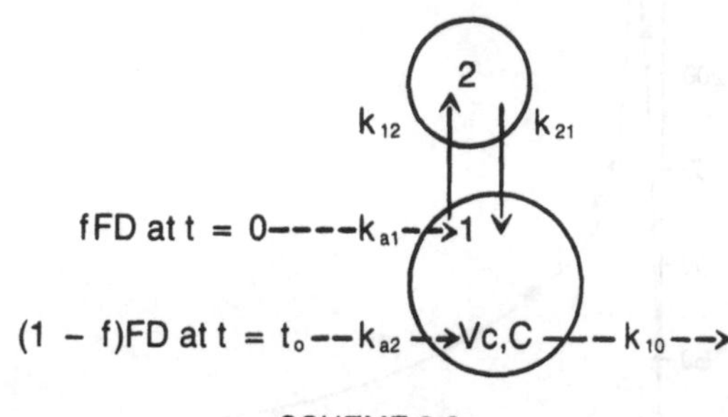

SCHEME 2.6.

In Scheme 2.6, F is the fraction of the dose, D, which is available. fF is the fraction of the dose available from the site with constant k_{a1} starting at time zero. $(1 - f)F$ is the fraction of the dose available from the site with constant k_{a2} starting at time t_o. k_{12} and k_{21} are disposition first-order rate constants and k_{10} is the first-order model elimination rate constant.

Equation (88) gives the concentration, C, in the central compartment at time t.

$$
\begin{aligned}
C = & \left\{\frac{k_{a1} fFD}{Vc}\right\} [(k_{21} - \lambda_1) \exp(-\lambda_1 t)/(k_{a1} - \lambda_1)(\lambda_2 - \lambda_1) \\
& + (k_{21} - \lambda_2) \exp(-\lambda_2 t)/(k_{a1} - \lambda_2)(\lambda_1 - \lambda_2) \\
& + (k_{21} - k_{a1}) \exp(-k_{a1} t)/(\lambda_1 - k_{a1})(\lambda_2 - k_{a1})] \\
& + \frac{k_{a2}(1 - f)FD}{Vc} [(k_{21} - \lambda_1) \exp(-\lambda_1 t)/(k_{a2} - \lambda_1)(\lambda_2 - \lambda_1) \\
& + (k_{21} - \lambda_2) \exp(-\lambda_2 t)/(k_{a2} - \lambda_2)(\lambda_1 - \lambda_2) \\
& + (k_{21} - k_{a2}) \exp(-k_{a2} t)/(\lambda_1 - k_{a2})(\lambda_2 - k_{a2})]
\end{aligned} \tag{88}
$$

The Exact Loo-Riegelman equation [10] [see Equation (55) under Model X] will give Equations (89a) and (89b) if this model is obeyed.

$$\text{FA} = f(1 - e^{-k_{a1} t}) \quad \text{when} \quad t < t_o \tag{89a}$$

$$\text{FA} = f\{1 - e^{-k_{a1} t}\} + (1 - f)\{1 - e^{-k_a (t - t_o)}\} \quad \text{when} \quad t > t_o \tag{89b}$$

This model can explain double peaks in plasma (blood or serum) concentration-time curves. An example is given below. However, not all data obeying the model show double peaks. Often data obeying the model appear very similar to data obeying Model X.

An example of data where there are double peaks is the data of subject #11, who was administered 100 mg of flurbiprofen in 40 ml of flurbiprofen oral solution (2.5 mg/ml) containing the sodium salt. The subject fasted overnight and for four hours after dosing [16]. Flurbiprofen was measured in plasma by a sensitive and specific HPLC method. The plasma concentrations of subject #11 were fitted to Equation (88) using the program MINSQ with $1/Y_i$ weighting. The fit is shown in Figure 2.13. The estimated parameters were: $f = 0.743$, $Vc/F = 5.78$ L, $k_{a1} = 6.92$ hr^{-1}, $k_{a2} = 102.5$ hr^{-1}, $k_{21} = 0.4152$ hr^{-1}, $\lambda_1 = 0.1095$ hr^{-1}, $\lambda_2 = 0.6634$ hr^{-1}, and $t_o = 1.244$ hr. Values derived from these parameters were: $k_{12} = 0.1827$ hr^{-1} and $k_{10} = 0.1750$ hr^{-1}.

The same subject #11 as above was administered one 100 mg flurbiprofen (acid) tablet under fasting conditions at a different time. Again plasma concentrations of flurbiprofen were measured at the same times as after the oral solution treatment. The Exact Loo-Riegelman method [see Equation (55) under Model X] was applied to these plasma concentrations and the fraction absorbed (FA) was determined as a function of time. These FA,t data

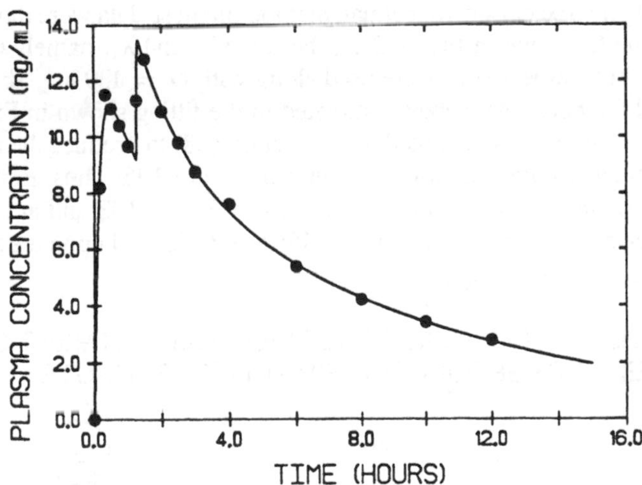

FIGURE 2.13. Example for Model XIII. Plasma concentrations of flurbiprofen in subject #11 following oral administration of 100 mg of flurbiprofen (as the sodium salt) in aqueous solution were fitted to Equation (88) using the program MINSQ [2] with $1/Y_i$ weighting. Estimated parameters are listed in the text.

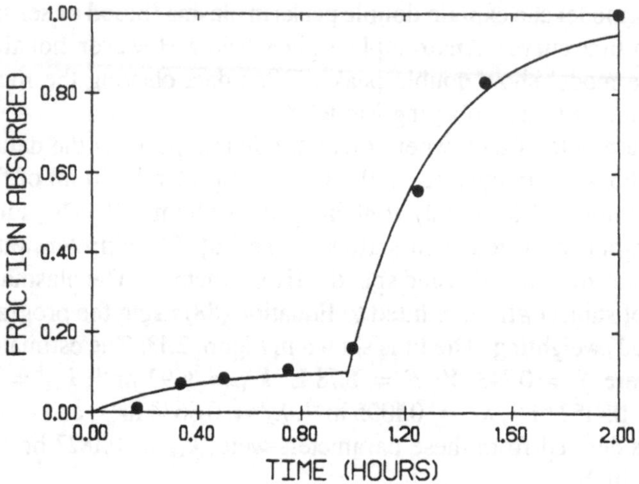

FIGURE 2.14. Exact Loo-Riegelman plot [10] derived from *C,t* data when subject #11 was administered 100 mg of flurbiprofen (as the acid) in a tablet. Points were fitted with Equations (89a) and (89b). Estimated parameters are shown in the text.

were fitted with Equations (89a) and (89b) and results are shown in Figure 2.14. The estimated parameters were: $f = 0.107$, $k_{a1} = 2.74$ hr^{-1}, $k_{a2} = 3.05$ hr^{-1}, and $t_o = 0.977$ hr.

A reconstruction was then performed. Since k_{12}, k_{21}, and k_{10} used in performing the Exact Loo-Riegelman method on these data were those derived from the fit shown in Figure 2.13, the same λ_1 and λ_2, namely 0.1095 and 0.6634 hr^{-1} respectively, were used along with $D = 100$ mg, $Vc/F = 5.78$ L, and the parameters above estimated in the fitting shown in Figure 2.14. These values were substituted into Equation (88) to produce the theoretical line drawn through the observed points in Figure 2.15. Thus, Equation (88) can sometimes produce double peaks as in Figure 2.13 and at other times produce a single peak as in Figure 2.15; results depend upon the absorption parameters and the lag time, t_o.

MODEL XIV. THE CLASSICAL TWO-COMPARTMENT OPEN MODEL WITH SEQUENTIAL FIRST-ORDER INPUTS

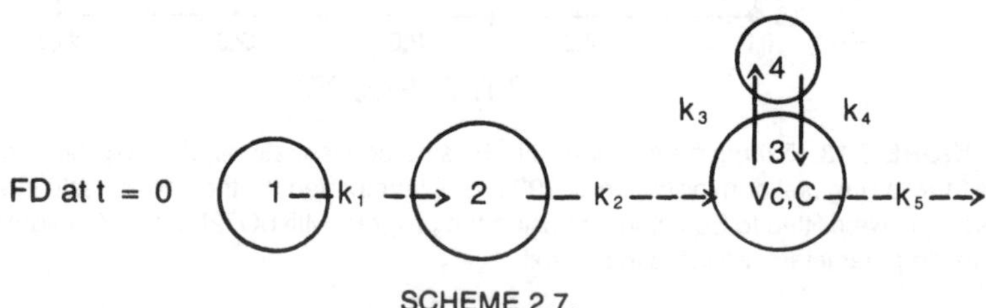

SCHEME 2.7.

Equation (90) gives the concentration, C, in the central compartment #3 at time t.

$$C = I_1 e^{-\lambda_1 t} + I_2 e^{-\lambda_2 t} + I_3 e^{-k_1 t} + I_4 e^{-k_2 t} \tag{90}$$

where

$$I_1 = (k_1 k_2 FD/Vc)[(k_4 - \lambda_1)/(\lambda_2 - \lambda_1)(k_1 - \lambda_1)(k_2 - \lambda_1)] \tag{91}$$

$$I_2 = (k_1 k_2 FD/Vc)[(k_4 - \lambda_2)/(\lambda_1 - \lambda_2)(k_1 - \lambda_2)(k_2 - \lambda_2)] \tag{92}$$

$$I_3 = (k_1 k_2 FD/Vc)[(k_4 - k_1)/(\lambda_1 - k_1)(\lambda_2 - k_1)(k_2 - k_1)] \tag{93}$$

$$I_4 = (k_1 k_2 FD/Vc)[(k_4 - k_2)/(\lambda_1 - k_2)(\lambda_2 - k_2)(k_1 - k_2)] \tag{94}$$

$$\lambda_1 + \lambda_2 = k_3 + k_4 + k_5 \tag{95}$$

$$\lambda_1 \lambda_2 = k_4 k_5 \tag{96}$$

If data were fitted to Equation (90) then the parameters of the model could be estimated with Equations (97) to (102) taken from Wagner [17]. In

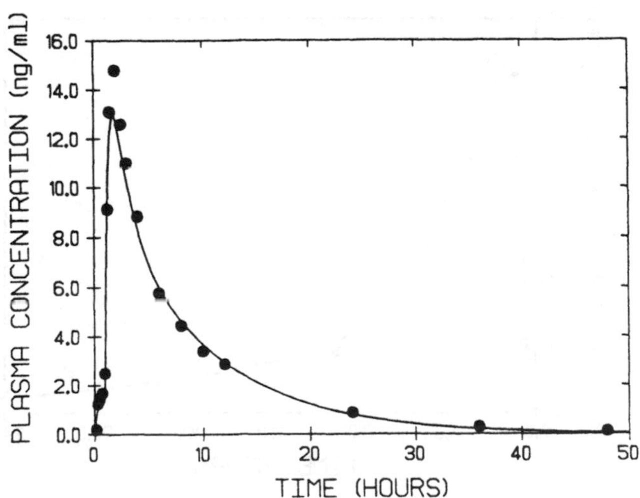

FIGURE 2.15. Flurbiprofen reconstruction based on Equation (88). See text for details.

such cases the largest two rate parameters are usually assumed to be equal to k_1 and k_2.

$$k_4 = \{k_1 k_2(\lambda_1 I_2 + \lambda_2 I_1) + \lambda_1 \lambda_2(k_2 I_3 + k_1 I_4)\}/k_1 k_2 C_o \qquad (97)$$

$$C_o = (k_1 - \lambda_2)(k_2 - \lambda_2)(k_1 - \lambda_1)(k_2 - \lambda_1)/\{k_{12}k_{22} - k_1 k_2 \lambda_1 \lambda_2\}$$

$$= FD/Vc \qquad (98)$$

$$\mathrm{AUC} = I_1/\lambda_1 + I_2/\lambda_2 + I_3/k_1 + I_4/k_2 \qquad (99)$$

$$k_5 = C_o/\mathrm{AUC} \qquad (100)$$

$$k_4 = \lambda_1 \lambda_2 / k_5 \qquad (101)$$

$$k_3 = \lambda_1 + \lambda_2 - k_4 - k_5 \qquad (102)$$

An example is the propoxyphene data of Wagner [17] and Wagner et al. [18,19]. Subject #1 received propoxyphene as single doses at three different times: treatment A was two commercial capsules containing 130 mg of drug; treatment B was the contents of two commercial capsules dissolved in Coca-Cola syrup; and treatment D was one commercial capsule containing 65 mg of drug. Propoxyphene plasma concentrations were measured at 0, 0.5, 1, 2, 3, 5, 7, 9, and 24 hours after dosing by a sensitive and specific GLC method after each treatment. The plasma concentrations were simultaneously fitted to Equations (90) to (96) by the nonlinear least squares method keeping k_2, k_3, k_4, k_5, λ_1, and λ_2 constant for each treatment but letting k_1 and C_o change with the treatment. Figure 2.16 shows the fit for

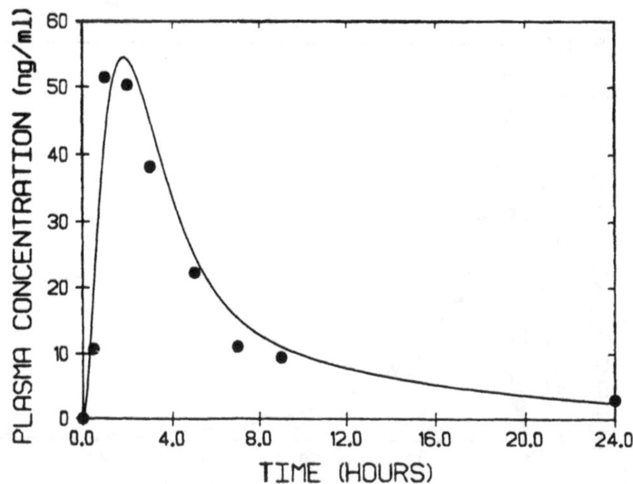

FIGURE 2.16. Fit of plasma concentrations of propoxyphene following treatment D (commercial capsule containing 50 mg of propoxyphene). See text for details.

treatment D, and the estimated parameters were: $k_1 = 0.699$ hr^{-1}, $k_2 = 2.73$ hr^{-1}, $k_3 = 0.413$ hr^{-1}, $k_4 = 0.203$ hr^{-1}, $k_5 = 0.464$ hr^{-1}, $C_o = 169$ ng/ml, $\lambda_1 = 0.0959$ hr^{-1}, and $\lambda_2 = 0.984$ hr^{-1}. More details of these fittings are given by Wagner [17].

Equation (103) would be obtained if the Exact Loo-Riegelman method [10,11] [see Equation (55) under Model X] was applied to Equations (90) to (96).

$$FA = 1 - \{(k_1 e^{-k_2 t} - k_2 e^{-k_1 t})/(k_1 - k_2)\} \tag{103}$$

MODEL XV. THE CLASSICAL TWO-COMPARTMENT OPEN MODEL WITH PARALLEL ZERO-ORDER AND FIRST-ORDER INPUTS

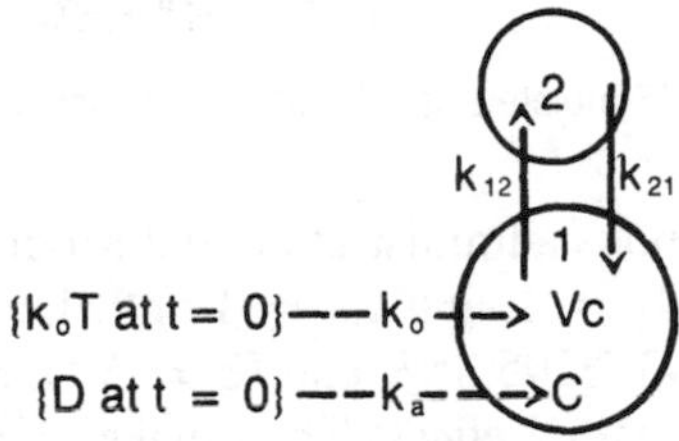

SCHEME 2.8.

In Scheme 2.8, D is the dose available at a first-order rate and $k_o T$ is the dose available at a zero-order rate; T is the duration of the zero-order rate. As before, k_{12} and k_{21} are first-order distribution rate constants and Vc is the volume of the central compartment #1, but the volume that would be estimated is Vc/F, where F is the bioavailability of the total dose, $D + k_o T$; but F is not in Equation (104) in order to avoid confusion.

$$\begin{aligned} C = {} & (k_a D/Vc)[(k_{21} - \lambda_1)e^{-\lambda_1 t}/(\lambda_2 - \lambda_1)(k_a - \lambda_1) \\ & + (k_{21} - \lambda_2)e^{-\lambda_2 t}/(k_a - \lambda_2)(\lambda_1 - \lambda_2) \\ & + (k_{21} - k_a)e^{-k_a t}/(\lambda_1 - k_a)(\lambda_2 - k_a)] \\ & + k_o(k_{21} - \lambda_1)(1 - e^{-\lambda_1 t})/(\lambda_1(\lambda_2 - \lambda_1)Vc) \\ & + k_o(k_{21} - \lambda_2)(1 - e^{-\lambda_2 t})/(\lambda_2(\lambda_1 - \lambda_2)Vc) \end{aligned} \tag{104}$$

Like Model VII this model may be used to simulate results that would be obtained with a sustained-release oral formulation in which part of the dose is available rapidly and the remainder of the dose is released at a zero-order rate.

A simulation was performed as an example. Suppose we had a drug for

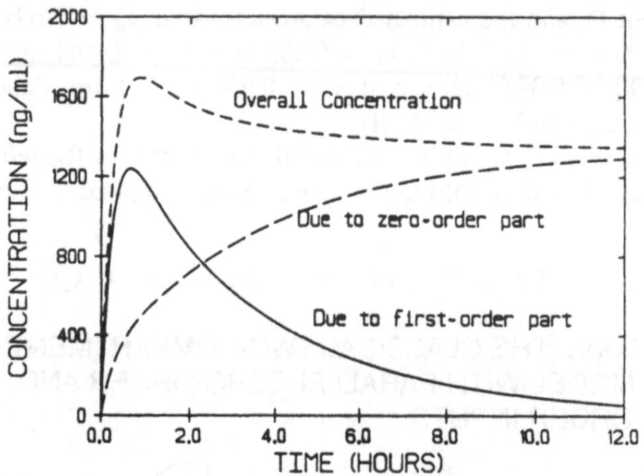

FIGURE 2.17. Simulated data for Model XV based on Equation (104).

which bolus intravenous administration and Model VIII had given the parameters $k_{12} = 2$, $k_{21} = 1$, and $k_{10} = 1$ hr^{-1}. Equations (24) to (26) gave $\lambda_1 = 0.2679$, $\lambda_2 = 3.73205$ hr^{-1}, and $Vc = 5$ L. Suppose a single dose of 25 mg produces a desired effect. Then make $D = 25$ mg $= 25{,}000$ μg, make k_oT be the dose released at a zero-order rate over T hours, and make $T = 12$ hr. Then, $k_oT = DT\lambda_1 = (25)(12)(0.26795) = 80$ mg $= 80{,}000$ μg. Hence, $k_o = 80{,}000/12 = 6666.67$ μg/hr. Let $k_a = 1.25$ hr^{-1} for reasonably rapid release. Substitution of these values into Equation (104) provides the data shown in Figure 2.17.

Figure 2.17 shows the contribution of the rapidly available dose that is absorbed with a first-order rate constant of 1.25 hr^{-1}, and the contribution of the zero-order release portion, as well as the overall concentration in the central compartment. The shape of the latter is readily changed by adjusting the parameter values. After twelve hours in this simulation, the concentration would fall off according to biexponential kinetics with the λ_1 and λ_2 values indicated above.

MODEL XVI. CLASSICAL TWO-COMPARTMENT OPEN MODEL WITH BOLUS LOADING DOSE AND ZERO-ORDER INPUT

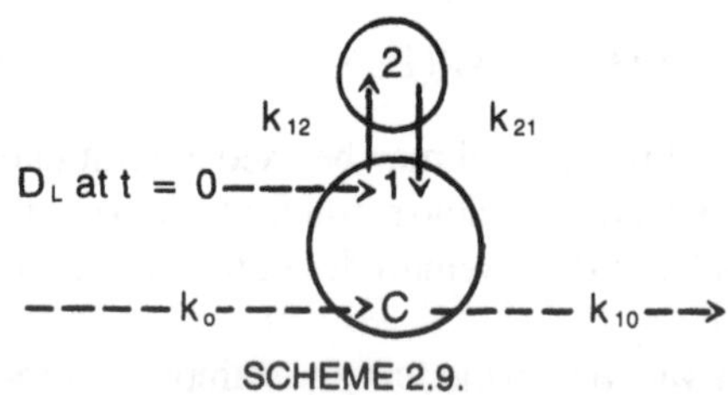

SCHEME 2.9.

The concentration, C, at time t in the central compartment is given by Equation (105).

$$C = k_o/Vck_{10} - [(D_L\lambda_2 - k_o)(k_{21} - \lambda_2)\exp(-\lambda_2 t)/\{(Vc\lambda_2(\lambda_2 - \lambda_1)\}$$

$$- (D_L\lambda_1 - k_o)(k_{21} - \lambda_1)\exp(-\lambda_1 t)/\{(Vc\lambda_1(\lambda_1 - \lambda_2)\}] \qquad (105)$$

where k_o is the zero-order input rate started at the same time as the bolus dose is given, D_L is the bolus loading dose put in the central compartment at zero time, k_{12} and k_{21} are distribution rate constants, and Vc is the volume of the central compartment. The hybrid rate constants, λ_1 and λ_2, are given by Equations (24) and (25) under Model VIII. The first term on the right-hand side of Equation (105) is equal to the steady-state concentration if the infusion was continued long enough and is independent of the loading dose D_L.

This model has been discussed by Mitenko and Ogilvie [20], Boyes et al. [21], and Wagner [17a]. The latter author showed that: with loading doses equal to k_o/k_{10}, as recommended by Boyes et al. [21], the initial concentration is equal to the desired steady-state concentration, then drops below it, and then there is a very slow rise to the steady-state concentration.

With a loading dose equal to k_o/λ_1, as recommended by Mitenko and Ogilvie [20], the initial concentration is very high and may be toxic, then falls rapidly to the steady-state concentration. Wagner [17a] shows plots as examples.

The method of administration discussed under Model XVII is more desirable than either of the above possibilities.

MODEL XVII. CLASSICAL TWO-COMPARTMENT OPEN MODEL WITH TWO CONSECUTIVE ZERO-ORDER INFUSION RATES

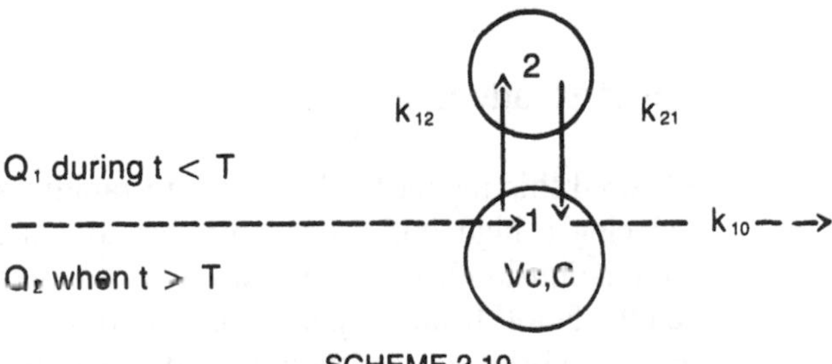

SCHEME 2.10.

In Scheme 2.10, Q_1 is the initial infusion rate (mass/time), Q_2 is the final infusion rate (mass/time), T is the duration of the first infusion and the time when the rate is instantaneously changed; k_{12}, k_{21}, k_{10}, Vc, and C have the same meanings as before.

During the first infusion at the rate Q_1 the concentration, C, in the central compartment at time t is given by Equation (106) or Equation (74) under Model XII.

$$C = (Q_1/Vck_{10})[1 - \{(k_{10} - \lambda_1)\exp(-\lambda_2 t)/(\lambda_2 - \lambda_1\} - \{(\lambda_2 - k_{10})\exp(-\lambda_1 t)/(\lambda_2 - \lambda_1)\}] \tag{106}$$

During the second infusion at the rate Q_2 the concentration is given by Equation (107), where $C_{ss} = Q_2/Vck_{10}$ and is the steady-state concentration.

$$C = C_{ss} + [\{(k_{21} - \lambda_2)(Q_2 - \lambda_2 A_1(T)) - \lambda_2 k_{21} A_2(T)\} \times \exp(-\lambda_2(t - T))]/\{\lambda_2(\lambda_2 - \lambda_1)Vc\} - [\{(k_{21} - \lambda_1)(Q_2 - \lambda_1 A_1(T)) - \lambda_1 k_{21} A_2(T)\} \times \exp(-\lambda_1(t - T)]/\{\lambda_1(\lambda_2 - \lambda_1)Vc\} \tag{107}$$

where

$$A_1(T) = Q_1[1/k_{10} + \{(k_{21} - \lambda_2)\exp(-\lambda_2 T)\}/\{\lambda_2(\lambda_2 - \lambda_1)\} - \{(k_{21} - \lambda_1)\exp(-\lambda_1 T)\}/\{\lambda_1(\lambda_2 - \lambda_1)\}] \tag{108}$$

$$A_2(T) = Q_1 k_{12}[\{1/k_{21}k_{10}\} + \{\exp(-\lambda_2 t)/\{\lambda_2(\lambda_2 - \lambda_1)\} - \{\exp(-\lambda_1 t)/\{\lambda_1(\lambda_2 - \lambda_1)\}] \tag{109}$$

Rapid Attainment of Steady State

Wagner [17b,22] used this model to show how to rapidly attain steady state without producing a high initial concentration as discussed under Model XVI. If the relationship between the two infusion rates is given by Equation (110) then the falloff from the peak concentration at time T is dependent on the larger hybrid rate constant λ_2 and is independent of the smaller hybrid rate constant λ_1.

$$Q_1 = \frac{Q_2}{1 - e^{-\lambda_1 T}} \tag{110}$$

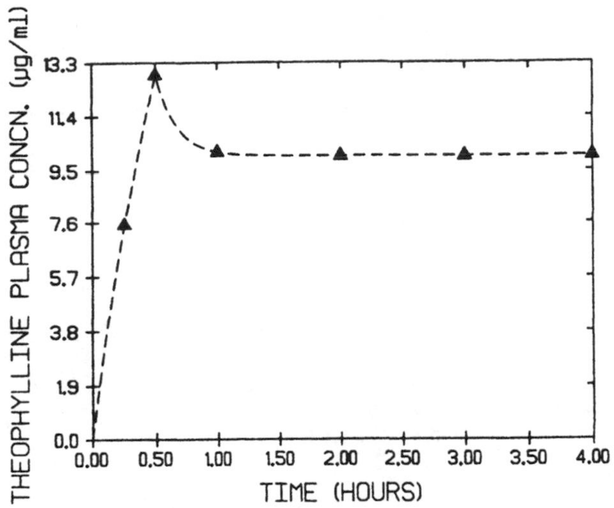

FIGURE 2.18. Simulated data for Model XVII based on Equations (106) to (109) and the parameter values given in the text.

The final rate, Q_2, is given by Equation (111), where CL is the clearance and is equal to Vck_{10}.

$$Q_2 = Vck_{10}C_{ss} = \mathrm{CL}\,C_{ss} \tag{111}$$

As an example, the mean parameter values of theophylline of seven asthmatics were used, namely: $\lambda_1 = 0.162\ \mathrm{hr}^{-1}$, $\lambda_2 = 5.99\ \mathrm{hr}^{-1}$, $Vc = 0.277$ L/kg, and $k_{10} = 0.312\ \mathrm{hr}^{-1}$ [20]. A steady-state target plasma concentration of 10 μg/ml was chosen as well as $T = 0.5$ hr for the duration of the first infusion. Hence, Equation (111) gives $Q_2 = (0.277)(0.310)(10) = 0.864$ mg/kg/hr. Equation (110) gives $Q_1 = 0.864/\{1 - e^{-(0.162)(0.5)}\} = 11.1$ mg/kg/hr. Substituting these values into Equations (106) to (109) gave the plot shown in Figure 2.18. The peak concentration at time T at the end of the initial infusion is 12.9 μg/ml and a steady-state plasma concentration of 10 μg/ml is reached at two hours. The peak concentration may be decreased by increasing the value of T, and Wagner [17b,22] gave some guidelines. Hence, the steps in applying the method are:

(1) Choose the target steady-state concentration, C_{ss}.
(2) Calculate the final infusion rate, Q_2, with Equation (111).
(3) Choose the time T over which to administer the initial infusion rate; usually 0.5 hours is satisfactory.
(4) Calculate the initial infusion rate, Q_1, with Equation (110).

To apply the method one must know the CL and λ_1.

FIGURE 2.15. Simulation [illegible] Model [illegible] Equations (III) [illegible] (II) and the [illegible] values given in the text.

The final rate, Q_f, is given by Equation (II), where Cl is the clearance and [illegible]

$$Q_f = [illegible] \qquad (III)$$

As an example, the [illegible] parameters [illegible] of several [illegible] were used, namely, [illegible] = 0.[illegible] hr[illegible], [illegible] = 0.[illegible] and k_{10} = 0.3[illegible] [illegible] 20 [illegible] steady state target plasma concentration of [illegible] was chosen as well as T = 0.5 hr for the duration of the first infusion [illegible]. Equation (II) gives [illegible] = (0.2[illegible]) [illegible] = [illegible] and [illegible] Equation (III) gives Q_0 = 0.[illegible] [illegible] [illegible] [illegible] [illegible] values [illegible] Equation [illegible] gave the [illegible] concentration [illegible] of the [illegible] and [illegible] at [illegible] reached at two hours. The plasma concentration [illegible] the [illegible] of [illegible] Figure 2.15 gave [illegible]. Here are the steps in applying the method:

(1) Choose the target steady state concentration [illegible]

(2) Calculate the final infusion rate Q_f with Equation (II).

(3) Choose the time T over which to administer the [illegible] (usually 0.5 hour is satisfactory).

(4) Calculate the initial infusion rate Q_0 with Equation (III).

To apply the method one must know the CL and [illegible].

CHAPTER 3

Complicated Linear Models

MODEL XVIII. FAMILY OF REVERSIBLE METABOLISM MODELS

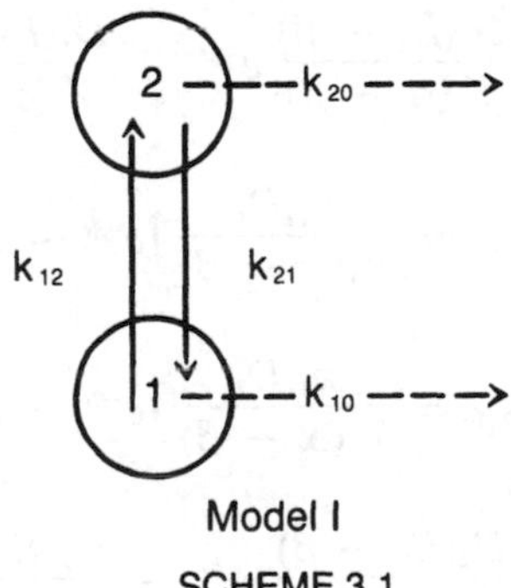

Model I

SCHEME 3.1.

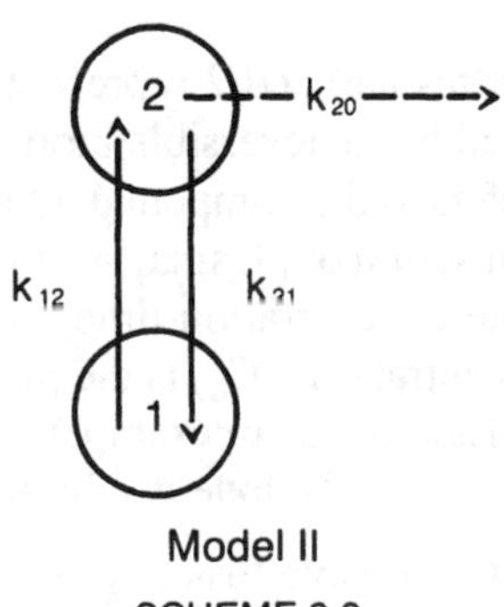

Model II

SCHEME 3.2.

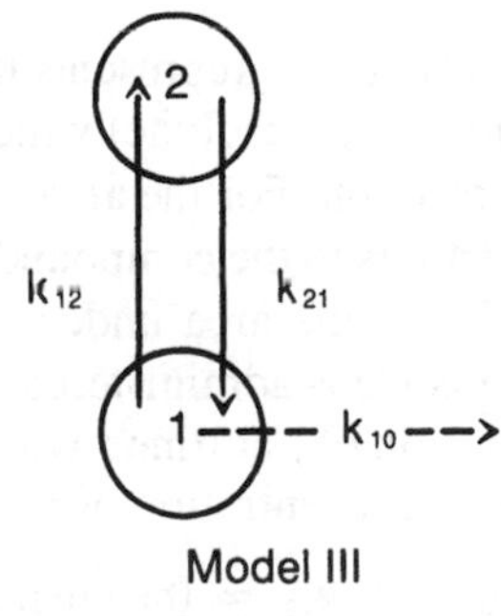

Model III

SCHEME 3.3.

TABLE 3.1. Requirements.

Model I	Model II	Model III
$AUC_{2,2} > AUC_{1,2}$	$AUC_{2,2} = AUC_{1,2}$	$AUC_{2,2} > AUC_{1,2}$
$AUC_{1,1} > AUC_{2,1}$	$AUC_{1,1} > AUC_{2,1}$	$AUC_{1,1} = AUC_{2,1}$

Models I through III are the simplest pharmacokinetic models which apply in cases of reversible metabolism at low doses when linear kinetics would be involved. Area under the curve (AUC) relationships are given in Table 3.1, which allow one to make a choice of which of the three models would apply in a particular case. Also, expressions for clearances and other pharmacokinetic parameters are given in Table 3.2.

Following intravenous administration, the time courses are given by Equations (112) through (115). It is interesting that such time courses are biexponential but distribution is not involved.

$$C_{1,1} = \frac{D_1(E_2 - \beta)}{V_1(\alpha - \beta)} e^{-\beta t} - \frac{D_1(E_2 - \alpha)}{V_1(\alpha - \beta)} e^{-\alpha t} \quad (112)$$

$$C_{1,2} = \frac{k_{12}D_1}{V_1(\alpha - \beta)} [e^{-\beta t} - e^{-\alpha t}] \quad (113)$$

$$C_{2,1} = \frac{k_{21}D_2}{V_1(\alpha - \beta)} (e^{-\beta t} - e^{-\alpha t}) \quad (114)$$

$$C_{2,2} = \frac{D_2(E_1 - \beta)}{V_2(\alpha - \beta)} e^{-\beta t} - \frac{D_2(E_1 - \alpha)}{V_2(\alpha - \beta)} e^{-\alpha t} \quad (115)$$

Symbolism

In general, 1 represents the drug (prednisone) and 2 represents the metabolite (prednisolone) which are involved in the reversible biotransformation reaction. For the areas, $AUC_{i,j}$, i refers to the compound administered and j refers to the compound measured in serum or plasma. As an example, $AUC_{1,1}$ is the area under the prednisone concentration-time curve when prednisone is administered. For the concentrations, $C_{i,j}$ is the plasma concentration of j at time t when i is administered. As an example, $C_{2,1}$ is the plasma concentration of prednisone when prednisolone is administered.

$CL_{10} = V_1k_{10}$ = the clearance for the irreversible loss of prednisone
$CL_{20} = V_2k_{20}$ = the clearance for the irreversible loss of prednisolone

TABLE 3.2. Expressions for Clearances and Other Parameters.

Parameter	Model I	Model II	Model III
CL_{10}	$\dfrac{(D_1 AUC_{2,2} - D_2 AUC_{1,2})}{(AUC_{1,1}AUC_{2,2} - AUC_{1,2}AUC_{2,1})}$	0	$D_1/AUC_{1,1} = D_2/AUC_{2,1}$
CL_{20}	$\dfrac{(D_2 AUC_{1,1} - D_1 AUC_{2,1})}{(AUC_{1,1}AUC_{2,2} - AUC_{1,2}AUC_{2,1})}$	$\dfrac{D_1}{(AUC_{1,2})} = \dfrac{D_2}{(AUC_{2,2})}$	0
CL_{12}	$CL_{10} \Big/ \left[\dfrac{(AUC_{2,2})}{(AUC_{1,2})} - 1\right]$	$\dfrac{D_2}{(AUC_{1,1} - AUC_{2,1})}$	$CL_{10} \Big/ \left[\dfrac{(AUC_{2,2})}{(AUC_{1,2})} - 1\right]$
CL_{21}	$CL_{20} \Big/ \left[\dfrac{(AUC_{1,1})}{(AUC_{2,1})} - 1\right]$	$CL_{20} \Big/ \left[\dfrac{(AUC_{1,1})}{(AUC_{2,1})} - 1\right]$	$\dfrac{D_1}{(AUC_{2,2} - AUC_{1,2})}$
CL_{t1}	$CL_{10} + CL_{12}\left[\dfrac{k_{20}}{k_{20} + k_{21}}\right]$	$CL_{12}\left[\dfrac{k_{20}}{k_{20} + k_{21}}\right]$	CL_{10}
CL_{t2}	$CL_{20} + CL_{21}\left[\dfrac{k_{10}}{k_{10} + k_{12}}\right]$	CL_{20}	$CL_{21}\left[\dfrac{k_{10}}{k_{10} + k_{12}}\right]$

(continued)

TABLE 3.2. *(continued)*.

Parameter	Model I	Model II	Model III
$\frac{(AUC_{1,1})}{(AUC_{1,2})}$	$\frac{CL_{20} + CL_{21}}{CL_{12}}$	$\frac{CL_{20} + CL_{21}}{CL_{12}}$	$\frac{CL_{21}}{CL_{12}}$
$\frac{(AUC_{1,2})}{(AUC_{2,2})}$	$\frac{k_{12}}{k_{10} + k_{12}}$	1	$\frac{k_{12}}{k_{10} + k_{12}}$
E_1	$k_{10} + k_{12}$	k_{12}	$k_{10} + k_{12}$
E_2	$k_{10} + k_{21}$	$k_{20} + k_{21}$	k_{21}
α,β	$k_{10}k_{20} + k_{12}k_{20} + k_{10}k_{21}$	$k_{12}k_{20}$	$k_{10}k_{21}$
$\alpha + \beta$	$k_{10} + k_{12} + k_{20} + k_{21}$	$k_{12} + k_{20} + k_{21}$	$k_{10} + k_{12} + k_{21}$

CL_{12} = V_1k_{12} = the clearance associated with the conversion of prednisone to prednisolone
CL_{21} = V_2k_{21} = the clearance associated with the conversion of prednisolone to prednisone
CL_{t1} = the total clearance of prednisone
CL_{t2} = the total clearance of prednisolone
D_1 = the dose of prednisone
D_2 = the dose of prednisolone
E_1 = the exit rate constant for prednisone
E_2 = the exit rate constant for prednisolone
k_{10} = first-order rate constant for irreversible elimination of prednisone
k_{12} = first-order rate constant for the conversion of prednisone to prednisolone
k_{21} = first-order rate constant for the conversion of prednisolone to prednisone
k_{20} = first-order rate constant for the irreversible elimination of prednisolone
V_1 = volume of distribution of prednisone
V_2 = volume of distribution of prednisolone

Wagner et al. [23] reported details of a human study with prednisone and prednisolone given orally in 5 mg doses. Area analysis indicated that Model III applied. The authors cautioned that thc same model may not apply to higher doses of these drugs.

MODEL XIX. THREE-COMPARTMENT OPEN MODEL WITH CENTRAL COMPARTMENT ELIMINATION AND BOLUS INTRAVENOUS ADMINISTRATION

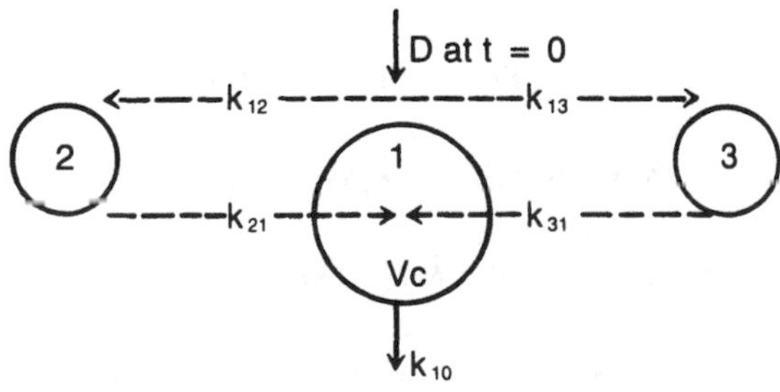

SCHEME 3.4.

Equations (116) through (126) apply to this model. Equation (116) gives the time course of the concentration in the central compartment #1.

$$C = D[(k_{21} - \lambda_1)(k_{31} - \lambda_1)e^{-\lambda_1 t}/(\lambda_2 - \lambda_1)(\lambda_3 - \lambda_1)$$
$$+ (k_{21} - \lambda_2)(k_{31} - \lambda_2)e^{-\lambda_2 t}/(\lambda_1 - \lambda_2)(\lambda_3 - \lambda_2)$$
$$+ (k_{21} - \lambda_3)(k_{31} - \lambda_3)e^{-\lambda_3 t}/(\lambda_1 - \lambda_3)(\lambda_2 - \lambda_3)]/Vc \quad (116)$$

where

$$\lambda_1 + \lambda_2 + \lambda_3 = k_{10} + k_{12} + k_{13} + k_{21} + k_{31} \quad (117)$$

$$\lambda_1\lambda_2 + \lambda_1\lambda_3 + \lambda_2\lambda_3 = k_{10}k_{21} + k_{13}k_{21} + k_{10}k_{21} + k_{21}k_{31} + k_{12}k_{31} \quad (118)$$

$$\lambda_1\lambda_2\lambda_3 = k_{10}k_{21}k_{31} \quad (119)$$

Equation (116) may also be written as Equation (120).

$$C = C_1e^{-\lambda_1 t} + C_2e^{-\lambda_2 t} + C_3e^{-\lambda_3 t} \quad (120)$$

After matching the coefficients in Equations (116) and (120) one can derive Equations (121) through (126) [24].

$$Vc = D/(C_1 + C_2 + C_3) \quad (121)$$

$$k_{21} = \lambda_2 + C_2(\lambda_1 - \lambda_2)(\lambda_2 - \lambda_3)/(\lambda_2 - k_{31})(C_1 + C_2 + C_3) \quad (122)$$

$$k_{31} = \lambda_3 + C_3(\lambda_1 - \lambda_3)(\lambda_2 - \lambda_3)/(k_{31} - \lambda_3)(C_1 + C_2 + C_3) \quad (123)$$

$$k_{10} = \lambda_1\lambda_2\lambda_3/k_{21}k_{31} \quad (124)$$

$$k_{12} = (\lambda_2\lambda_3 + \lambda_1\lambda_2 + \lambda_1\lambda_3) - k_{21}(\lambda_1 + \lambda_2 + \lambda_3)$$
$$- k_{10}k_{31} + k_{21}^2/(k_{31} - k_{21}) \quad (125)$$

$$k_{13} = \lambda_1 + \lambda_2 + \lambda_3 - (k_{10} + k_{12} + k_{21} + k_{31}) \quad (126)$$

An example is the diazepam data of Kaplan et al. [25]. Four subjects were

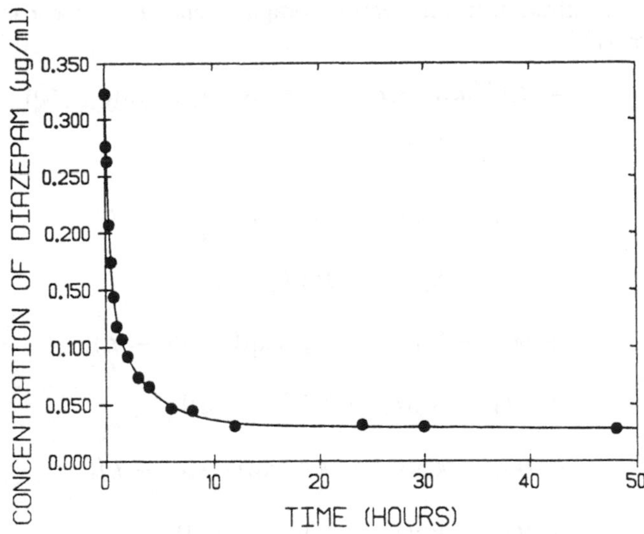

FIGURE 3.1. Example of Model XIX. Plasma concentrations of diazepam following a 10 mg dose of diazepam by bolus I.V. injection in subject #1 were fitted with Equation (127).

administered 10 mg bolus intravenous doses of diazepam and plasma was sampled at various times after administration. Plasma was then assayed for diazepam. Data of subject #1 are shown in Figure 3.1 and were fitted with the triexponential equation shown as Equation (127).

$$C = 0.03187e^{-0.02294t} + 0.1081e^{-0.3000t} + 0.1953e^{-2.8405t} \quad (127)$$

Application of Equations (121) to (126) gave the parameters: $Vc = 29.8$ L, $k_{10} = 0.0200\ \text{hr}^{-1}$, $k_{12} = 1.202\ \text{hr}^{-1}$, $k_{13} = 0.4769\ \text{hr}^{-1}$, $k_{21} = 1.336\ \text{hr}^{-1}$, and $k_{31} = 0.0722\ \text{hr}^{-1}$. Data for the other subjects were well-fitted similar to subject #1.

MODEL XX. THREE-COMPARTMENT OPEN MODEL WITH CENTRAL COMPARTMENT ELIMINATION AND FIRST-ORDER ABSORPTION

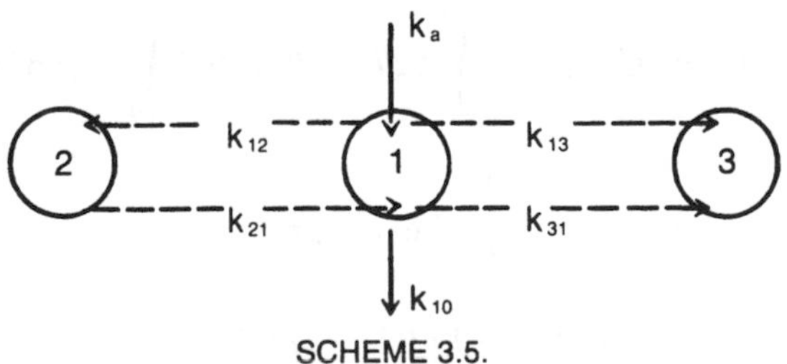

SCHEME 3.5.

The concentration in the central compartment #1 at time t is given by Equation (128).

$$
\begin{aligned}
C = {} & k_a C_o[(k_{21} - \lambda_1)(k_{31} - \lambda_2) \exp\{-\lambda_1(t - t_o)\} \\
& / \{(\lambda_2 - \lambda_1)(\lambda_3 - \lambda_1)(k_a - \lambda_1)\} \\
& + (k_{21} - \lambda_2)(k_{31} - \lambda_2) \exp\{-\lambda_2(t - t_o)\} \\
& / \{(\lambda_1 - \lambda_2)(\lambda_3 - \lambda_2)(k_a - \lambda_2)\} \\
& + (k_{21} - \lambda_3)(k_{31} - \lambda_3) \exp\{-\lambda_3(t - t_o)\} \\
& / \{(\lambda_1 - \lambda_3)(\lambda_2 - \lambda_3)(k_a - \lambda_3)\} \\
& + (k_{21} - k_a)(k_{31} - k_a) \exp\{-k_a(t - t_o)\} \\
& / \{(\lambda_1 - k_a)(\lambda_2 - k_a)(\lambda_3 - k_a)\}]
\end{aligned}
\tag{128}
$$

In Equation (128), k_a is the first-order absorption rate constant, $C_o = FD/Vc$, where FD is the amount absorbed and Vc is the volume of the central compartment #1, the $k_{i,j}$s are first-order distribution rate constants and the λ_is are hybrid rate constants defined by Equations (117) to (119) under Model XIX.

An example is the flunarizine data of Wagner et al. [13]. Thirty milligram doses of flunarizine were administered to five epileptic patients and plasma samples were collected over a seventy-day period. The plasma samples were assayed for flunarizine at N.I.H. One of the subjects was designated as MO3PS. Downslope data from 6 to 433 hours for this subject were fitted to a triexponential equation and disposition parameters for Model XX of Scheme 3.5 were derived as described in a later chapter. The Exact Loo-Riegelman method [10,11] for triexponential disposition [Equation (129)] was applied to all of the C, t data for this subject and produced the points shown in Figure 3.2. These points were fitted with the first-order Equation (130).

$$
\begin{aligned}
A_T/Vc = {} & C_T + k_{10} \int_0^T C dt + k_{12} e^{-k_{21}T} \int_0^T C \exp(k_{21}t) dt \\
& + k_{13} e^{-k_{31}T} \int_0^T C \exp(k_{31}t) dt
\end{aligned}
\tag{129}
$$

$$
A_T/V_c = 173[1 - e^{-0.893(t-0.556)}] \tag{130}
$$

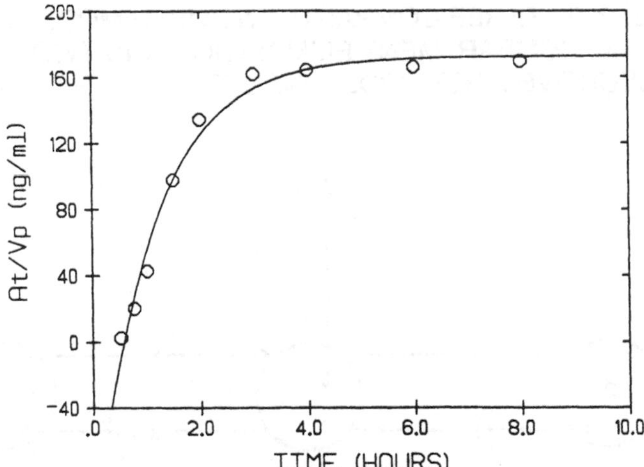

FIGURE 3.2. Exact Loo-Riegelman plot [10,11] fitted with Equation (130) and derived from plasma concentrations of flunarizine after oral administration of 30 mg of flunarizine to subject MO3PS.

The disposition parameters derived from downslope data for subject MO3PS were: $k_{10} = 0.0612$, $k_{12} = 0.0745$, $k_{13} = 0.107$, $k_{21} = 0.0564$, $k_{31} = 0.00162$ hr^{-1}, $C_o = 173$ ng/ml, and $t_o = 0.556$ hr from Equation (130). These parameter values were substituted into Equation (128) to produce the reconstructed curve fitting the data points and shown in Figure 3.3.

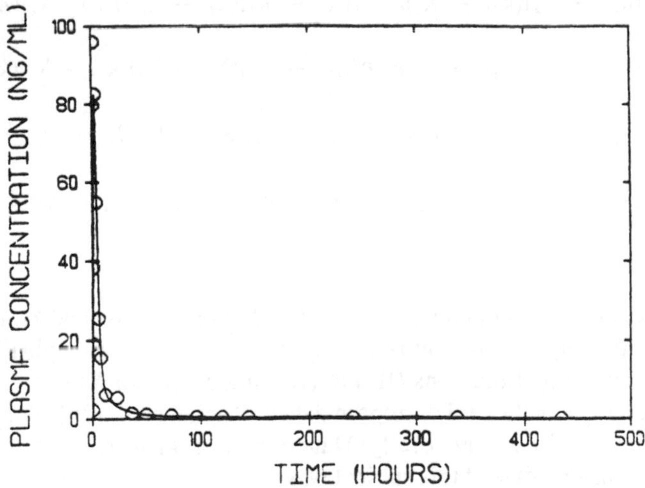

FIGURE 3.3. Reconstruction of flunarizine C,t data of subject MO3PS based on Equation (128) and the parameters derived as discussed in the text.

MODEL XXI. THREE-COMPARTMENT OPEN MODEL WITH CENTRAL COMPARTMENT ELIMINATION AND TWO CONSECUTIVE FIRST-ORDER INPUTS

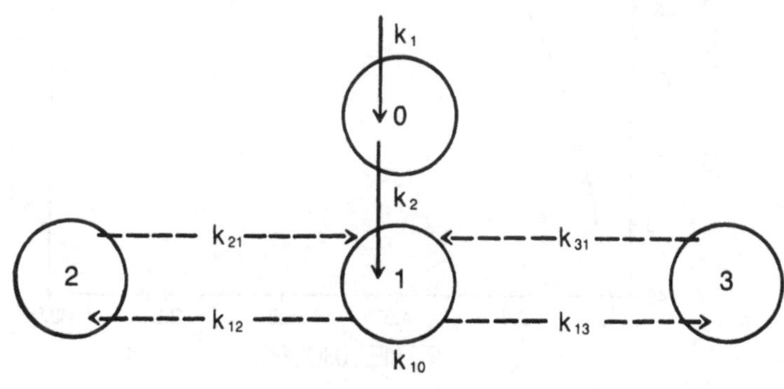

SCHEME 3.6.

Equation (131) gives the concentration in the central compartment #1 at time t.

$$
\begin{aligned}
C = k_1k_2C_o[&(k_{21} - \lambda_1)(k_{31} - \lambda_1)e^{-\lambda_1 t} \\
&/ (\lambda_2 - \lambda_1)(k_1 - \lambda_1)(k_2 - \lambda_1)(\lambda_3 - \lambda_1) \\
&+ (k_{21} - \lambda_2)(k_{31} - \lambda_2)e^{-\lambda_2 t}/(\lambda_1 - \lambda_2)(k_1 - \lambda_2)(k_2 - \lambda_2)(\lambda_3 - \lambda_2) \\
&+ (k_{21} - \lambda_3)(k_{31} - \lambda_3)e^{-\lambda_3 t}/(\lambda_1 - \lambda_3)(\lambda_2 - \lambda_3)(k_1 - \lambda_3)(k_2 - \lambda_3) \\
&+ (k_{21} - k_1)(k_{31} - k_1)e^{-k_1 t}/(\lambda_1 - k_1)(\lambda_2 - k_1)(\lambda_3 - k_1)(k_2 - k_1) \\
&+ (k_{21} - k_2)(k_{31} - k_2)e^{-k_2 t}/(\lambda_1 - k_2)(\lambda_2 - k_2)(\lambda_3 - k_2)(k_1 - k_2)]
\end{aligned} \tag{131}
$$

In Equation (131) k_1 and k_2 are input rate constants, k_{21} and k_{31} are microscopic (model) distribution rate constants, and the λ_is are hybrid rate constants defined by Equations (117) to (119) under Model XIX.

An example is the oral diazepam data of Kaplan et al. [25]. Subjects were given 10 mg of diazepam orally. Data of subject #1 were used as the example. Downslope data of this subject was fitted with a triexponential equation and disposition parameters of Model XX of Scheme 3.5 were derived as described in a later chapter. Parameters estimated were: $k_{10} = 0.0327\ \text{hr}^{-1}$, $k_{12} = 2.24\ \text{hr}^{-1}$, $k_{13} = 0.657\ \text{hr}^{-1}$, $k_{21} = 0.612\ \text{hr}^{-1}$, $k_{31} = 0.0481\ \text{hr}^{-1}$,

$C_o = 0.718\ \mu g/ml$, $\lambda_1 = 0.00173\ hr^{-1}$, $\lambda_2 = 0.162\ hr^{-1}$, and $\lambda_3 = 3.44\ hr^{-1}$.

$$FA = 1 - [(k_1 e^{-k_2 t} - k_2 e^{-k_1 t})/(k_1 - k_2)] \tag{132}$$

The Exact Loo-Riegelman method [Equation (129) under Model XX] was applied to the C,t data of this subject to give the points shown in Figure 3.4. The A_{max}/Vc value, equivalent to the C_o parameter of Equation (131), was 0.718 μg/ml. FA values were then calculated with Equation (56) (under Model X). These FA,t points were fitted with Equation (132), which is applicable to two consecutive first-order inputs, with rate constants k_1 and k_2. The estimated rate constants were $k_1 = 0.554\ hr^{-1}$, and $k_2 = 1.87\ hr^{-1}$. The above parameters, k_1, k_2, and C_o, were then used as initial values with Equation (131), keeping all other parameters as constants (values above), and a least squares fit performed with MINSQ [2]. The final estimated values were $k_1 = 0.875\ hr^{-1}$, $k_2 = 0.863\ hr^{-1}$, and $C_o = 0.700\ \mu g/ml$. The least squares fit of the data is shown in Figure 3.5.

This rather complicated procedure was necessary in order to get good initial and final estimates of the parameters. The steps taken to get Figure 3.5 were:

(1) Downslope data from two to forty-eight hours were fitted by the method of least squares to the triexponential Equation (120) shown under Model XIX.

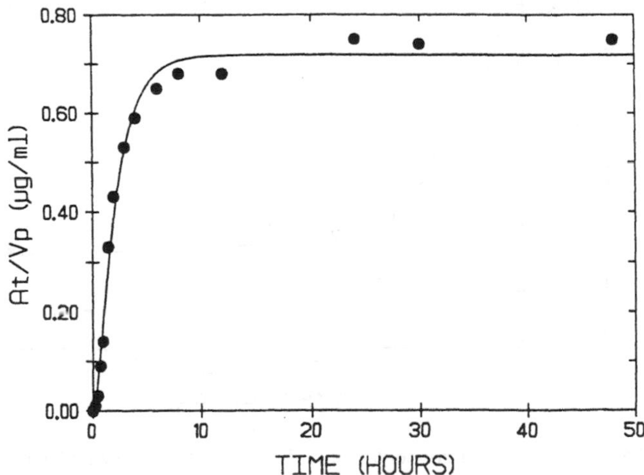

FIGURE 3.4. Example of Model XXI. Exact Loo-Riegelman plot [10,11] based on plasma diazepam concentrations measured in subject #1 given 10 mg of diazepam orally [25].

(2) The same data were then fitted by the method of least squares to an equation given in the next chapter with λ_1, λ_2, and λ_3 held constant and the parameters k_{10}, and sum $k_{21} + k_{31}$ and a time shift, t_s, estimated.

(3) The microscopic rate constants, k_{ij}s of Model XX, were calculated as described later.

(4) The Exact Loo-Riegelman method [Equation (129)] was then applied to all of the C, t data.

(5) A_{max} was calculated as the mean of the A_T/Vc values for those points used in the original triexponential fit.

(6) FA values were calculated with Equation (56) under Model X.

(7) The FA,t values were fitted by the method of least squares to Equation (132) in order to estimate k_1 and k_2.

(8) All the C, t data were then fitted by the method of least squares to Equation (132) and the result is shown in Figure 3.5.

One can see in Figure 3.5 that the theoretical line does not reach the observed peak concentrations. This is a common problem in pharmacokinetics, particularly with more complicated linear models as this Model XXI. This may sometimes be corrected by employing zero-order input rather than first-order input; but this is not always so. In this particular case, Figure 3.4 indicates that absorption was not zero order.

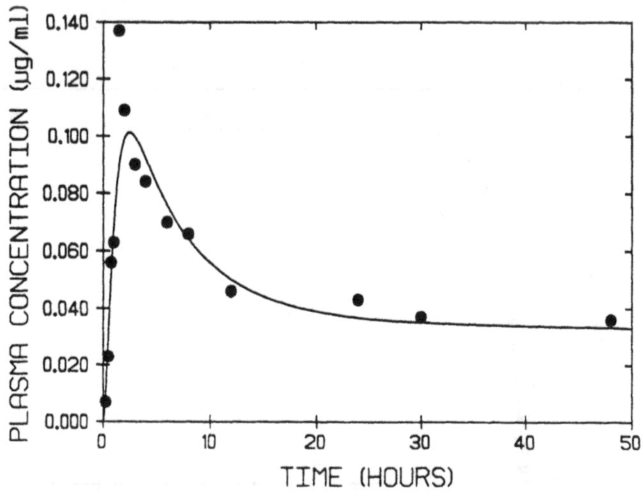

FIGURE 3.5. Fit of plasma concentrations of diazepam in subject #1 following oral administration of 10 mg of diazepam to the five-term polyexponential Equation (131) using the procedure described in the text.

MODEL XXII. THREE-COMPARTMENT OPEN MODEL WITH CENTRAL COMPARTMENT ELIMINATION AND ZERO-ORDER INPUT

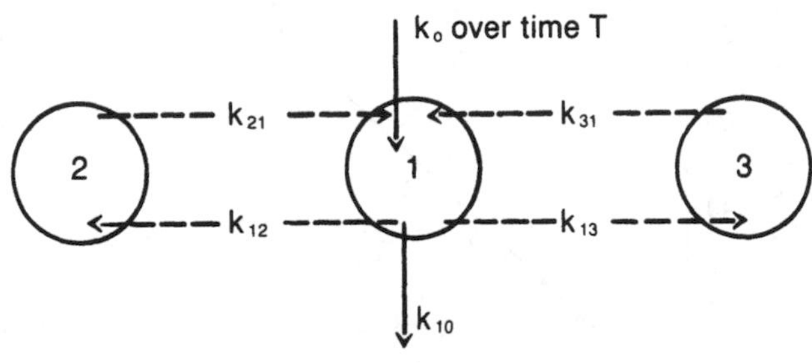

SCHEME 3.7.

The concentration, *C*, in the central compartment at time *t* during the zero-order input is given by Equation (133).

$$C = \frac{k_o}{Vc}[(k_{21} - \lambda_1)(k_{31} - \lambda_1)\{e^{+\lambda_1(t-t_o)} - 1\}e^{-\lambda_1(t-t_o)}/\lambda_1(\lambda_2 - \lambda_1)(\lambda_3 - \lambda_1)$$
$$+ (k_{21} - \lambda_2)(k_{31} - \lambda_2)\{e^{+\lambda_2(t-t_o)} - 1\}e^{-\lambda_2(t-t_o)}/\lambda_2(\lambda_1 - \lambda_2)(\lambda_3 - \lambda_2)$$
$$+ (k_{21} - \lambda_3)(k_{31} - \lambda_3)\{e^{+\lambda_3(t-t_o)} - 1\}e^{-\lambda_3(t-t_o)}/\lambda_3(\lambda_1 - \lambda_3)(\lambda_2 - \lambda_3)] \tag{133}$$

The concentration, *C*, in the central compartment at time *t* after the zero-order input has ceased at time *T* is given by Equation (134).

$$C = \frac{k_o}{Vc}[(k_{21} - \lambda_1)(k_{31} - \lambda_1)\{e^{+\lambda_1(T-t_o)} - 1\}e^{-\lambda_1(t-t_o)}/\lambda_1(\lambda_2 - \lambda_1)(\lambda_3 - \lambda_1)$$
$$+ (k_{21} - \lambda_2)(k_{31} - \lambda_2)\{e^{+\lambda_2(T-t_o)} - 1\}e^{-\lambda_2(t-t_o)}/\lambda_2(\lambda_1 - \lambda_2)(\lambda_3 - \lambda_2)$$
$$+ (k_{21} - \lambda_3)(k_{31} - \lambda_3)\{e^{+\lambda_3(T-t_o)} - 1\}e^{-\lambda_3(t-t_o)}/\lambda_3(\lambda_1 - \lambda_3)(\lambda_2 - \lambda_3)] \tag{134}$$

An example is subject MO2SL, one of the five epileptic patients given a single dose of 30 mg of flunarizine. This patient's *C*, *t* data were treated as described for subject MO3PS under Model XX above. In this case, however, the Exact Loo-Riegelman method [Equation (129) under Model XX] gave a linear plot (see Figure 3.6) rather than a first-order plot (see Figure

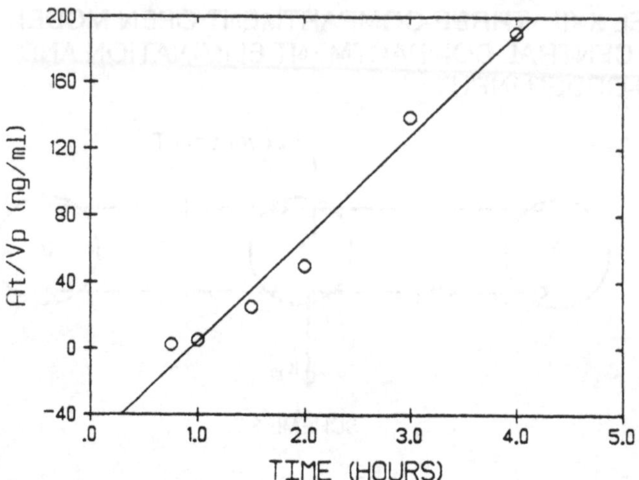

FIGURE 3.6. Zero-order absorption of flunarizine in subject MO2SL given a single oral dose of 30 mg of flunarizine as indicated by this Exact Loo-Riegelman plot [10,11].

3.2 under Model XX). The slope of the straight line in Figure 3.6 is $k_o/Vc = 61.6$ ng/(ml × hr).

Since the linear plot in Figure 3.6 indicated zero-order absorption in this patient, the estimated parameters were substituted into Equations (133) and (134) and a reconstruction made for all the C, t data. The result is shown in Figure 3.7 but only the first thirty-seven hours are plotted as a result of the great range of all the data.

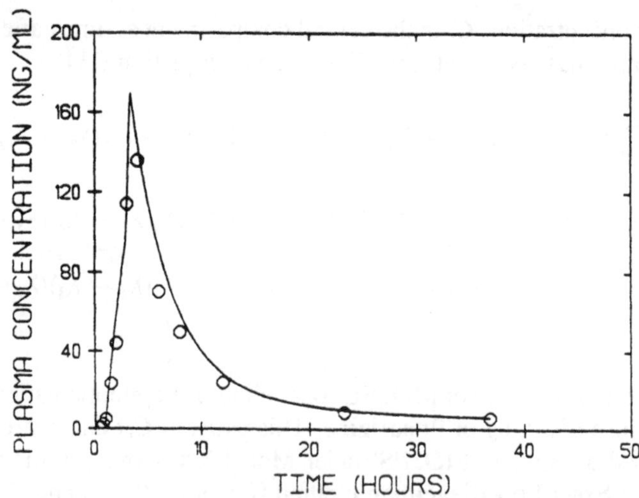

FIGURE 3.7. Reconstruction of the flunarizine plasma concentrations of subject MO2SL who exhibited zero-order absorption. Line is based on Equations (133) and (134).

MODEL XXIII. FIRST-PASS THREE-COMPARTMENT OPEN MODEL WITH BOLUS INTRAVENOUS ADMINISTRATION

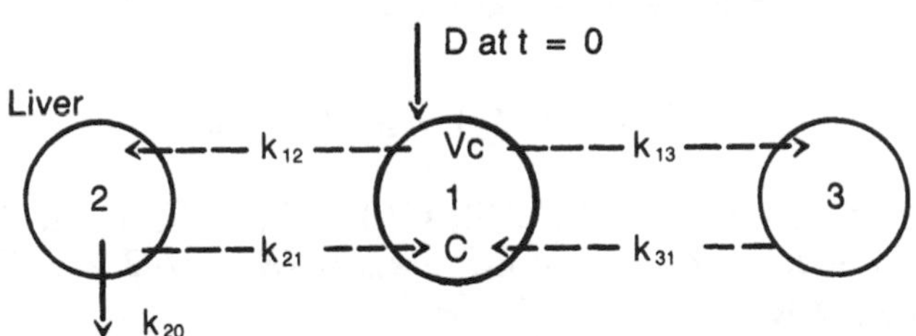

SCHEME 3.8.

In Model XIX, the drug was eliminated only from the central compartment #1. In Model XXIII, the drug is eliminated only from compartment #2. One assumes that the liver is part or all of compartment #2. When the drug is administered intravenously, as in this model, the drug is put into compartment #1 at time zero. If the drug were given orally, as in Model XXIV, the drug is put into compartment #2. Hence, this is a first-pass model, since when the drug is given intravenously it must go from compartment #1 to compartment #2 before it can get out of the body, hence part of the dose is protected. But when the drug is given orally, as in Model XXIII, all of the drug is exposed to eliminating mechanisms.

The concentration, C, in the central compartment #1 at time t is given by Equation (135) where $E_1 = k_{12} + k_{13}$, $E_2 = k_{20} + k_{21}$, and $E_3 = k_{31}$.

$$C = \frac{D}{Vc}[(E_2 - \lambda_1)(E_3 - \lambda_1)\exp(-\lambda_1 t)/\{(\lambda_2 - \lambda_1)(\lambda_3 - \lambda_1)\}$$
$$+ (E_2 - \lambda_2)(E_3 - \lambda_2)\exp(-\lambda_2 t)/\{(\lambda_1 - \lambda_2)(\lambda_3 - \lambda_2)\}$$
$$+ (E_2 - \lambda_3)(E_3 - \lambda_3)\exp(-\lambda_3 t)/\{(\lambda_2 - \lambda_3)(\lambda_1 - \lambda_3)\}] \tag{135}$$

In Equation (135),

$$G_1 = \lambda_1 + \lambda_2 + \lambda_3 = E_1 + E_2 + E_3 \tag{136}$$

$$G_2 = k_{12}(k_{20} + k_{31}) + k_{20}(k_{13} + k_{31}) + k_{21}(k_{13} + k_{31}) \tag{137}$$

$$G_3 = k_{12}k_{20}k_{31} \tag{138}$$

Estimation of Model Parameters

To estimate model parameters, first fit data to the triexponential Equation (139).

$$C = C_1e^{-\lambda_1 t} + C_2e^{-\lambda_2 t} + C_3e^{-\lambda_3 t} \tag{139}$$

Let $X = C_1/C(0)$, $Y = C_2/C(0)$, $C(0) = C_1 + C_2 + C_3$, $K_1 = k_{12} + k_{13} = G_1 - K_2 - K_{31}$, $K_2 = k_{21} + k_{20} = (R + Q)/2$ and $K_3 = k_{31} = (R - Q)/2$ where

$$R = K_2 + K_3 = X(\lambda_3 - \lambda_1) + Y(\lambda_3 - \lambda_2) + \lambda_1 + \lambda_2 \tag{140}$$

$$Q = K_2K_3 = [\{X(\lambda_3 - \lambda_1) + Y(\lambda_3 - \lambda_2) + \lambda_1 - \lambda_2\}^2 + 4Y(\lambda_2 - \lambda_1)(\lambda_3 - \lambda_2)]^{1/2} \tag{141}$$

Then,

$$k_{13} = \{G_3 - K_3(G_2 - K_1K_3 - K_2K_3)\}/\{K_3(K_3 - K_2)\} \tag{142}$$

$$k_{12} = K_1 - k_{13} \tag{143}$$

$$k_{21} = \{K_1K_2 + K_2K_3 + K_3K_1 - G_2 - k_{13}K_3\}/k_{12} \tag{144}$$

$$k_{20} = K_2 - k_{21} \tag{145}$$

Nagashima et al. [26] published the above equations and an example is taken from their article on bishydroxycoumarin. A single dose of 150 mg of bishydroxycoumarin as the disodium salt in aqueous solution (5 mg/ml) was injected intravenously over a five-minute period and the drug was measured in plasma by a sensitive and specific procedure. Zero time was taken as the time when the injection was completed. They fitted the data with the triexponential Equation (146).

$$C = 14.5e^{-0.085t} + 10e^{-0.46t} + 22e^{-3.5t} \tag{146}$$

Matching coefficients in Equations (139) and (146) gives $C_1 = 14.5$, $C_2 = 10$, and $C_3 = 22$. Hence $C(0) = 14.5 + 10 + 22 = 46.5$, $X = 14.5/46.5 = 0.3118$, and $Y = 10/46.5 = 0.2151$. Equation (140) gives $R = 2.2637$ and Equation (141) gives $Q = 1.6692$. Thus, $K_2 = 1.966$ and $k_{31} = 0.297$. Equations (142) to (145) give $k_{13} = 0.22$ hr^{-1}, $k_{12} = 1.56$

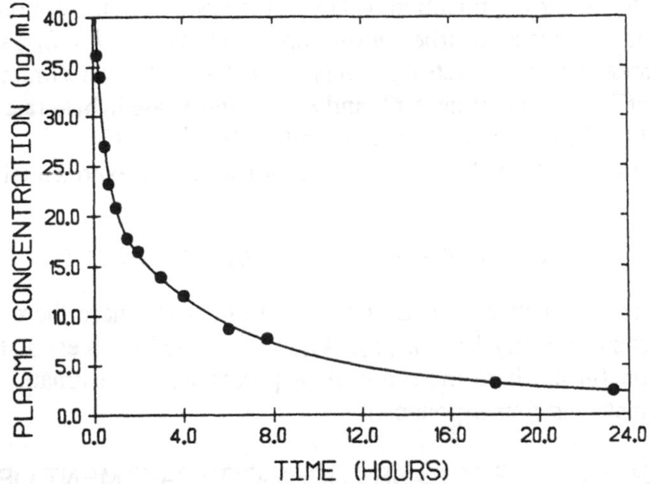

FIGURE 3.8. Example of Model XXIII. Plasma bishydroxycoumarin concentrations following a single 150 mg dose of bishydroxycoumarin injected I.V. over a five-minute period fitted to the triexponential Equation (146).

hr^{-1}, k_{21} = 1.67 hr^{-1}, and k_{20} = 0.295 hr^{-1}. Figure 3.8 shows the fit of the bishydroxycoumarin data with Equation (146).

MODEL XXIV. FIRST-PASS THREE-COMPARTMENT OPEN MODEL WITH FIRST-ORDER ABSORPTION

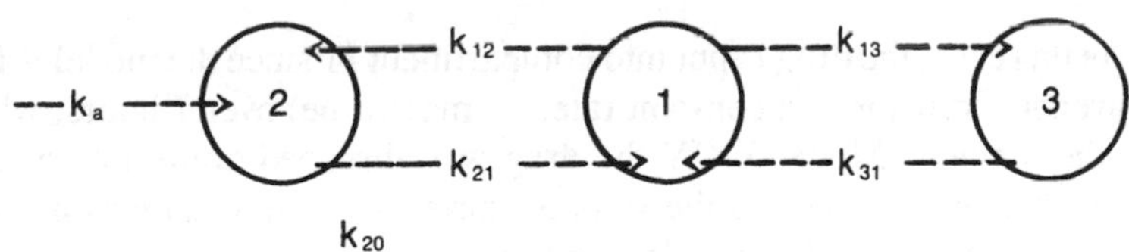

SCHEME 3.9.

The concentration, C, in the central compartment #1 at time t is given by Equation (148).

$$C = \frac{FDk_a k_{21}}{Vc}[(k_{31} - \lambda_1)e^{-\lambda_1 t}/\{(\lambda_2 - \lambda_1)(\lambda_3 - \lambda_1)(k_a - \lambda_1)\}$$

$$+ (k_{31} - \lambda_2)e^{-\lambda_2 t}/\{(\lambda_1 - \lambda_2)(\lambda_3 - \lambda_1)(k_a - \lambda_2)\}$$

$$+ (k_{31} - \lambda_3)e^{-\lambda_3 t}/\{(\lambda_1 - \lambda_3)(\lambda_2 - \lambda_3)(k_a - \lambda_3)\}$$

$$+ (k_{31} - k_a)e^{-k_a t}/\{(\lambda_1 - k_a)(\lambda_2 - k_a)(\lambda_3 - k_a)\}] \qquad (147)$$

In Scheme 3.9 and Equation (147) k_a is the first-order absorption rate constant and the drug is absorbed into compartment #2, k_{12}, k_{21}, k_{13}, and k_{32} are first-order distribution rate constants, k_{20} is the first-order elimination rate constant from compartment #2, and λ_1, λ_2, and λ_3 are hybrid rate constants defined by Equations (136) to (138) under Model XXIII.

Equation (147) may be written as the four-term polyexponential Equation (148).

$$C = B_1e^{-\lambda_1 t} + B_2e^{-\lambda_2 t} + B_3e^{-\lambda_3 t} + B_4e^{-k_a t} \tag{148}$$

However, it is better to fit data to Equation (147) where there are seven parameters—namely Vc/F, k_a, k_{21}, k_{31}, λ_1, λ_2, and λ_3—to estimate, than to Equation (148) where there are eight parameters to estimate (i.e., four coefficients and four exponents).

MODEL XXV. FIRST-PASS THREE-COMPARTMENT OPEN MODEL WITH CONSTANT RATE INFUSION

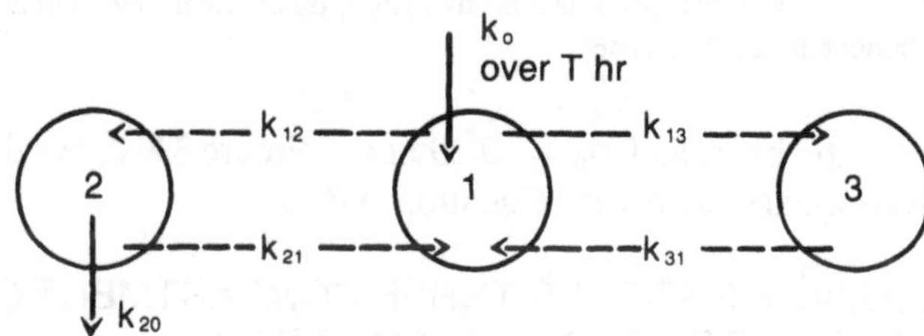

SCHEME 3.10.

In this case, the drug is put into compartment #1 since the model is for intravenous infusion at a constant rate, k_o (mass/time) over T hours, whereas in the previous Model XXIV the drug was absorbed orally into compartment #2. In Scheme 3.10 the symbols have the same meanings as before. The concentration, C, in the central compartment #1 at time t is given by Equation (149). During the infusion $b = t$ and after the infusion has ceased, $b = T$. In Equation (149), $E_2 = k_{20} + k_{21}$.

$$\begin{aligned} C = \frac{k_o}{Vc}[&(E_2 - \lambda_1)(k_{31} - \lambda_1)(1 - e^{+\lambda_1 b})e^{-\lambda_1 t}/\{-\lambda_1(\lambda_2 - \lambda_1)(\lambda_3 - \lambda_1)\} \\ &+ (E_2 - \lambda_2)(k_{31} - \lambda_2)(1 - e^{+\lambda_2 b})e^{-\lambda_2 t}/\{-\lambda_2(\lambda_1 - \lambda_2)(\lambda_3 - \lambda_2)\} \\ &+ (E_2 - \lambda_3)(k_{31} - \lambda_3)(1 - e^{+\lambda_3 b})e^{-\lambda_3 t}/\{-\lambda_3(\lambda_2 - \lambda_3)(\lambda_1 - \lambda_3)\}] \end{aligned} \tag{149}$$

C, t data may be fitted to Equation (149) and the parameters Vc, E_2, k_{31}, λ_1, λ_2, and λ_3 estimated.

CHAPTER 4

Linear Multicompartment Disposition Parameters from Downslope Extravascular *C, t* Data

Classically, when disposition is monoexponential, one estimates the first-order elimination rate constant from post-absorptive concentration-time data and then applies the Wagner-Nelson method [6] to data in the absorptive phase to obtain values of A_T/V (amount of drug available to time T divided by the volume of distribution) as a function of time. Frequently, but not always, these values can be fitted to a function in order to estimate one or more absorption rate constants. As a result of intrasubject variation of the elimination rate constant of many drugs, it is the author's opinion that this parameter should not be estimated from data derived from a different treatment than the data being analyzed for absorption. Thus, one goes to downslope data to estimate the elimination rate constant and then uses this value in applying the Wagner-Nelson method [6] to the data at the front end of the curve to estimate absorption.

Classically, disposition parameters of multicompartment models have been determined by analysis of C, t data following either bolus intravenous injection or constant rate intravenous infusion. Classically, to obtain absorption parameters from extravascular C, t data it has been assumed that the disposition parameters remain constant (i.e., there is no intrasubject variation in disposition) and the same disposition parameter values used in various types of deconvolution equations [10,27,28], such as the Exact Loo-Riegelman equations [10], the Loo-Riegelman method [27], or numerical deconvolution (Reference [28] for example).

There are two major problems with the above approach:

(1) There *is* intrasubject variation in multicompartment disposition parameters (see Tables 4.1 to 4.6) and it should not be ignored.

(2) About 60% of the drugs on the U.S. market do not have a commercially available intravenous formulation.

Hence, there is a great need for methods to derive multicompartment disposition parameters from terminal extravascular data so that these parameters may be used in deconvolution equations [10,27,28] applied to all of the C,t data and absorption parameters estimated from data between zero time and some time after the peak concentration has been reached. Thus, a set of extravascular C,t data can be analyzed sequentially to obtain the complete compartment model, which explains the data.

INTRASUBJECT VARIATION OF DISPOSITION

Table 4.1 indicates a suitable way of estimating intra- and intersubject coefficients of variation. The table is arranged with the subjects in rows and the replicates of some measurement (in this case oral clearance of verapamil) in columns. In this case there were four replicates corresponding to days 2, 8, 16, and 21. The intrasubject coefficient of variation in percent (C.V. = 100 × S.D./subject mean) was calculated from the four clearances in each row. On the far right of the table the lowest, highest, and median C.V.s are indicated. In the sixth column the subject means are shown and the grand mean and intersubject C.V. were calculated from these subject means and are shown at the bottom of this sixth column. The relative magnitude of intra- and intersubject C.V.s is indicated by comparing median intrasubject C.V. with the intersubject C.V. The extremes of intrasubject C.V.s are indicated by the lowest and highest values. References for these verapamil clearances and data from which values in Tables 4.2 and 4.3

TABLE 4.1. Example of Method of Calculation of Intra- and Intersubject Coefficients of Variation.

	Oral Clearance of Verapamil (L/min)						
Subject	Day 2	Day 8	Day 16	Day 21	Subject Mean	Intrasubject C.V. (%)	
1	2.53	2.42	3.01	2.31	2.57	12.0	Lowest
2	5.22	4.06	4.69	10.43	6.10	48.0	Highest
3	3.71	4.51	2.72	3.15	3.52	22.0	
4	2.09	2.38	2.57	1.74	2.20	16.5	Median
5	4.17	3.76	3.14	3.18	3.56	13.9	
		Grand mean			3.59		
		Intersubject C.V. (%)			42.4		

TABLE 4.2. Comparison of Intra- and Intersubject Coefficients of Variation of Oral Clearances of Several Drugs.

Drug	Units	No. Subjects	No. Doses	Grand Mean	Inter-Subject C.V. (%)	Intrasubject C.V. (%)		
						Lowest	Median	Highest
Furosemide	L/min	8	2*	0.139	16.5	0.8	4.17	13.0
Flurbiprofen	L/hr	15	4	1.31	22.5	2.37	6.55	14.6
Verapamil	L/min	5	4	3.59	42.4	12.0	16.5	48.0
Acetaminophen	L/min	10	2	0.479	23.6	1.29	9.12	20.1
Phenacetin	L/min	7	5	8.74	119.	34.7	50.1	109.
Prednisone	L/hr	12	2	17.9	19.5	1.50	5.25	32.9
Flurbiprofen	L/hr	12	3	1.41	20.0	0.35	5.58	11.85
Cimetidine	L/min	24	2	0.973	23.9	0.62	10.0	46.4
Carbamazepine	L/hr	6	3	1.23	19.7	1.25	13.6	18.0
Prednisolone	L/hr	12	4	8.05	8.5	2.9	7.80	21.5
Prednisolone	L/hr	12	4	8.16	13.6	13.1	22.1	34.3
Tolmetin	L/min	24	3	0.313	22.8	17.9	10.4	24.1
Penicillamine	L/min	4	4	0.847	30.5	11.3	27.1	64.7
Free Ibuprofen	L/min	15	4	10.7	15.4	7.4	10.2	19.4

*These were intravenous doses.

TABLE 4.3. Comparison of Intra- and Intersubject Coefficients of Variation of Elimination Rate Constants of Several Drugs.

Drug	No. Subjects	No. Doses	Grand Mean (1/hr)	Inter-Subject C.V. (%)	Intrasubject C.V. (%)		
					Lowest	Median	Highest
Bumetanide	12	3	0.847	19.7	2.8	25.1	76.3
Clindamycin	10	3	0.159	27.0	3.3	15.8	29.8
Lincomycin	6	6	0.136	10.7	18.9	22.5	49.2
Warfarin	4	5	0.0161	11.2	25.9	30.8	37.9
Verapamil	5	4	0.172	12.2	6.3	15.4	19.0
Phenacetin	7	5	0.116	119.	24.1	86.4	173.
Phenylbutazone	7	5	0.205	35.6	21.3	71.8	110.
p-Aminosal. acid	8	3	1.25	30.2	19.9	45.0	81.1
Phenytoin	8	4	0.0531	19.3	11.1	35.8	59.0
Acetaminophen	10	2	0.210	23.7	0.2	8.81	25.8
Prednisone	9	3	0.297	8.11	1.4	27.9	43.5
Prednisone	12	4	0.285	10.1	5.0	11.2	26.4
Prednisone	12	4	0.202	11.8	5.4	16.2	33.0
Prednisolone	12	4	0.270	9.6	0.5	10.8	23.6
Prednisolone	12	4	0.313	8.8	4.8	13.8	25.4
Tolmetin	24	3	0.321	23.0	4.4	24.4	71.3
Penicillamine	4	4	0.333	10.4	19.2	19.6	33.6
Ibuprofen	15	4	0.310	10.5	8.8	17.9	32.1
Flurbiprofen	15	4	0.0966	14.8	3.4	11.8	19.2
Flurbiprofen	12	3	0.106	15.1	2.75	8.45	21.3
Theophylline	7	5	0.0891	18.8	5.1	20.5	50.5
Theophylline	12	6	0.0840	16.0	3.5	9.99	21.4

were estimated were either given by Wagner et al. [13] or calculated from studies carried out by the author of this book. These were crossover studies in which statistical analysis indicated there were no significant differences in treatment averages; hence, statistically the results were the same as if only one treatment had been administered more than one time.

Table 4.2 lists a comparison of oral clearances of several drugs, excluding the first row which was estimated from replicate intravenous clearances of furosemide. The relative magnitudes of intra- and intersubject C.V.s depends upon the drug. The ratio of median intrasubject C.V. to intersubject C.V. for the fourteen rows of Table 4.2 are: 0.25, 0.29, 0.39, 0.39, 0.42, 0.27, 0.28, 0.42, 0.69, 0.92, 1.63, 0.46, 0.89, and 0.66. Hence, obviously, there is considerable intrasubject variation of oral clearance.

Table 4.3 summarizes similar data where the disposition parameter is the elimination rate constant obtained from terminal C, t data for several drugs. Here the median intrasubject C.V. is often of similar magnitude or larger than the intersubject C.V.

The author tried to obtain additional disposition data after intravenous administration. There are published articles that refer to studies where subjects were dosed more than once intravenously. However, such articles usually report only mean parameter values and not individual subject values. The author wrote to several researchers requesting the individual subject data but no answers to the requests were obtained.

Lee [29] did an extensive study with cimetidine in the dog. Four adult male beagle dogs were each administered two single bolus intravenous doses of 10 mg/kg of cimetidine and plasma samples were obtained from thirteen to nineteen times after injection over a six-hour period. The plasma samples were analyzed for cimetidine by an HPLC procedure. Each set was fitted by nonlinear least squares to a biexponential equation with $1/Y_i$ weighting using the program RSTRIP [2]. The intravenous clearance was estimated as dose/AUC, where AUC was obtained by integrating the biexponential equation from zero to infinity. The disposition parameters k_{12}, k_{21}, and k_{10} of Model VIII were also obtained with Equations (26) through (29). These four disposition parameters were then analyzed by the method shown in Table 4.1 and results are shown in Table 4.4. The individual clearances are listed in Table 4.5 along with the analysis of intra- and intersubject C.V.s. Note that in Table 4.4 that median intrasubject C.V. of 9.94% for CL is essentially the same as the intersubject C.V. of 10.4%. For k_{10} and k_{12} the median intrasubject C.V.s are 16.6% and 15.7%, respectively, compared with the intersubject C.V.s of 31.9% and 50.8%, respectively. However, for k_{21} the median intrasubject C.V. of 55.2% is much larger than the intersubject C.V. of 13.4%.

*TABLE 4.4. Disposition of Cimetidine after Intravenous Injection.**

Disposition Parameter	Grand Mean	Intersubject C.V. (%)	Intrasubject C.V. (%)		
			Lowest	Highest	Median
CL (L/kg/hr)	0.628	10.4	2.85	47.8	9.94
Vc (L/kg)**	0.612	27.9	9.13	20.6	13.0
k_{10} (1/hr)**	1.090	31.9	4.09	30.1	16.6
k_{12} (1/hr)**	3.185	50.8	3.04	73.5	15.7
k_{21} (1/hr)**	2.270	13.4	2.51	55.2	37.7

*There were four dogs with two studies in each of three dogs and three studies in one dog.
**Parameters of Model VIII based on biexponential fits of plasma concentration-time data.

METHODS OF ESTIMATING DISPOSITION PARAMETERS WHEN DISPOSITION IS BIEXPONENTIAL

The numbering of the methods for estimating disposition parameters is the same as that of Wagner et al. [14].

Method I

This is the classical method where Model VIII of Chapter 2 is assumed and Equations (26) through (29) are utilized to estimate the disposition parameters Vc, k_{12}, k_{21}, and k_{10} of the two-compartment open model with central compartment elimination from the two coefficients and two exponents of the biexponential equation. Note that the parameters of this model (and not the Rowland Model IX) must be used in applying the Exact Loo-Riegelman method [10,11].

TABLE 4.5. Intravenous Clearances of Cimetidine in the Dog.

Dog	CL (L/kg/hr)			Mean	Intrasubject C.V. (%)	
	Trial 1	Trial 2	Trial 3			
1	0.632	0.607	–	0.620	2.85	Low
2	0.509	0.608	–	0.559	12.52	
3	0.723	1.047	0.377	0.716	47.80	High
4	0.583	0.647	–	0.615	7.36	
	Mean			0.628	Intrasubject	
	Intersubject C.V. (%)			10.38	median = 9.94%	

Method II

This is the MRT method using Model XII. It may be used with intravenous infusion or extravascular data if it is known that input is zero order and the time during which zero-order input occurs is known. The method is clearly explained by Equations (78) to (86) and was illustrated by intestinal infusion of labetalol under Model XII. This method is independent of input rate.

Method IIIA

Equation (85) may be written as Equation (150).

$$C_o = \frac{FD}{Vc} = k_{10}(\text{AUC}) \tag{150}$$

If you replace D/Vc of the bolus I.V. Equation (20) with the far right-hand side of Equation (150) then use the identity $k_{21} = \lambda_1\lambda_2/k_{10}$ and add a time shift, t_s, you convert the I.V. equation to the extravascular Equation (151). This is explained in Figure 4.1.

$$C = \text{AUC}[\lambda_1(\lambda_2 - k_{10})e^{-\lambda_1(t-t_s)} + \lambda_2(k_{10} - \lambda_1)e^{-\lambda_2(t-t_s)}] \tag{151}$$

To apply Method IIIA one fits post-absorptive extravascular data (i.e., downslope data post-peak) to the biexponential Equation (152).

$$C = B_1e^{-\lambda_1 t} + B_2e^{-\lambda_2 t} \tag{152}$$

Then using the λ_1 and λ_2 estimated in this fitting as constants one fits the same post-absorptive data to Equation (151) and estimates k_{10} and t_s. Then one uses Equations (28) and (29) to estimate k_{21} and k_{12}.

There are no "rules" as to choice of first data point to use of the downslope data, but the second or third point past the peak is a good start.

AUC in Equation (151) is estimated with Equation (153).

$$\text{AUC} = \text{Trapezoidal AUC } 0\text{–}T + C_T/\lambda_1 \tag{153}$$

where T is the last observed time of observation and C_T is the estimated concentration at that time, which may be obtained from the biexponential Equation (152).

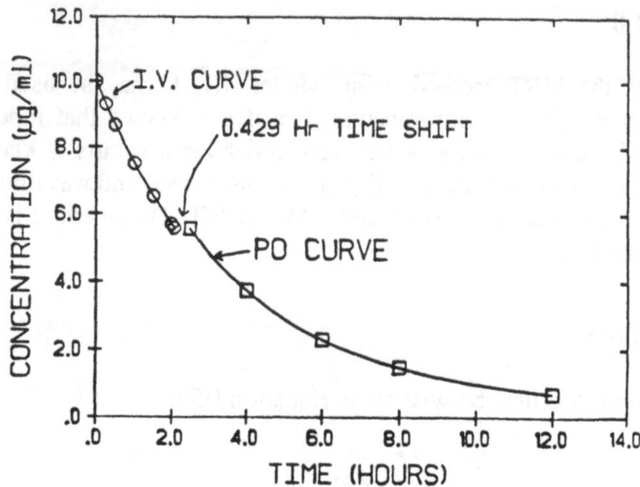

FIGURE 4.1. Explanation of Method IIIA. I.V. data were simulated with Equation (20). The po data was simulated using Equation (45) under Model X. The parameters were λ_1 = 0.109 hr^{-1}, λ_2 = 0.331 hr^{-1}, k_{10} = 0.240 hr^{-1}, k_{12} = 0.050 hr^{-1}, k_{21} = 0.150 hr^{-1}, k_{10} (AUC) = 10 µg/ml, and k_a = 2.50 hr^{-1}. Fitting post-absorptive C,t data from 2.5 to 32 hr to Equation (152) gave λ_1 = 0.109 hr^{-1} and λ_2 = 0.340 hr^{-1}. The same data set was then fitted to Equation (151), holding λ_1 and λ_2 constant, and estimating k_{10} = 0.242 hr^{-1} and t_s = 0.429 hr (from Reference [13], with permission from Plenum Press).

EXAMPLES OF APPLICATION OF METHOD IIIA

Lee [29] gave four adult female mongrel dogs (18.2–22.7 kg) with surgically implanted duodenal fistulas (15 cm distal to the pyloric sphincter) duodenal infusions of 20 mg/kg cimetidine delivered in buffered, isotonic solutions through the duodenal fistulas. Solutions were prepared at pH 4, 6, and 8 and targeted at 4.27 mg/ml, thus the 20 mg/kg doses ranged from 85 to 106 ml of solution. Solutions were delivered via a syringe pump (Model 22, Harvard Apparatus). The pump was interfaced with a microcomputer (PS/2 Model 55 IBM) programmed to change the pump delivery rate at five-second intervals to mimic first-order delivery with rate 0.17/min which was equivalent to the mean gastric emptying rate in previous 40 and 200 ml volume dog gastric emptying studies. The drug was delivered for about four half-lives or 16.3 min, with a final flow rate of about 1.1 ml/min. Blood was sampled on the same schedule as in the I.V. studies discussed above. Plasma was assayed by an HPLC method as in the I.V. studies.

Despite the fact that the dogs given the drug I.V. were different dogs than those given the drug by duodenal infusion, Lee chose to analyze the data by using the Exact Loo-Riegelman method [10,11] and using the I.V.-derived parameters to evaluate the duodenal infusion data. This author repeated this procedure for comparison purposes, but also analyzed the duodenal infusion data by Method IIIA and used the resulting parameters also in applying the Exact Loo-Riegelman method [10,11].

When the FA, t values were obtained by both methods described above it was found that each set was best fitted by the biexponential Equation (103) shown under Model XIV. For each data set a reconstruction was then made based on Equations (90) to (96) and the original C, t data.

Figure 4.2(a) shows results obtained in fitting the post-absorptive data to the biexponential Equation (152). The same data were then fitted to Equation (151) and the $k_{10} = 1.03$ hr^{-1} was estimated, then the k_{12} and k_{21} values shown in the second column of Table 4.6 were estimated. The Exact Loo-Riegelman method [10,11] was then applied and the FA, t values plotted in Figure 4.2(b) were obtained and fitted to Equation (103) to yield k_1 and k_2 values also shown in the second column of Table 4.6. The C_o value of 19.3 μg/ml was obtained as the A_{max}/Vc value calculated from the A_T/Vc values obtained in application of the Exact Loo-Riegelman method.

A reconstruction of the original C, t data was then made by substituting the parameters shown in the second column of Table 4.6 into Equations (90) to (94). This trend line is shown drawn through the original data in Figure 4.2(c).

Figure 4.3(a) shows the fit of one set of pH 6 dog cimetidine duodenal infusion FA,t data to Equation (103) and Figure 4.3(b) shows the reconstruction; the parameters used are shown in column 4 of Table 4.6. Figure 4.3(c) shows the reconstruction for the other pH 4 data set [see the first one in Figure 4.2(c)] and the parameters are shown in column 3 of Table 4.6.

Figure 4.4 shows one of the reasons that use of Method I (I.V. parameters) gives poor answers. Figure 4.4(a) shows the reconstruction of a pH 8 data set, based on Method IIIA parameters shown in column 7 of Table 4.6. Figure 4.4(b) shows the reconstruction based on the Method I (I.V.) parameters; obviously Method IIIA is preferred to Method I. The same data set was fitted by the nonlinear least squares method to Equations (90) to (94) using the program MINSQ (2) using both Method IIIA and Method I parameters as initial estimates. Results are shown in Figure 4.5. The fits are very comparable, but the least squares estimates of the parameters, which are compared in Table 4.7 are not. The k_{21} value of 5.14 hr^{-1} arising from the I.V. parameters is significantly out of line with values determined by Method IIIA (Table 4.6).

Method IIIB

This is the same as Method IIIA except that there is no initial fit to the biexponential Equation (153). Instead the post-absorptive C, t data is fitted

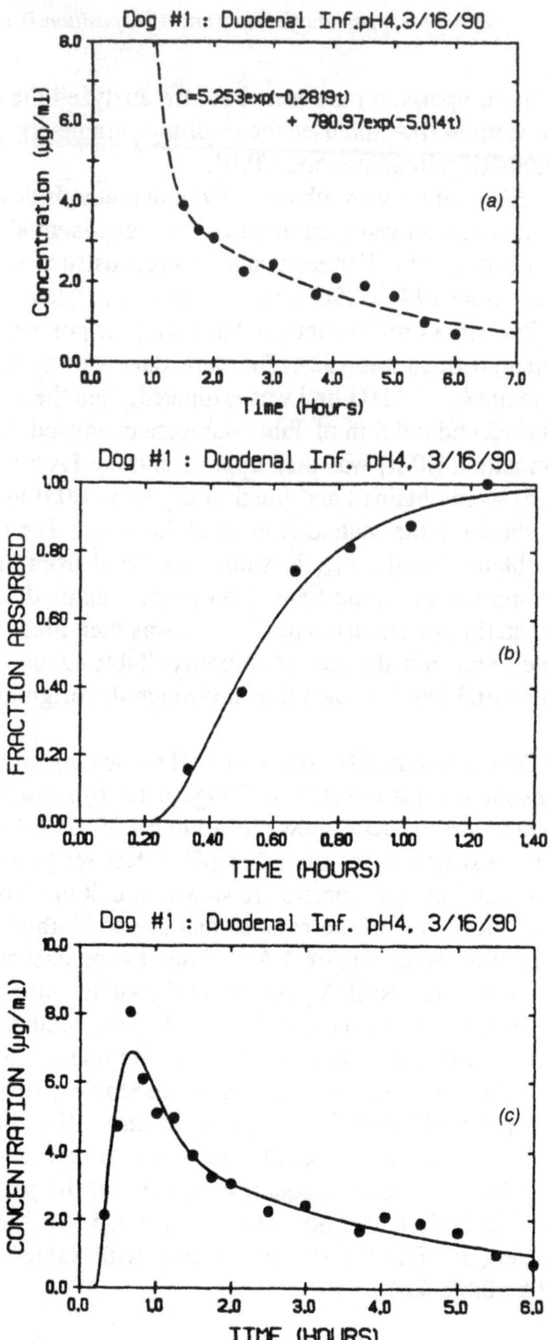

FIGURE 4.2. (a) Method IIIA applied to cimetidine plasma concentrations in dog #1 following a duodenal infusion of 20 mg/kg at pH 4 in study. (b) Fit of Exact Loo-Riegelman FA,*t* data to Equation (103) (see Model XIV). (c) Reconstruction of original cimetidine *C,t* data based on Equations (90) to (96) (see Model XIV) and the estimated parameters listed in Table 4.6 for study #1.

TABLE 4.6. Estimated Parameters of Cimetidine in Dog #1.

	Method IIIA after Duodenal Infusion						Method I: I.V.	
	pH 4		pH 6		pH 8			
Parameter	Study 1	Study 2	Study 1	Study 2	Study 1	Study 2	Study 1	Study 2
λ_1 (1/hr)	0.282	0.395	0.491	0.348	0.370*	0.239	0.321	0.348
λ_2 (1/hr)	5.01	3.16	6.18	2.11	–	0.528	2.56	6.34
k_{10} (1/hr)	1.03	0.538	1.83	1.12	–	0.421	0.713	0.842
k_{12} (1/hr)	2.89	0.697	3.19	0.681	–	0.0461	1.02	3.23
k_{21} (1/hr)	1.38	2.32	1.66	0.654	–	0.299	1.15	2.63
k_1 (1/hr)	4.34	7.05	2.11	2.53	2.88**	2.45	–	–
k_2 (1/hr)	6.43	7.05	3.21	1.55	31.4**	2.44	–	–
C_o (μg/ml)	19.3	17.3	32.7	16.5	12.8**	11.2	11.3	13.9

*This is K of the one-compartment open model.
**From fitting of Wagner-Nelson FA,t data.

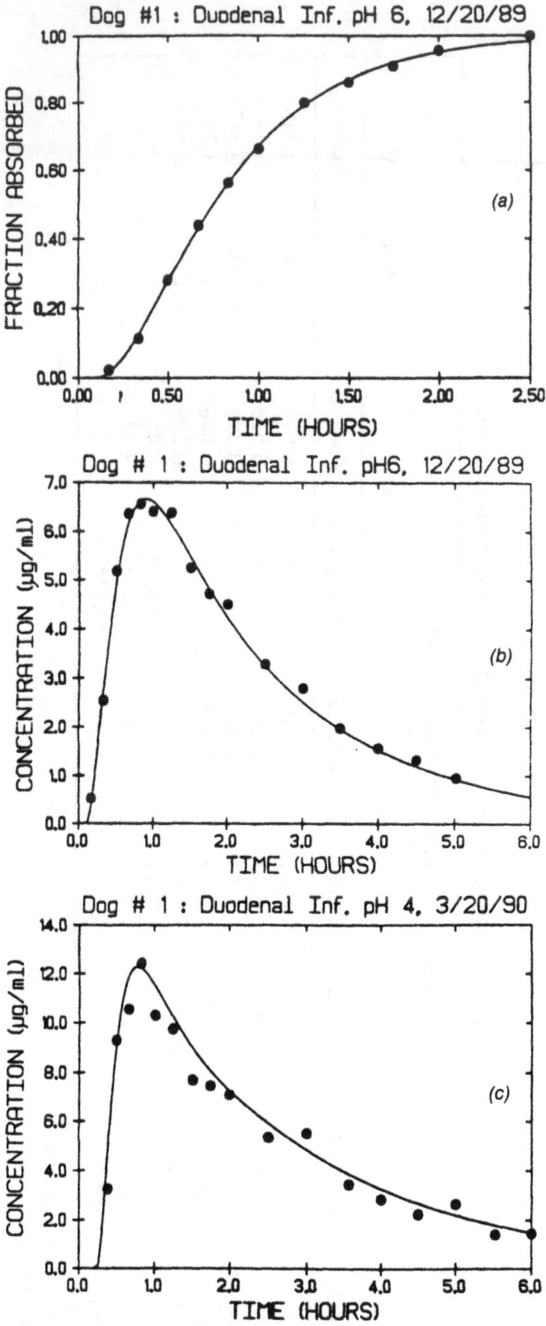

FIGURE 4.3. (a) Fit of Exact Loo-Riegelman FA,*t* data to Equation (103) (see Model XIV) derived from cimetidine plasma concentrations following a duodenal infusion of 20 mg/kg at pH 6 in dog #1. (b) Reconstruction of *C,t* data for dog #1 given a duodenal infusion of 20 mg/kg at pH 6 based on Equations (90) to (96) and the estimated parameters shown in Table 4.6. (c) Reconstruction of *C,t* data of dog #1 after duodenal infusion of 20 mg/kg at pH 4 in study #2.

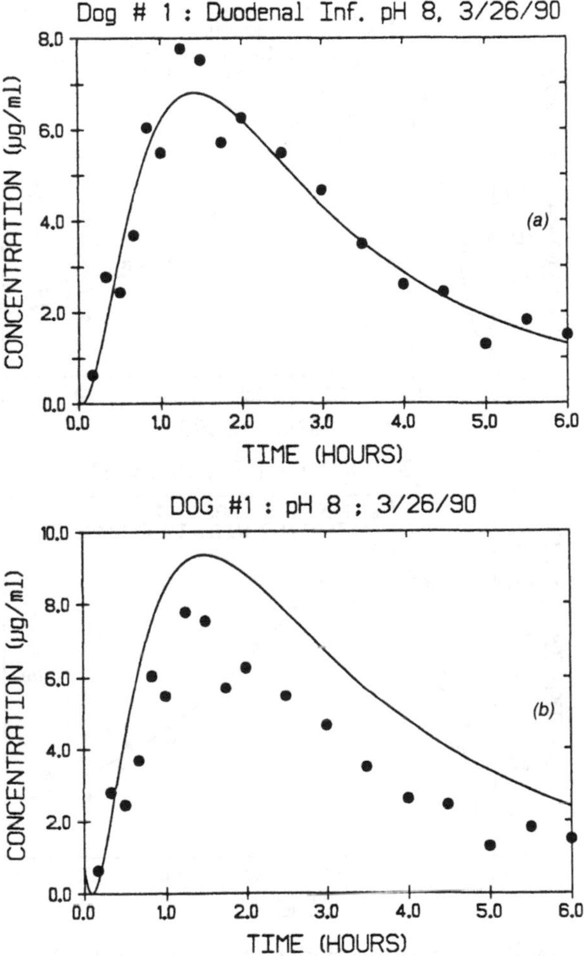

FIGURE 4.4. (a) Reconstruction of *C,t* data based on Method IIIA parameters in Table 4.6 for dog #1 given an intestinal infusion at pH 8. (b) Reconstruction of the same *C,t* data based on Method I (i.e., I.V. parameters shown in Table 4.4).

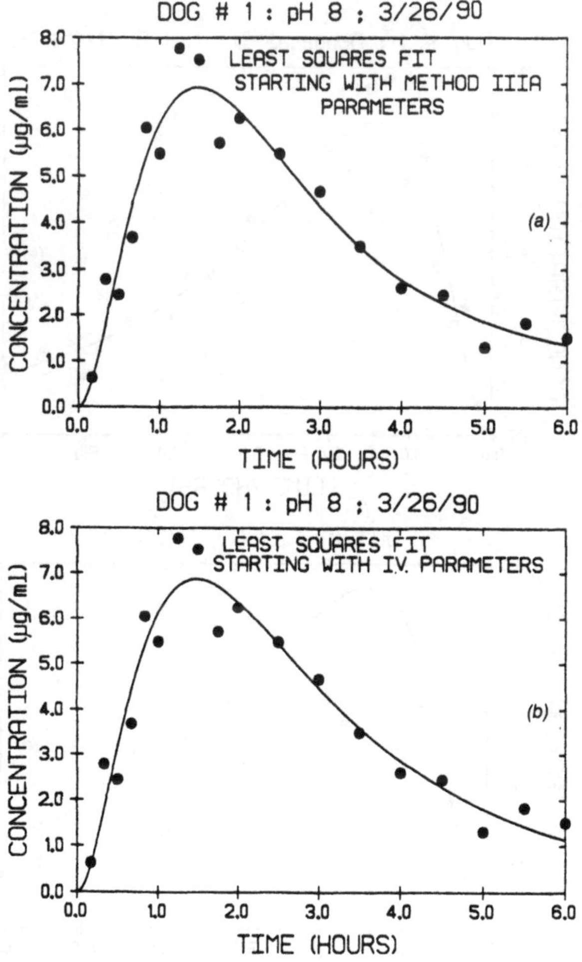

FIGURE 4.5. Least squares fits of cimetidine plasma concentrations in dog #1 after a pH 8 infusion. (a) Initial estimates were obtained by Method IIIA; (b) initial estimates were the I.V. parameters. The fits were very comparable but not the least squares parameter estimates; these are compared in Table 4.7.

TABLE 4.7. Estimated Parameters for Fits Shown in Figure 4.5.

Parameter	Method IIIA	Method I
k_1 (1/hr)	2.04	1.78
k_2 (1/hr)	1.93	3.68
k_{21} (1/hr)	0.0935	5.14
C_o (μg/ml)	13.7	8.26
t_o (hr)	0	0.0027
λ_1 (1/hr)	0.0422	0.473
λ_2 (1/hr)	0.686	3.60

by nonlinear least squares directly to Equation (152) and four parameters—namely λ_1, λ_2, k_{10}, and t_s—are estimated. Method IIIA works better than Method IIIB, but the former requires more work.

Wagner et al. [13] describe two other methods, designated Methods IV and V, but they are not improvements on Method IIIA and hence are not discussed here.

In his article on system analysis, Veng-Pedersen [30] stated: "Unfortunately, it is sometimes difficult and even impossible to reconstruct the basic drug level response from the parameters presented in pharmacokinetic studies." As indicated above, with cimetidine in dog #1, reconstructions were excellent; and results in the other three dogs studied by Lee [29], with data evaluation by the author, were also excellent. Previously, Wagner et al. [13] determined disposition parameters from fifty sets of extravascular concentration-time data and successfully made reconstructions using the estimated parameters.

Wagner et al. [13] also showed that Methods IIIA and IIIB are influenced by the rate of absorption. The two methods only give good parameter estimates when absorption is rapid, such as from immediate-release dosage forms, and are not effective for sustained-release dosage forms. This was illustrated by performing simulations with Model X and calculating the ratio shown as Equation (155) when the model equation is written as Equation (154).

$$C = B_1 e^{-\lambda_1 t} + B_2 e^{-\lambda_2 t} - B_3 e^{-k_a t} \tag{154}$$

where $B_3 = B_1 + B_2$.

$$\text{Fractional area} = \frac{B_3/k_a}{B_1/\lambda_1 + B_2/\lambda_2 - B_3/k_a} \tag{155}$$

Methods IIIA and IIIB work satisfactorily when the fractional area values are about 0.1.

METHODS OF ESTIMATING DISPOSITION PARAMETERS WHEN DISPOSITION IS TRIEXPONENTIAL

Method I

This is the classical method where Model XIX is assumed to apply. Data are fitted to the triexponential Equation (120), then the parameters Vc, k_{21}, k_{31}, k_{10}, k_{12}, and k_{13} are estimated with Equations (121) through (126).

Method II

A method based on MRT and triexponential disposition, analogous to Method II when disposition is biexponential, has not been satisfactorily worked out.

Method IIIA

Substituting k_{10} (AUC) for D/Vc in Equation (116) (under Model XIX) and letting $z = k_{21} + k_{31}$, followed by algebraic manipulation, gives Equation (156).

$$C = \mathrm{AUC}[(k_{10}(\lambda_1^2 - z\lambda_1) + \lambda_1\lambda_2\lambda_3)e^{-\lambda_1(t-t_s)}/\{(\lambda_2 - \lambda_1)(\lambda_3 - \lambda_1)\}$$

$$+ (k_{10}(\lambda_2^2 - z\lambda_2) + \lambda_1\lambda_2\lambda_3)e^{-\lambda_2(t-t_s)}/\{(\lambda_1 - \lambda_2)(\lambda_3 - \lambda_2)\}$$

$$+ (k_{10}(\lambda_3^2 - z\lambda_3) + \lambda_1\lambda_2\lambda_3)e^{-\lambda_3(t-t_s)}/\{(\lambda_1 - \lambda_3)(\lambda_2 - \lambda_3)\}] \quad (156)$$

Hence, like Equation (152), Equation (156) may be used to fit postabsorptive (downslope) extravascular data. Such data are initially fitted to Equation (157) and the three coefficients and three exponents are estimated.

$$C = B_1e^{-\lambda_1 t} + B_2e^{-\lambda_2 t} + B_3e^{-\lambda_3 t} \quad (157)$$

Then, holding the above values of λ_1, λ_2, and λ_3 as constants, the same postabsorptive data are fitted to Equation (156) and k_{10}, z, and t_s are estimated.

The other disposition parameters are then estimated with Equations (158) through (162).

$$k_{21}k_{31} = \lambda_1\lambda_2\lambda_3/k_{10} \quad (158)$$

$$k_{21} = 0.5[z + (z^2 - 4k_{21}k_{31})^{1/2}] \quad (159)$$

$$k_{31} = 0.5[z - (z^2 - 4k_{21}k_{31})^{1/2}] \quad (160)$$

$$k_{12} = [(\lambda_2\lambda_3 + \lambda_1\lambda_2 + \lambda_1\lambda_3) - k_{21}(\lambda_1 + \lambda_2 + \lambda_3)$$

$$- k_{10}k_{31} + k_{21}^2]/(k_{31} - k_{21}) \qquad (161)$$

$$k_{13} = \lambda_1 + \lambda_2 + \lambda_3 - (k_{10} + k_{12} + k_{12} + k_{31}) \qquad (162)$$

The above disposition parameters [Equations (158) to (162)] may then be substituted into the Exact Loo-Riegelman equation [10,11] appropriate for triexponential disposition to obtain A_T/Vc as a function of time; by averaging those A_T/Vc values that correspond to the time points used for downslope fitting to Equations (156) and (157), one obtains A_{max}/Vc. Then one obtains FA values using Equation (56) under Model X. The FA, t data may be fitted to the appropriate equation to estimate one or more absorption rate constants. The A_{max}/Vc value is equivalent to $C_o = FD/Vc$, and may be used in reconstruction or least squares data fitting.

EXAMPLES OF APPLICATION OF METHOD IIIA

The oral diazepam data of Kaplan et al. [25] was treated by Methods I and IIIA. The estimated disposition parameters were substituted into the appropriate Exact Loo-Riegelman equation [10,11] and the FA,t data fitted to Equation (132) (shown under Model XXI). Hence, Equation (131) would fit the complete C, t data set in each case, except for the fact that when Method I was applied to subject #4 I.V. data it was found to be fitted to a biexponential rather than a triexponential equation. Results are summarized in Table 4.8.

For diazepam subject #4, disposition was biexponential after I.V. administration but triexponential after po administration. Hence disposition was not constant in this subject. In the other three subjects, disposition was reasonably constant intrasubject and this represented the most constant of drugs studied by Wagner et al. [13]. Table 4.9 summarizes intra- and intersubject variation of diazepam disposition in subjects #1 to #3. In one case, the three values of λ_1, λ_2, and λ_3 (λ_i exponents)–a total of nine values–were analyzed, and in the other case the five values of k_{10}, k_{12}, k_{13}, k_{21}, and k_{31} (k_{ij} rate constants)–a total of fifteen values–were analyzed by the method outlined in Table 4.1. In Figure 4.6 the nine values of λ_1, λ_2, and λ_3 obtained from po data are correlated on an ln-ln plot with the nine values obtained from I.V. data. The least squares line shown in the figure has an intercept of -0.134 and a slope of 1.10 with $r = 0.989$. In Figure 4.7 the fifteen values the k_{ij}s obtained from po data are correlated on an ln-ln plot with the fifteen values obtained from I.V. data. The least squares line shown in the figure has an intercept of -0.071 and a slope of 0.873 with $r = 0.840$.

TABLE 4.8. Diazepam Parameters in Four Subjects by Methods I and IIIA.

	Subject #1		Subject #2		Subject #3		Subject #4	
Parameter	I	IIIA	I	IIIA	I	IIIA	I*	IIIA
k_{10} (1/hr)	0.0256	0.0327	0.128	0.134	0.0647	0.263	0.200	0.185
k_{12} (1/hr)	1.21	2.24	0.888	2.02	0.516	1.62	2.57	2.22
k_{13} (1/hr)	0.499	0.657	0.533	0.536	0.0839	0.275	–	0.272
k_{21} (1/hr)	1.31	0.612	1.84	0.879	1.07	0.374	1.46	0.761
k_{31} (1/hr)	0.0637	0.0481	0.173	0.0634	0.0466	0.0282	–	0.0643
k_1 (1/hr)	0.505	0.554	2.91	2.22	3.20	0.394	3.22	1.15
k_2 (1/hr)	3.81	1.87	2.87	2.31	3.18	24.9	3.24	4.66
A_{max}/Vc (μg/ml)	0.410	0.718	0.314	0.512	0.237	1.09	0.405	0.653
λ_1 (1/hr)	0.0025	0.0017	0.0252	0.0100	0.0154	0.0118	0.0703	0.0247
λ_2 (1/hr)	0.300	0.162	0.541	0.219	0.127	0.0958	4.17	0.114
λ_3 (1/hr)	2.81	0.344	3.00	3.40	1.64	2.45	–	3.23

*Biexponential and not a triexponential fit.

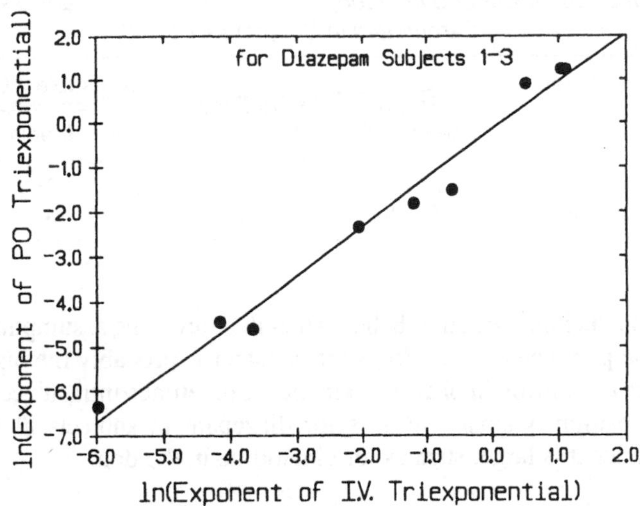

FIGURE 4.6. Correlation of the nine values of λ_1, λ_2, and λ_3 for diazepam from oral data with the corresponding nine values from I.V. data on an ln-ln plot. The slope of the line is 1.10 and r = 0.989. See text for details.

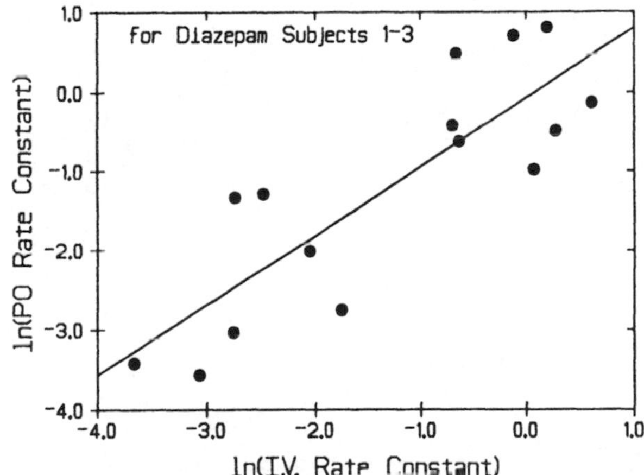

FIGURE 4.7. The fifteen values of k_{ij}s for diazepam obtained from oral data are correlated on an ln-ln plot with corresponding values obtained from I.V. data. The slope of the line is 0.873 and r = 0.840. See text for details.

TABLE 4.9. Intra- and Intersubject Variation of Diazepam Disposition Parameters in Subjects #1 to #3.

Data	Grand Mean (hr^{-1})	Intersubject C.V. (%)	Intrasubject C.V. (%)		
			Low	Median	High
3 λ_i exponents	1.01	136.	8.44	26.6	61.1
5 k_{ij} rate constants	0.608	95.3	0.40	42.2	85.6

In conclusion, the author believes that the prevailing assumption that disposition parameters of all drugs are constant is probably the biggest error that has been made in pharmacokinetics. Sometimes intrasubject variation in disposition is small, such as for diazepam in subjects #1 to #3, but sometimes it is large, such as for cimetidine in the dog.

CHAPTER 5

Noncompartmental and System Analysis

THE n-COMPARTMENT OPEN MAMMILLARY MODEL

The basic physical model for noncompartmental analysis is shown in Scheme 5.1 [31].

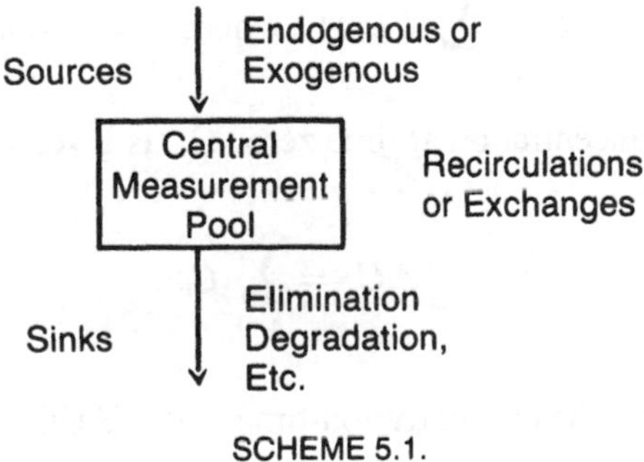

SCHEME 5.1.

The terms "model-independent" or "model free" are undesirable since a definite model, with a definite structure, is implied in noncompartmental analysis. This basic structure is shown in Scheme 5.1.

In many pharmacokinetic applications, Scheme 5.1 may be written as in Scheme 5.2, where there is a central compartment into which input occurs and from which elimination occurs, as well as $n - 1$ peripheral compartments; the drug that moves to a peripheral compartment from the central compartment must return to the central compartment to be eliminated from the "body" but the rate constants for these transfers are not specified. For linearity the elimination rate constant or parallel elimination rate constants from the central compartment #1 are first order.

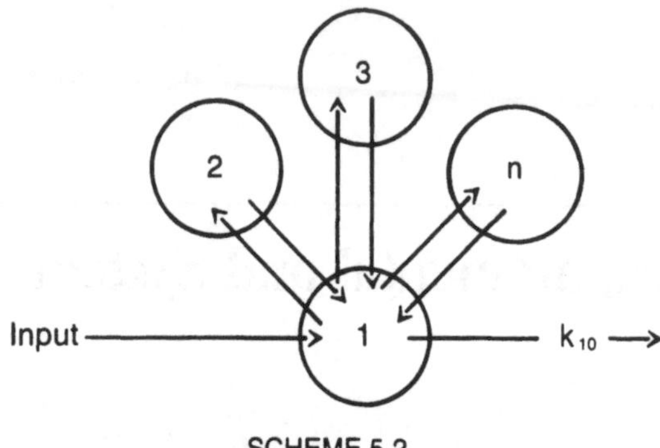

SCHEME 5.2.

The parameters defined below for the n-compartment open mammillary model are frequently used to describe the pharmacokinetics of a drug.

The parameters are derived from the bolus intravenous Equation (163), where C is the plasma (serum) or whole blood drug concentration, C_i is the coefficient, and λ_i is the exponent of the ith exponential term.

$$C = \sum_{n=1}^{n} C_i e^{-\lambda_i t} \quad \text{where } i = 1 \text{ to } n \tag{163}$$

The initial concentration at time zero, C_o, is given by Equation (164).

$$C_o = \sum_{n=1}^{n} C_i \tag{164}$$

The area under the concentration-time curve, AUC, is given by Equation (165).

$$\text{AUC} = \sum_{i=1}^{n} C_i/\lambda_i \tag{165}$$

The volume of the central compartment #1 is given by Equation (166), where D is the dose administered.

$$Vc = D/C_o \tag{166}$$

The clearance, CL, is given by Equation (167).

$$\text{CL} = D/\text{AUC} \tag{167}$$

The elimination half-life, $t_{1/2}$, is given by Equation (168), where λ_1 is the smallest exponent (i.e., $\lambda_1 < \lambda_2 < \lambda_n$) and is the apparent elimination rate constant corresponding to the very tail end of the concentration-time curve.

$$t_{1/2} = 0.693/\lambda_1 \tag{168}$$

The mean residence time, MRT, is the average time a drug molecule spends in the "body" and is given by Equation (169).

$$\text{MRT} = \left[\sum_{i=1}^{n} C_i/\lambda_i^2\right] \Big/ \text{AUC} \tag{169}$$

It should be noted that the MRT depends upon the site of input and the site of elimination [32], and Equation (169) assumes central compartment input and elimination.

The mean residence time in the systemic circulation or central compartment, MRTC, is given by Equation (170).

$$\text{MRTC} = \frac{\text{AUC}}{C_o} = \frac{1}{k_{10}} \tag{170}$$

The mean residence time in the peripheral compartment(s), MRTP, is the difference MRT − MRTC.

The volume of distribution steady state, V_{ss}, multiplied by the average steady-state drug concentration is the average amount of drug in the body if the model of Scheme 5.2 applies. The formula for V_{ss} is given as Equation (171).

$$V_{ss} = \text{CL MRT} = D\left[\sum_{i=1}^{n} C_i/\lambda_i^2\right] \Big/ \left[\sum_{i=1}^{n} C_i/\lambda_i\right]^2 \tag{171}$$

The volume of distribution area or beta, V_β, multiplied by the concentration, C, in the terminal log-linear phase, is the amount of drug in the body if the model of Scheme 5.2 applies. This volume is given by Equation (172).

$$V_\beta = \frac{\text{CL}}{\lambda_1} = \frac{D}{\lambda_1(\text{AUC})} \tag{172}$$

The extrapolated volume of distribution, V_{ext}, is the one-compartment volume and is given by Equation (173).

$$V_{ext} = \frac{D}{C_1} \tag{173}$$

Note that the order of magnitudes of the volumes is:

$$V_{ext} > V_{\beta} > V_{ss} > Vc$$

INFUSIONS

The equation corresponding to Equation (163) for the time during infusion is Equation (174), where the X_is are different than the C_is. By relating the coefficients of Equations (22) and (23) with those of Equation (74) or the coefficient of Equation (116) with that of Equation (134), we obtain the relationship shown in Equation (175).

$$C = \sum_{i=1}^{n} X_i(1 - e^{-\lambda_i t}) \tag{174}$$

$$C_i = \lambda_i T X_i \tag{175}$$

Hence, if *during* infusion, data are fitted to Equation (174), one can use Equation (175) to convert the X_is to C_is then use Equations (164) through (173) to estimate n-compartment mammillary model parameters. Here T = infusion time.

Analogously, if post-infusion data are fitted to Equation (176), one can use Equation (177) to convert the Y_is to C_is and then apply Equations (164) through (173).

$$C = \sum_{n=1}^{n} Y_i e^{-\lambda_i t} \tag{176}$$

$$C_i = \frac{\lambda_i T Y_i}{e^{+\lambda_i T} - 1} \tag{177}$$

When there is central compartment input and elimination, the same MRT as estimated by Equation (169) may also be estimated by Equation (178), where AUMC is the area under the moment curve.

$$\text{MRT} = \frac{\text{AUMC}}{\text{AUC}} = \int_0^T tCdt \bigg/ \int_0^T Cdt \tag{178}$$

Although many authors have termed AUMC the area under the first moment curve, Veng-Pedersen wrote [33]: "It is inappropriate to denote AUMC as the first statistical moment or to denote AUC as the 'zero-th' statistical moment, since $C(t)$ is not a pdf" (probability density function).

Example

A dose of 600 mg of a drug was administered by bolus intravenous injection to a woman weighing 50 kg. The plasma concentration-time data were computer-fitted to the equation:

$$C\ (\mu\text{g/ml}) = 10e^{-0.1t} + 20e^{-1t} \quad (t \text{ in hr}) \tag{179}$$

If $n = 2$ in Equation (163), it becomes Equation (180).

$$C = C_1 e^{-\lambda_1 t} + C_2 e^{-\lambda_2 t} \tag{180}$$

Since the smallest exponent in Equation (179) is 0.1 then $\lambda_1 = 0.1$; since the coefficient corresponding to 0.1 is 10 then $C_1 = 10$. Then $\lambda_2 = 1$ and $C_2 = 20$.

The equation being applied is shown on the left below and the particular parameter value derived from Equation (179) is shown beside it. Note that $D = 600/50 = 12$ mg/kg.

Equation	Parameter	
(164)	$C_o = 10 + 20 = 30\ \mu\text{g/ml}$	(181)
(165)	$\text{AUC} = \frac{10}{0.1} + \frac{20}{1} = 120\ (\mu\text{g} \times \text{hr})/\text{ml}$	(182)
(166)	$Vc = 600/30 = 20$ L or 0.4 L/kg	(183)
(167)	$\text{CL} = 600/120 = 5$ L/hr or 0.1 L/(kg × hr)	(184)
(168)	$t_{1/2} = 0.693/0.1 = 6.93$ hr	(185)
(169)	$\text{MRT} = \left[\frac{10}{(0.1)^2} + \frac{20}{(1)^2}\right] \Big/ (120) = 8.5$ hr	(186)
(170)	$\text{MRTC} = \frac{120}{30} = 4$ hr	(187)
(171)	$V_{ss} = 5 \times 8.5 = 42.5\ \text{L} = 0.85$ L/kg	(188)
(172)	$V_\beta = \frac{5}{0.1} = 5$ L or 1 L/kg	(189)
(173)	$V_{ext} = 600/10\ 60$ L or 1.2 L/kg	(190)

DIRECT METHOD OF CALCULATION

One may also calculate the n-compartment mammillary model parameters directly from the concentration-time data without fitting to a polyexponential equation as shown below. Equation (180) was used to calculate the time and concentration values shown in columns 1 and 2 (Table 5.1), and the product tC is shown in the third column.

First one obtains the terminal ln C *versus* t line. Using the data in the six- to twelve-hour range, one obtains Equation (191), where the λ_1 estimate is 0.1016 hr^{-1}.

$$\ln C = 2.3193 - 0.1016t \tag{191}$$

Since $e^{2.3193} = 10.17$, Equation (191) may be written as Equation (192).

$$C = 10.17e^{-0.1016t} \tag{192}$$

Since the last sampling time, T, is twelve hours, then

$$C_T = C_{12\,\text{hr}} = 10.17e^{-(0.1016)(12)} = 3.005\ \mu\text{g/ml}$$

AUC is given by Equation (193).

$$\text{AUC} = \int_0^T C dt + \frac{C_T}{\lambda_1} = 91.065 + \frac{3.004}{0.1016} = 120.6 \tag{193}$$

where the integral was estimated by trapezoidal rules. AUMC is given by

TABLE 5.1. Data for Direct Calculation Example.

t (hr)	C (μg/ml)	tC
0	30	0
0.25	25.3	6.325
0.5	21.6	10.8
0.75	18.7	14.025
1	16.4	16.4
1.5	13.1	19.65
2	10.9	21.8
4	7.07	28.28
6	5.54	33.24
8	4.50	36.00
10	3.68	36.8
12 = T	3.01	36.12

Equation (194) where the integral is estimated from the t and C,t values in the first and third columns of Table 5.1 by trapezoidal rules.

$$\text{AUMC} = \int_0^T tCdt + \frac{C_T T}{\lambda_1} + \frac{C_T}{\lambda_1^2}$$

$$= 355.7725 + \frac{(3.004)(12)}{0.1016} + \frac{3.004}{(0.1016)^2}$$

$$= 355.7725 + 354.803 + 291.013 = 1002 \qquad (194)$$

Hence

$$\text{MRT} = \frac{1002}{120.6} = 8.31 \text{ hr} \qquad (195)$$

$$\text{CL} = \frac{600}{120.6} = 4.975 \text{ L/hr} = 0.0995 \text{ L/(kg} \times \text{hr)} \qquad (196)$$

$$V_{ss} = (0.0995)(8.31) = 0.827 \text{ L/kg} \qquad (197)$$

$$V_\beta = \frac{0.0995}{0.1016} = 0.979 \text{ L/kg} \qquad (198)$$

$$V_{ext} = 600/10.17 = 59.0 \text{ L} = 1.18 \text{ L/kg} \qquad (199)$$

$$t_{1/2} = 0.693/0.1016 = 6.82 \text{ hr} \qquad (200)$$

Note that MRTC was not estimated since C_o was unknown.

EXTRAVASCULAR DATA

Equations (164) to (173) or (193) to (200) may also be used with extravascular data, but in this case the absorbed dose is *FD*, where *F* is the bioavailability, and since one uses the dose in the calculations, the parameters estimated are Vc/F, CL/F, V_{ss}/F, V_β/F, and V_{ext}/F. Note that $t_{1/2}$, MRT, and MRTC are independent of bioavailability. Note also that when using extravascular polyexponential equations with one or more negative coefficients, the signs must be obeyed in the calculations. For example, AUC from the equation:

$$C = 10e^{-0.1t} + 20e^{-1t} - 30e^{-4.5t} \text{ is } \frac{10}{0.1} + \frac{20}{1} - \frac{30}{4} = 112.5$$

MEAN INPUT TIME (MIT)

The MRT estimated from extravascular data or from I.V. infusion data involves a contribution for the mean transit time for input. This contribution is termed the mean input time, symbolized by MIT. Table 5.2 gives the value of MIT for various types of input.

MIT is the mean transit time for input and is given by Equation (201), where X is amount of drug going in and the integral is the total area under the amount *versus* time curve for input [34].

$$\text{MIT} = \int_0^{\text{Inf}} X dt/\text{Dose} \tag{201}$$

Hence, the MRT estimated from any data other than bolus I.V. data is given by Equation (202).

$$\text{MRT estimated from data other than bolus I.V. data} = \text{MRT} + \text{MIT} \tag{202}$$

where MRT is obtained from bolus I.V. data.

TABLE 5.2. MIT as a Function of Type of Input [34].

Mode of administration	MIT
Intravenous bolus dose	0
Intravenous infusion	$\tau/2$ (τ = input time)
First-order input	$1/k_a$
Bolus + infusion (Models IV and XVI)	$R_o\tau^2/(R_o\tau + D)$ (R_o = infusion rate)
Two consecutive infusions (Model XVII)	$\dfrac{(R_o\tau^2/2)_1 + (R_o\tau^2/2)_2}{(R_o\tau)_1 + (R_o\tau)_2}$

As indicated above under Equation (170):

$$\text{MRTP} = \text{MRT} - \text{MRTC} \tag{203}$$

where MRTP is the MRT of the peripheral compartment(s), MRT is the system MRT, and MRTC is the MRT of the central compartment. Veng-Pedersen and Gillespie [35] point out that:

$$Vc = \text{MRTC CL} \tag{204}$$

and

$$V_{ss} = \text{MRT CL} \tag{205}$$

hence

$$\frac{\text{MRT}}{\text{MRTC}} = \frac{V_{ss}}{Vc} \tag{206}$$

Wagner [36] also showed that

$$\frac{V_{ss}}{Vc} - 1 = \sum_{\substack{i=1 \\ j=1}}^{n} k_{ij}/k_{ji} \tag{207}$$

Thus when $n = 2$, the right-hand side of Equation (205) is equal to k_{12}/k_{21} and when $n = 3$ it becomes $k_{12}/k_{21} + k_{13}/k_{31}$.

Wagner [36] also showed that for an infusion to steady state the amount of drug in the body, $A_b = V_{ss}C_{ss}$, and is given by Equation (208). Thus, the ratio V_{ss}/V_β has a significance also.

$$V_{ss}C_{ss} = \left[\frac{V_{ss}}{V_\beta}\right]\left[\frac{R_o}{\lambda_1}\right] \tag{208}$$

SYSTEM ANALYSIS

System analysis mathematically describes a general property of a pharmacokinetic system without modelling in specific terms the kinetic processes responsible for the general property considered [37]. For example, the use of the superposition principle for drug level predictions is considered a system approach under the above definition. The following example of superposition is taken from Wagner [17c].

The superposition or overlaying principle provides a method of predicting multiple-dose drug concentrations in linear systems from drug concentrations after a single dose, and also provides a simple method for estimating a loading dose to allow rapid achievement of steady state. Single-dose data were generated with Equation (3) under Model II with $FD/V = 100$, $k_a = 1.0455\ \text{hr}^{-1}$, $K = 0.17425\ \text{hr}^{-1}$, and $t_o = 0$. The non-bracketed numbers under dose 1 of Table 5.3 were obtained by substituting those parameter values into Equation (3). This particular set of C, t data gave a linear semilogarithmic plot in the six- to twelve-hour time period. The least squares line is shown as Equation (209).

$$\ln C = 4.7793 - 0.1735t \tag{209}$$

This equation may also be written as Equation (210).

$$C = 119.0e^{-0.1735t} \tag{210}$$

Using Equation (210) the bracketed numbers in Table 5.3 were calculated. This simulates the usual extrapolation of single-dose C, t data beyond the last sampling time. In this example it is assumed that the same dose used to obtain the single-dose C, t data is administered every six hours. In Table 5.3 one copies the concentrations in column 3 into column 4 but starts dosing at six hours indicated by the 0 at six hours under dose 2. Then one copies the concentrations in column 3 into column 5 under dose 3, putting the 0 at twelve hours this time. In the last column of the table one adds all concentrations in each row; this is headed total $= Cn(t)$.

The table is made long enough to show that steady state is essentially reached when the pattern of Cns in each six-hour dosing interval is about the same. Figure 5.1 is a plot of $Cn(t)$ *versus* time.

In linear kinetics each dose acts independently, which provides for the superposition principle. One writes the contribution of each dose as in Table 5.3 and then adds the contributions to obtain $Cn(t)$, which is the predicted multiple dose concentration.

If one administers an appropriate loading dose—i.e., an initial dose higher than the maintenance doses—one can attain steady state rapidly. There are several methods to estimate the loading dose, as follows:

(1) $$\frac{\text{Loading dose}}{\text{Maintenance dose}} = \frac{\text{Peak concentration after sixth dose}}{\text{Peak concentration after first dose}}$$

$$= \frac{115.6}{69.86} = 1.65$$

TABLE 5.3. Superposition Principle for Prediction of Drug Concentrations after Multiple Doses in a Linear System.

Dose No.	Time (hrs)	Dose 1	Dose 2	Dose 3	Dose 4	Dose 5	Dose 6	Total $Cn(t)$
1	0	0						0
	0.5	38.84						38.84
	1	58.63						58.63
	2	69.86						69.86
	3	65.93						65.93
	4	57.94						57.94
	5	49.57						49.57
2	6	41.96	0					41.96
	6.5	[38.53]	38.84					77.37
	7	35.36	58.63					93.99
	8	29.74	69.86					99.60
	9	25.00	65.93					90.93
	10	21.01	57.94					78.95
	11	17.65	49.57					67.22
3	12	14.83	41.96	0				56.79
	12.5	[13.61]	38.53	38.84				90.98
	13	[12.48]	35.36	58.63				106.5
	14	[10.49]	29.74	69.86				110.1
	15	[8.82]	25.00	65.93				99.75
	16	[7.41]	21.01	57.94				86.36
	17	[6.23]	17.65	49.57				73.45
4	18	[5.24]	14.83	41.96	0			62.03
	18.5	[4.81]	13.61	38.53	38.84			95.79
	19	[4.41]	12.48	35.36	58.63			110.9
	20	[3.70]	10.49	29.74	69.86			113.8
	21	[3.11]	8.82	25.00	65.93			102.9
	22	[2.62]	7.41	21.01	57.94			88.98
	23	[2.20]	6.23	17.65	49.57			75.65
5	24	[1.85]	5.24	14.83	41.96	0		63.88
	24.5	[1.70]	4.81	13.61	38.53	38.84		97.49
	25	[1.56]	4.41	12.48	35.36	58.63		112.4
	26	[1.31]	3.70	10.49	29.74	69.86		115.1
	27	[1.10]	3.11	8.82	25.00	65.93		104.0
	28	[0.92]	2.62	7.41	21.01	57.94		89.90
	29	[0.78]	2.20	6.23	17.65	49.57		76.43
6	30	[0.65]	1.85	5.24	14.83	41.96	0	64.53
	30.5	[0.60]	1.70	4.81	13.61	38.53	38.84	98.09
	31	[0.55]	1.56	4.41	12.48	35.36	58.63	113.0
	32	[0.46]	1.31	3.70	10.49	29.74	69.86	115.6
	33	[0.39]	1.10	3.11	8.82	25.00	65.93	104.4
	34	[0.33]	0.92	2.62	7.41	21.01	57.94	90.23
	35	[0.27]	0.78	2.20	6.23	17.65	49.57	76.70
	36	[0.23]	0.65	1.85	5.24	14.83	41.96	64.76

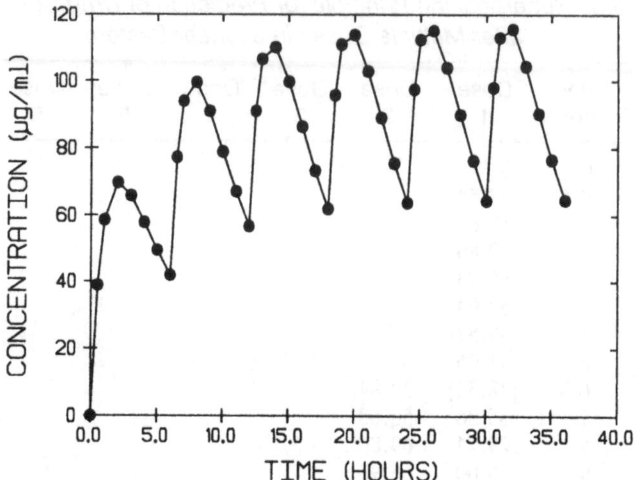

FIGURE 5.1. Plot of $Cn(t)$ from Table 5.3 *versus* time showing the approach to steady state.

$$(2)\ \frac{\text{Minimum steady-state concentration}}{\text{Minimum concentration after first dose}} = \frac{64.76}{41.96} = 1.54$$

$$(3)\ \frac{\text{AUC during dosage interval at steady state}}{\text{AUC during dosage interval after first dose}} = \frac{568.6}{412.1} = 1.38$$

Method (1) was used to produce Figure 5.2. The concentrations under

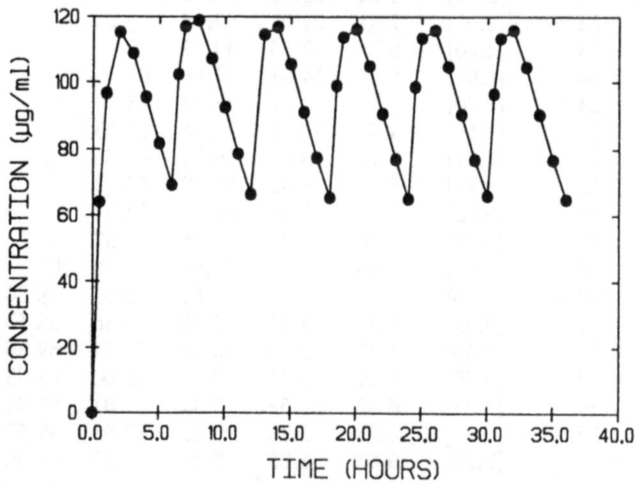

FIGURE 5.2. Plot of $Cn(t)$ *versus* time after a loading dose 1.65 times the maintenance dose. The steady state was reached after the loading dose.

dose 1 in Table 5.3 were multiplied by 1.65 to simulate giving a loading dose 1.65 times the maintenance dose to start a new table (not shown). The same numbers as in Table 5.3 under doses 2 through 6 were copied into the new table and each row summed as before to produce a new set of $Cn(t)$ values. These new $Cn(t)$ values were plotted *versus* t to produce Figure 5.2. Note that the steady state was reached almost immediately. Since Methods (2) and (3) give lower ratios than Method (1), they would not provide peak concentrations after the first (loading) dose as high as in Figure 5.2. Hence steady state would not be reached quite as rapidly.

THE SUPERPOSITION PRINCIPLE IN MATHEMATICAL FORM

Wagner [17c] derived Equation (211), which provides the superposition principle in mathematical form.

$$Cn(t') = C_1(t) + C_1(t)\left[\frac{1 - e^{-(n-1)\beta\tau}}{1 - e^{-\beta\tau}}\right] e^{-\beta t'} \tag{211}$$

where t' = time after administration of the nth dose, t = time after administration of the first dose, n = the dose number, and τ = the dosage interval. Parameter values used for the Table 5.3 example are β = 0.1735 hr^{-1} [from Equation (209)] and τ = 6 hours. If we estimate for dose n = 6, then substitution into Equation (211) gives:

$$Cn(t') = C_1(t) + 41.96\left[\frac{1 - e^{-(5)(0.1735)(6)}}{1 - e^{-(0.1735)(6)}}\right] e^{-0.1735t'}$$

$$= C_1(t) + 41.96\left[\frac{0.9945}{0.6469}\right] e^{-0.1735t'}$$

$$= C_1(t) + 64.51e^{-0.1735t'} \tag{212}$$

Table 5.4 gives the $Cn(t')$ values calculated with Equation (212) and compares them with those from Table 5.3.

Note that in applying the superposition principle, τ must be chosen so that it is greater than or equal to the time when the log-linear phase is established after the first dose.

Note also that although the example used above involved first-order input, the superposition method works when absorption is not first order, and is even irregular providing the overall data are linear.

TABLE 5.4. Sixth Dose Cn(t′) *Values Estimated with Equation (211) Compared with Those from Table 5.3.*

t' (hr)	t (hr)	$Cn(t')$ Equation (211)	$Cn(t)$ Table 5.2
0	30	64.51	64.53
1	31	112.9	113.0
2	32	115.5	115.6
3	33	104.3	104.4
4	34	90.17	90.23
5	35	76.66	76.70
6	36	64.74	64.76

DECONVOLUTION

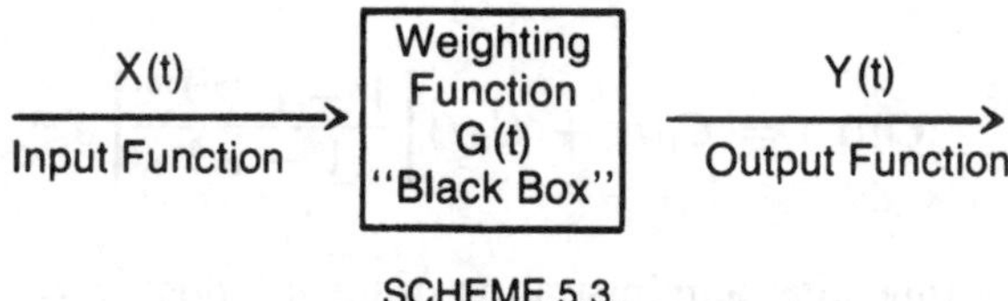

SCHEME 5.3.

A black box is a physical system that transforms an input into an output. The input and output are given by continuous or sectionally continuous functions. If $X(t)$ is the input function (i.e., the rate of appearance of drug in the "black box") and $Y(t)$ is the output function of a linear black box, then there is a function, $G(t)$, called the weighting function of the black box, such that:

$$Y(t) = \int_0^t G(t - T)X(T)dT \tag{213}$$

Hence, it is said that $Y(t)$ is given by the convolution between $G(t)$ and $X(t)$ where $0 < T < t$. Transformation of Equation (213) to the Laplacian domain gives:

$$y(s) = g(s)x(s) \tag{214}$$

where $y(s)$, $g(s)$, and $x(s)$ are the Laplace transforms of $Y(t)$, $G(t)$, and $X(t)$, respectively. Also, $g(s)$ is called the transfer function and its matrix is called the transfer matrix of the black box.

As usually applied in pharmacokinetics, $Y(t)$ is the function describing the plasma (serum or whole blood) concentration-time curve following extravascular administration, $G(t)$ is the function describing the concentration-time curve following bolus intravenous (or impulse) administration, and $X(t)$ is the function describing input.

Deconvolution is the inverse of convolution, and, for example, from Equation (213) would involve obtaining $X(t)$ from $Y(t)$ and $G(t)$. It could also involve obtaining $G(t)$ from $Y(t)$ and $X(t)$.

One deconvolution method is the point-area method, in which one uses areas under the intravenous C, t curve and single concentration values of the extravascular curve. The weighting function $[G(t)]$ curve is decomposed into a series of rectangles having equal base sizes, so the method requires sampling at equal time intervals during the absorption phase.

Operations are carried out as in ordinary multiplication.

	G_1	G_2	G_3	....	
	X_1	X_2	X_3	....	
	X_1G_1	X_1G_2	X_1G_3		
		X_2G_1	X_2G_2	X_2G_3	
			X_3G_1	X_3G_2	X_3G_3
Sum =	Y_1	Y_2	Y_3	....	

Hence,

$$Y_1 = X_1G_1 \quad \text{and} \quad X_1 = Y_1/G_1$$

$$Y_2 = X_1G_2 + X_2G_1 \quad \text{and} \quad X_2 = (Y_2 - X_1G_2)/G_1$$

$$Y_3 = X_1G_3 + X_2G_2 + X_3G_1 \quad \text{and} \quad X_3 = (Y_3 - X_1G_3 - X_2G_2)/G_1$$

Similarly,

$$X_4 = (Y_4 - X_3G_2 - X_2G_3 - X_1G_4)/G_1$$

$$X_5 = (Y_5 - X_4G_2 - X_3G_3 - X_2G_4 - X_1G_5)/G_1$$

In general,

$$X_n = \left(Y_n - \sum_{i=2}^{n} G_iX_{n-i+1}\right)\Bigg/G_1 \tag{215}$$

The point-area method is illustrated with the same simulated extravascular data as used in the superposition example above. The intravenous data were simulated with Equation (216) below.

$$C_{\text{I.V.}} = 100e^{-0.17425t} \tag{216}$$

The areas under different segments of the intravenous C,t curve were obtained with Equation (217).

$$\int_0^{t_i} C_{\text{I.V.}}\,dt = \frac{100}{0.17425}[1 - e^{-0.17425t_i}] = 573.888[1 - e^{-0.17425t_i}] \tag{217}$$

Table 5.6 summarizes the absorption data for the deconvolution example of Table 5.5.

Note that in making the deconvolution calculations in Table 5.5 and below it is the area in the previous segment of the $C_{\text{I.V.}},t$ curve (G_i) which goes with the C_{po} point (Y_i)–i.e., values in the third column of Table 5.5 are differences.

The accuracy in estimating the absorption rate constant and fractions remaining at the absorption site (Table 5.6) by this point-area deconvolution

TABLE 5.5. Illustration of the Point-Area Deconvolution Method.

t_i	$\int_0^{t_i} C_{\text{I.V.}}\,dt$	G_i	$Y_i = C_{po}$
0	0		
1	91.71	91.77	58.63
2	168.87	77.10	69.86
3	233.63	64.76	65.93
4	288.04	54.41	57.94
5	333.75	45.71	49.57

$X_1 = 58.63/91.77 = 0.6389$

$X_2 = [69.86 - (77.10)(0.6389)]/91.77 = 0.2245$

$X_3 = [65.93 - (77.10)(0.2245) - (64.76)(0.6389)]/91.77 = 0.07896$

$X_4 = [57.94 - (77.10)(0.07896) - (64.76)(0.2245) - (54.41) \times (0.6389)]/91.77 = 0.0278$

$X_5 = [49.57 - (77.10)(0.0278) - (64.76)(0.07896) - (54.41) \times (0.2245) - (45.71)(0.6389)]/91.77 = 0.00974$

TABLE 5.6. Summary of Absorption Data for Deconvolution Example.

t_{i-1}	X_i	$G_i/0.6389$[a]	$e^{-1.0.455t_i-1}$[b]
0	0.6389	1.0000	1.0000
1	0.2245	0.3514	0.3515
2	0.07896	0.1236	0.1236
3	0.0278	0.0435	0.0434
4	0.00974	0.01524	0.01527

[a]Equivalent to fraction remaining at the absorption site. Estimated k_a from these values was 1.0456 hr^{-1} compared with actual value of 1.0455 hr^{-1}.
[b]Actual fraction remaining at the absorption site.

method is the result of using single exponential disposition [Equation (216)] and error-free C_{po}, t data (fourth column of Table 5.5). Benet and Chiang [38] showed that if there is biexponential disposition one obtains a log-linear relationship for G_i *versus* t but, unfortunately, the slope of the log-linear plot is not the value of the absorption rate constant unambiguously. Usually, the point-area method gives a low estimate of the absorption rate constant. They also stated that deconvolution by the point-area method is error-sensitive when errors exist in the $C_{I.V.}$ and C_{po} data.

There are methods of deconvolution which are satisfactory [28,39–46] and the reader is referred to those references for further details.

In disposition decomposition analysis, disposition kinetics are decomposed into an elimination function, $q(c)$, and a distribution function, $h(t)$, via deconvolution [47,48].

One should not forget that deconvolution as usually carried out involves the assumption that there is no intrasubject variation in disposition, i.e., that disposition remains constant when the drug is administered orally and intravenously to the same person on different occasions. There is a method [48], however, which allows correction of clearance based on the terminal elimination rate constant, λ_1.

MEAN RESIDENCE TIME REVISITED

A more general definition of mean residence time (MRT) based on a single dose is given as Equation (218).

$$\text{MRT} = \left[\int_0^{\text{Inf}} A_b dt\right] \Big/ D \tag{218}$$

where A_b is the amount of drug in the body at time t. For the Rowland two-compartment open model (Model IX) this is equal to:

$$\text{MRT} = \left[\int_0^{\text{Inf}} A_1 dt + \int_0^{\text{Inf}} A_2 dt\right] \Big/ D = \frac{(k_{20} + k_{21})}{k_{12}k_{20}} + \frac{1}{k_{20}}$$

$$= \frac{1}{k_{20}}\left[1 + \frac{(k_{20} + k_{21})}{k_{12}}\right] \tag{219}$$

where A_1 and A_2 are the amounts in compartments #1 and #2, respectively. The rate constant terms on the right-hand side of Equation (218), which are equal to the areas indicated by the integrals, were obtained by means of Laplace transforms [49].

There is another way of obtaining the areas and MRT. In general, for a 2×2 square matrix, A [Equation (220)], the determinant, Δ, is given by Equation (221). The negative of the inverted matrix, $-A^{-1}$, is given by Equation (222),

$$A = \begin{bmatrix} a_{11} & a_{12} \\ a_{21} & a_{22} \end{bmatrix} \tag{220}$$

$$\Delta = a_{11}a_{22} - a_{12}a_{21} \tag{221}$$

$$-A^{-1} = \begin{bmatrix} a_{22}/\Delta & a_{12}/\Delta \\ a_{21}/\Delta & a_{11}/\Delta \end{bmatrix} \tag{222}$$

where the elements, a_{ij}, refer to the original, not the inverted matrix.

The differential equations for Model IX are as follows:

$$\frac{dA_1}{dt} = -k_{12}A_1 + k_{21}A_2 \tag{223}$$

$$\frac{dA_2}{dt} = k_{12}A_1 - (k_{20} + k_{21})A_2 \tag{224}$$

where A_1 and A_2 are the amounts in compartments #1 and #2, respectively, at time t.

The coefficient matrix for Equations (223) and (224) is Equation (225).

$$A = \begin{bmatrix} -k_{12} & k_{21} \\ k_{12} & -(k_{20} + k_{21}) \end{bmatrix} \tag{225}$$

The determinant of the square matrix [from Equation (221)] is:

$$\Delta = k_{12}(k_{20} + k_{21}) - k_{12}k_{21} = k_{12}k_{20} \tag{226}$$

Based on Equation (222) the negative of the inverted matrix is given by Equation (227).

$$-A^{-1} = \begin{vmatrix} \dfrac{(k_{20} + k_{21})}{k_{12}k_{20}} & \dfrac{k_{21}}{k_{12}k_{20}} \\ \\ \dfrac{k_{12}}{k_{12}k_{20}} & \dfrac{k_{12}}{k_{12}k_{20}} \end{vmatrix} = \begin{vmatrix} \dfrac{k_{20} + k_{21}}{k_{12}k_{20}} & \dfrac{k_{21}}{k_{12}k_{20}} \\ \\ \dfrac{1}{k_{20}} & \dfrac{1}{k_{20}} \end{vmatrix} \begin{matrix} \leftarrow \text{MRT of compartment \#1} \\ \\ \leftarrow \text{MRT of compartment \#2} \end{matrix}$$

↑ When dosing is into compartment #1 — ↑ When dosing is into compartment #2

(227)

Hence, when dosing is into compartment #1 the MRT of Model IX is given by Equation (228):

$$\text{MRT of Model IX} = \frac{1}{k_{20}}\left[1 + \frac{(k_{20} + k_{21})}{k_{12}}\right] \tag{228}$$

and when dosing is into compartment #2:

$$\text{MRT of Model IX} = \frac{1}{k_{20}}\left[1 + \frac{k_{21}}{k_{12}}\right] \tag{229}$$

The methods of obtaining MRTs described in this section are appropriate when input is not into, or elimination is not from, the central compartment, and in such models the MRT is not given by AUMC/AUC, as formerly described.

Unlike the classical Model VIII, Model IX has the above MRT larger than AUMC/AUC, as indicated by Equation (230).

$$\text{MRT (above)} - \frac{\text{AUMC}}{\text{AUC}} = \frac{1}{(k_{20} + k_{21})} \tag{230}$$

For Model IX:

$$V_{ss} = \text{MRT CL} \tag{231}$$

where the MRT is as given by Equation (228), and

$$V_{ss} \text{ is not equal to } \left\{\frac{\text{AUMC}}{\text{AUC}}\right\} \text{CL} \tag{232}$$

CHAPTER 6

Linear Multiple Dose Equations

MODEL I MULTIPLE DOSE

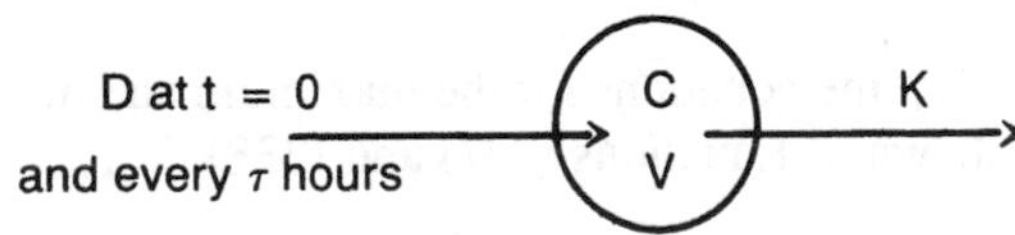

SCHEME 6.1.

Assume that a dose of size D is placed in the single compartment at zero time and every τ hours, where τ is the dosage interval or time between doses.

Let $(A_1)_{min},(A_2)_{min},\ldots,(A_n)_{min}$ be the minimum amount of drug in the "body" or compartment at time t hours after the first, second, . . ., and nth dose, respectively. Let $(A_1)_{max},(A_2)_{max},\ldots,(A_n)_{max}$ be the maximum amount of drug in the "body" or compartment immediately after administration of the first, second, . . ., and nth dose, respectively. Then,

$$(A_1)_{max} = D$$

$$(A_1)_{min} = De^{-K\tau}$$

$$(A_2)_{max} = D + De^{-K\tau} = D(1 + e^{-K\tau})$$

$$(A_2)_{min} = D(1 + e^{-K\tau})e^{-K\tau} = D(e^{-K\tau} + e^{-2K\tau})$$

Hence there is an infinite geometric series:

$$(A_n)_{max} = D(1 + e^{-K\tau} + e^{-2K\tau} + \ldots + e^{-(n-1)K\tau})$$

$$\text{Let } S = 1 + e^{-K\tau} + e^{-2K\tau} + \ldots + e^{-(n-1)K\tau} \tag{233}$$

Then:

$$e^{-K\tau}S = e^{-K\tau} + e^{-2K\tau} + \ldots + e^{-nK\tau} \tag{234}$$

Subtracting Equation (234) from Equation (233) gives Equation (235).

$$S(1 - e^{-K\tau}) = 1 - e^{-nK\tau} \tag{235}$$

Whence:

$$S = \sum_{n=1}^{n} e^{-nK\tau} = \frac{1 - e^{-nK\tau}}{1 - e^{-K\tau}} \tag{236}$$

Since $A_n = VC_n$, the equations for the maximum and minimum concentrations are as shown in Equations (237) and (238).

$$(C_n)_{max} = \left(\frac{D}{V}\right)\left(\frac{1 - e^{-nK\tau}}{1 - e^{-K\tau}}\right) \tag{237}$$

$$(C_n)_{min} = \left(\frac{D}{V}\right)\left(\frac{1 - e^{-nK\tau}}{1 - e^{-K\tau}}\right) e^{-K\tau} \tag{238}$$

At any time after the nth dose the concentration, C_n, is given by Equation (239).

$$C_n = \left(\frac{D}{V}\right)\left(\frac{1 - e^{-nK\tau}}{1 - e^{-K\tau}}\right) e^{-Kt} \tag{239}$$

After a number of doses that make $e^{-nK\tau}$ essentially equal to zero, a steady state is reached and the concentrations are given by Equations (240) and (241).

$$(C_{ss})_{max} = \left(\frac{D}{V}\right)\left(\frac{1}{1 - e^{-K\tau}}\right) \tag{240}$$

$$(C_{ss})_{min} = \left(\frac{D}{V}\right)\left(\frac{e^{-K\tau}}{1 - e^{-K\tau}}\right) \tag{241}$$

Note that for this Model I we have:

$$(C_{ss})_{max} - (C_{ss})_{min} = \frac{D}{V} = C_o \tag{242}$$

which was obtained by subtracting Equation (241) from Equation (240). This result is independent of both K and τ.

If one substitutes $nt_{1/2}$ for t (where $t_{1/2} = 0.693/K$), $\ln 2/t_{1/2}$ for K and let $\epsilon = t/t_{1/2}$, then Equation (239) becomes:

$$C_n = \left(\frac{D}{V}\right)\left(\frac{1 - 2^{-n\epsilon}}{1 - 2^{-\epsilon}}\right)2^{-n} \tag{243}$$

In a similar manner by substituting into Equations (240) and (241), one obtains:

$$(C_{ss})_{max} = \left(\frac{D}{V}\right)\left(\frac{1}{1 - 2^{-\epsilon}}\right) \tag{244}$$

$$(C_{ss})_{min} = \left(\frac{D}{V}\right)\left(\frac{2^{-\epsilon}}{1 - 2^{-\epsilon}}\right) \tag{245}$$

Hence, the ratio of maximum to minimum concentrations at steady state is given by Equation (246). Hence, when $\epsilon = 1$ (i.e., when $t = t_{1/2}$) the ratio of maximum to minimum steady-state concentrations is equal to 2.

$$\frac{(C_{ss})_{max}}{(C_{ss})_{min}} = \frac{1}{2^{-\epsilon}} = 2^{\epsilon} \tag{246}$$

Thus, for this simplest of all models, when the dosage interval is made equal to the elimination half-life the maximum plasma concentration is exactly twice the minimum plasma concentration and the amount of drug in the "body" fluctuates between the dose and twice the dose. Hence, according to this model, an appropriate dosage regimen is a loading dose equal to twice the maintenance dose, and maintenance doses are given at intervals equal to the elimination half-life.

MULTIPLE DOSING FUNCTION FOR EQUAL TIME INTERVALS

For the approach to steady state each term of a polyexponential equation for a single dose is multiplied by the term shown below, but here λ_i represents not only a hybrid rate parameter as formerly but also any model rate constant.

$$\frac{1 - e^{-n\lambda_i t}}{1 - e^{-\lambda_i t}}$$

To convert a single-dose polyexponential equation to one for the steady state, one multiplies each term of the single-dose equation by the term:

$$\frac{1}{1 - e^{-\lambda_i t}}$$

where again λ_i represents any hybrid or model rate constant.

Examples

Suppose the single-dose equation is Equation (247).

$$C(\mu g/ml) = 120\,(e^{-0.17425t} - e^{-1.0455t}) \tag{247}$$

If the same dose as the single dose is administered every six hours until steady state is attained, then the steady-state equation is Equation (248), where t is the time after dosing at steady state. In this example the λ_i values

$$C_{ss} = 120\left[\frac{e^{-0.17425t}}{1 - e^{-(0.17425)(6)}} + \frac{e^{-1.0455t}}{1 - e^{-(1.0455)(6)}}\right] \tag{248}$$

are $K = 0.17425\ hr^{-1}$ and $k_a = 1.0455\ hr^{-1}$ of Model II.

In this case one can readily find the maximum and minimum steady-state concentrations also. Remember in calculus that when there is a maximum of a function then the derivative of the function is equal to zero. If we calculate the values of the coefficients of Equation (248) we get Equation (249).

$$C_{ss}(\mu g/ml) = 185e^{-0.17425t} - 121e^{-1.0455t} \tag{249}$$

The derivative of Equation (249) is Equation (250) and this is set equal to zero.

$$\frac{dC_{ss}}{dt} = -(0.17425)(185)e^{-0.17425t_{max}} + (1.0455)(121)e^{-1.0455t_{max}} \tag{250}$$

Simplification of Equation (250) gives Equation (251).

$$32.24e^{-0.17425t_{max}} = 126.5e^{-1.0455t_{max}} \tag{251}$$

Rearrangement of Equation (251) gives Equation (252).

$$\frac{e^{-0.17425t_{max}}}{e^{-1.0455t_{max}}} = \frac{126.5}{32.24} = 3.924 \tag{252}$$

Taking natural logarithms of both sides of Equation (252) gives:

$$(-0.17425 + 1.0455)t_{max} = 0.87125t_{max} = \ln 3.924 = 1.367 \tag{253}$$

Thus:

$$t_{max} = \frac{1.367}{0.87125} = 1.57 \text{ hours} \tag{254}$$

Then substituting $t_{max} = 1.57$ for t in Equation (249) gives:

$$(C_{ss})_{max} = 185e^{-(0.17425)(1.57)} - 121e^{-(1.0455)(1.57)}$$

$$= 141 - 23 = 118\ \mu\text{g/ml} \tag{255}$$

The minimum steady-state concentration is obtained by substituting $\tau = 6$ hr for t in Equation (249):

$$(C_{ss})_{min} = 185e^{-(0.17425)(6)} - 121e^{-(1.0455)(6)}$$

$$= 65.0 - 0.2 = 64.8\ \mu\text{g/ml} \tag{256}$$

The area within a dosage interval at steady state is equal to the area from zero to infinity after a single dose; hence, one can estimate the average steady-state concentration by integrating the single-dose Equation (247) as shown below.

$$\int_0^{\text{Inf}} Cdt = 120\left(\frac{1}{0.17425} - \frac{1}{1.0455}\right) = 689 - 115 = 574 \tag{257}$$

Hence the average steady-state concentration, $(C_{ss})_{av}$, is this area divided by the dosage interval τ as indicated below.

$$(C_{ss})_{av} = \frac{574}{6} = 95.7\ \mu\text{g/ml} \tag{258}$$

What is an appropriate loading dose in this example? If we perform the same operations as indicated by Equations (250) through (254) on the single-dose Equation (247), we calculate a peak time of 2.057 hr after the single dose. If we substitute this time in Equation (247) we obtain a peak concentration of 69.9 μg/ml after the single dose. Hence as shown in Chapter 5 under superposition, we estimate:

$$\frac{\text{Loading dose}}{\text{Maintenance dose}} = \frac{\text{Peak concentration at steady state}}{\text{Peak concentration after single dose}}$$

$$= \frac{118}{69.9} = 1.69 \tag{259}$$

The superposition principle gave a ratio of 1.65 which is the same within error.

UNEQUAL DOSES AND/OR UNEQUAL DOSAGE INTERVALS WITHIN A DAY

In a hospital situation, it is often inconvenient to have equal dosage intervals if dosing occurs three or four times daily. There are several methods in the literature [50–52] which provide methods to make calculations in such cases. A method which provides for unequal doses and unequal dosage intervals is shown below.

Example

The doses and dosing times for this example are shown in Table 6.1.

Assuming Model I the amount of drug in the "body" is equal to the sum of the amounts that remain from each administered dose. The elimination half-life of the drug is five hours hence $K = 0.693/5 = 0.139\ \text{hr}^{-1}$. What is

TABLE 6.1. Doses and Dosing Times for the Example.

No.	Dose Amount (mg)	Day	Time	From First Dose (hr)
1	200	Day 1	9:00 A.M.	0
2	100		1:00 P.M.	4
3	100		7:00 P.M.	10
4	100	Day 2	9:00 A.M.	24
5	200		3:00 P.M.	30
6	300		6:00 P.M.	33
7	100	Day 3	11:00 A.M.	50

the amount of drug in the "body" at 8:00 A.M. on day 2, which is twenty-three hours after the first dose is administered?

$$A_b = D_1e^{-Kt_1} + D_2e^{-Kt_2} + D_3e^{-Kt_3}$$

$$= 200e^{-0.139(23-0)} + 100e^{-0.139(23-4)} + 100e^{-0.139(23-10)}$$

$$= 8.18 + 7.13 + 16.4 = 31.7 \text{ mg} \tag{260}$$

One can treat each term of a polyexponential equation in the same manner for each dose, as indicated above for the monoexponential equation.

The method of Ng [53] is the same as the superposition method described and illustrated in Chapter 5. The concentrations are derived from Model II using population parameters for theophylline. The equal doses are given at 8 A.M., 12 NOON, 6 P.M., and 10 P.M. When $n = 1$, four sets of C,t data are generated using four different equations—the same as Equation (3) under Model II, but the starting times are 0, 4, 10, and 14 hr and they have added multiple dosing functions. Then the four concentrations are added for each time as in the superposition method. When $n = 2$ (second set of four doses) four new sets of C,t data are generated but the residues from the first four doses are also added in to produce the overall sums. The author [53] gives a complete program for the Hewlett-Packard 41C calculator to do all calculations and provide the predicted C,t data. In this example, $t = 24$ hr in all equations since each dose is given only once a day at a specific time.

CHAPTER 7

Simple Nonlinear Models

MICHAELIS-MENTEN ELIMINATION KINETICS

The Michaelis-Menten equation [54] is Equation (261).

$$v = -\frac{dA}{dt} = -V\frac{dC}{dt} = \frac{V_m C}{K_m + C} \tag{261}$$

In Equation (261), v is velocity of elimination (mass/time), V is volume of distribution, C is the concentration at site of elimination at time t, V_m is the maximal velocity of elimination, and K_m is the Michaelis constant. If we let $V'_m = V_m/V$, then:

$$-\frac{dC}{dt} = \frac{V'_m C}{K_m + C} \tag{262}$$

Integration of Equation (262) with $C = C_o$ at $t = 0$ and $C = C$ at time t gives Henri's [55] equation, which is Equation (263).

$$C_o - C + K_m \ln (C_o/C) = V'_m t \tag{263}$$

The hyperbolic Michaelis-Menten equation has a shape as illustrated in Figure 7.1, which was obtained by substituting $V'_m = 0.22$ mg/(ml × hr) and $K_m = 0.1$ mg/ml into Equation (262) and simulating with the program MINSQ [2]; these parameter values are about the average ones for ethanol in man. Note that the K_m value is equal to the concentration when $-dC/dt = V'_m/2 = 0.22/2 = 0.11$.

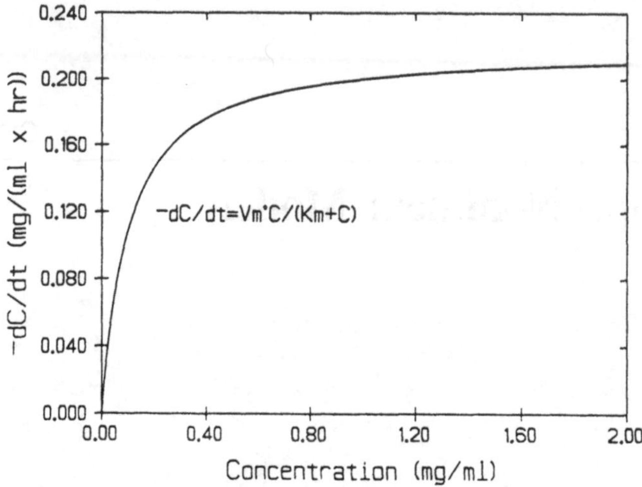

FIGURE 7.1. Simulated $-dC/dt, C$ data based on Equation (262) with $V_m' = 0.22$ and $K_m = 0.1$.

Figure 7.2 shows the "hockey-stick" shape of the integrated form [Equation (263)] with the same V_m' and K_m values and four C_o values, namely $C_o = 2$, 1, 0.5, and 0.25 mg/ml. These are rectilinear or Cartesian coordinate plots. When the same data are plotted on semilogarithmic graph paper, one obtains plots as shown in Figure 7.3. Note the inward curvature of the plots, with the bend in the upper region being greater the larger the initial C_o value. Note also that in Figure 7.2 there is a reasonably linear por-

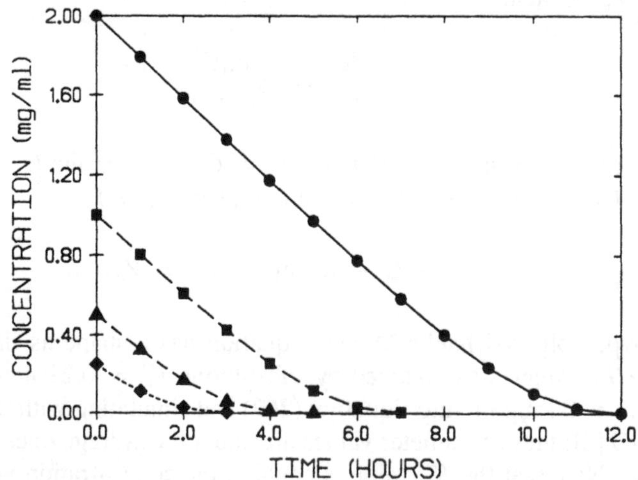

FIGURE 7.2. Simulated C,t data based on Equation (263) with $V_m' = 0.22$ and $K_m = 0.1$ and for different initial concentrations, C_o.

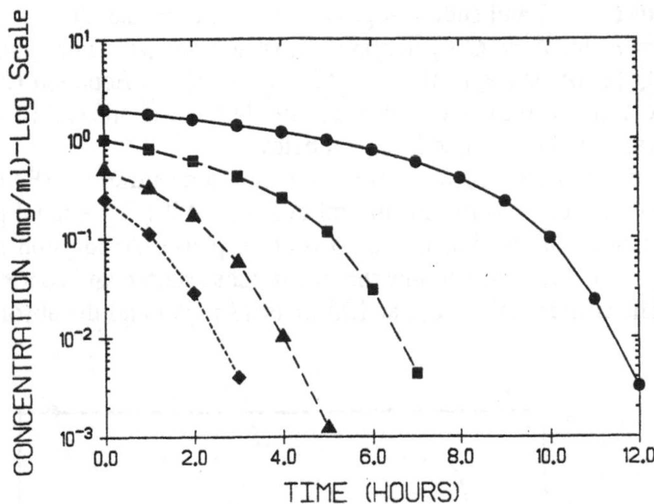

FIGURE 7.3. Semilogarithmic plot of same C,t data as in Figure 7.2.

tion which constitutes the "handle of each hockey stick"; this pseudolinear portion extends from C_o to $(1 - 1/e)C_o = 0.632\ C_o$, where e is the base of natural logarithms $= 2.7183$ [56]. The absolute value of the slope of this linear region, k_o, is not V'_m, but rather is approximately the value given by Equation (264).

$$k_o = \frac{0.632C_oV'_m}{K_m + 0.632C_o} \tag{264}$$

Hence the oft-repeated statement that the Michaelis-Menten function is zero order at the top and first order at the bottom is not really true. If it was zero order at the top, one would expect the slope of the pseudolinear portion to be equal to V'_m, but this is only true as the concentration approaches infinity in Figure 7.1 and C_o approaches infinity in Figure 7.2. Similarly if it was first order at the bottom one would expect the slope of the terminal linear portion in Figure 7.3 to be equal to $-V'_m/2.303K_m$, but the slopes are always less than this value.

METHODS OF ESTIMATING V'_m and K_m

Method IA. From the Initial Zero-Order Slopes, k_o

If three or more single doses are administered and the pseudolinear slopes are obtained after distribution and/or absorption have ceased, starting with $C = C_o$ and ending approximately at $C = 0.632C_o$, one then fits the $Y = k_o$ and $X = C_o$ by nonlinear least squares with a program such as MINSQ [2] or by the method of Wilkinson [63] to Equation (264). The method is applicable to data collected after both intra- and extravascular administrations. An example is shown below.

Figure 7.4 shows capillary blood ethanol concentrations (BAL) *versus* time for subject #5 following an oral dose of 0.451 g/kg ethanol [62]. The early dotted line portion is the absorption phase. Absorption ceased at $t(C_o) = 1.75$ hr, at which time the linear phase started and continued until 4 hr. The concentration, C_o, at 1.75 hr is 48 mg/dl and the absolute value

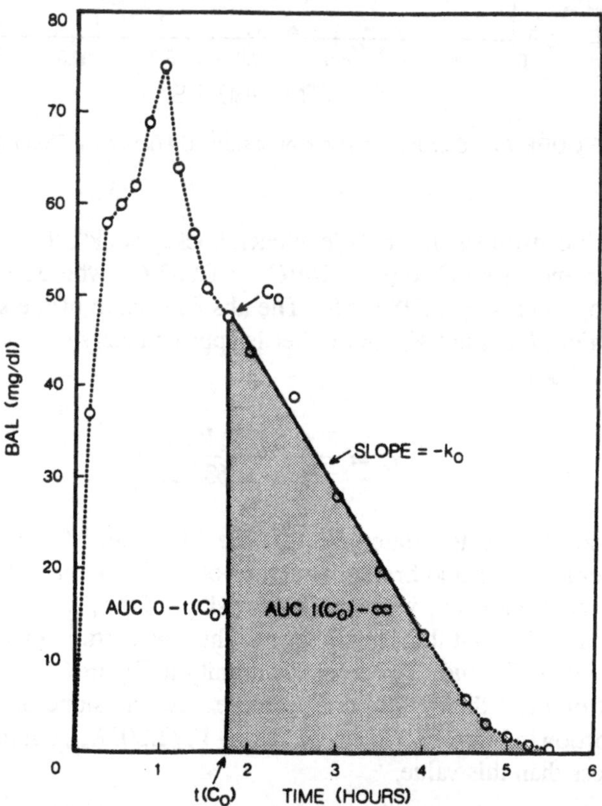

FIGURE 7.4. Example showing how k_o is estimated. Data are capillary blood alcohol concentrations (BAL) for subject #5 following an oral dose of 0.451 g/kg ethanol. Early data points are in the absorption phase. Absorption ceased at 1.75 hr where the vertical solid line is shown. The concentration, C_o, at 1.75 hr is 48 mg/dl. The solid slanting line has a negative slope of 15.9 mg/dl/hr and was obtained by least squares regression using the six points about the line.

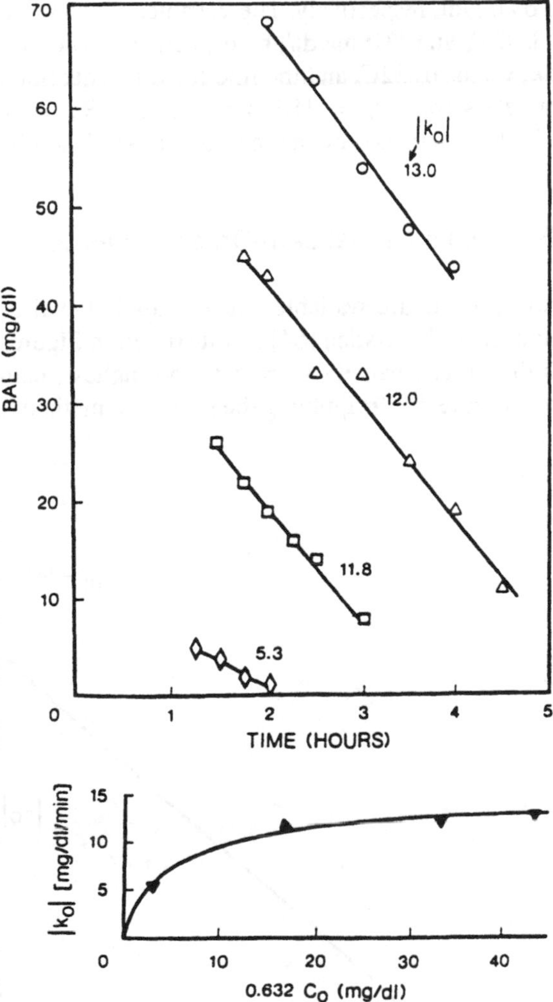

FIGURE 7.5. *Top:* shows how k_o was estimated for subject #7 following four oral doses of alcohol. *Bottom:* fit of absolute value of k_o *versus* 0.632 C_o to Equation (264) when $V_m' =$ 14.6 mg/dl/hr and K_m = 5.05 mg/dl were estimated.

of the slope of the solid line is 15.9 mg/dl/hr and was determined by linear least squares regression from the six points along the line with $r = -0.996$. The two areas shown in the figure will be discussed later.

Figure 7.5 at the top shows the k_o values and the C, t data from which they were estimated for subject #7 given fasting oral doses of 0.142, 0.283, 0.424, and 0.566 g/kg of ethanol. The first data point or C_o values are 4.9,

26, 45, and 69 mg/dl, respectively. The k_o values are shown on the plots and are 5.3, 11.8, 12.0, and 13.0 mg/dl/hr, respectively. The lower panel shows the plot of k_o versus $0.632C_o$ and the line fitted to Equation (264); the estimated parameters were $V'_m = 14.6$ mg/dl/hr and $K_m = 5.05$ mg/dl with $r^2 = 0.983$ when estimated by the method of Wilkinson [63].

Method IB. From the Initial Zero-Order Slopes, k_o

If only two k_o values are available, one can apply the direct linear plot of Eisenthal and Cornish-Bowden [64] as illustrated in Figure 7.6. The example utilizes the lowest, namely $k_o = 5.3$, and highest, namely $k_o = 13.0$, data shown in Figure 7.5. In applying the graphical method the k_o values are

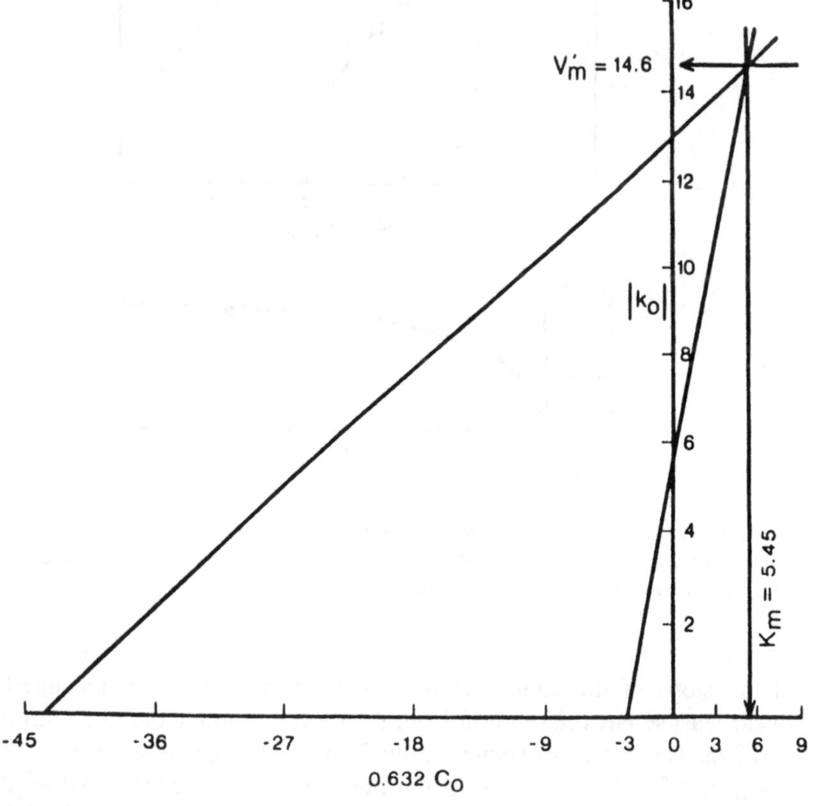

FIGURE 7.6. Example of application of plot of Eisenthal and Cornish-Bowden [64] (Method IB).

plotted vertically and values of $0.632C_o$ (3.10 corresponding to $k_o = 5.3$ and 43.6 corresponding to $k_o = 13.0$) are plotted horizontally. One joins the point pair with a line and extends the line into the top right quadrant. The two lines intersect in this quadrant. Going from this point of intersection, parallel to the X-axis, to the vertical k_o axis, one finds $V'_m = 14.6$. Going from the point of intersection, parallel to the vertical axis, to the $0.632C_o$ axis, one finds $K_m = 5.45$. These values are essentially identical to those obtained by Method IA.

The equivalent of the direct linear plot is found in Equations (265) and (266), which may be used instead of the direct linear plot.

$$V'_m = \frac{(0.632C_o)_2 - (0.632C_o)_1}{\dfrac{(0.632C_o)_2}{(k_o)_2} - \dfrac{(0.632C_o)_1}{(k_o)_1}} = \frac{43.6 - 3.10}{\dfrac{43.6}{13.0} - \dfrac{3.10}{5.3}} = 14.6 \quad (265)$$

$$K_m = \frac{(k_o)_2 - (k_o)_1}{\dfrac{(k_o)_1}{(0.632C_o)_1} - \dfrac{(k_o)_2}{(0.632C_o)_2}} = \frac{13.0 - 5.3}{\dfrac{5.3}{3.10} - \dfrac{13.0}{43.6}} = 5.45 \quad (266)$$

If there are three or more values of k_o and $0.632C_o$ then in Figure 7.6 one will have several lines crossing like a sheaf of wheat, and one takes the median values to obtain V'_m and K_m.

Method IIIA. Parabolic Area Equation

If the drug is given by bolus intravenous injection and the one-compartment open model with Michaelis-Menten elimination applies then C, t data are described by Equation (263) where $C_o = D/V$. If one integrates again one will obtain an expression for area under the C, t curve as in Figure 7.2. Wagner [65] did such an integration and reported Equation (267).

$$\text{AUC} = \frac{C_o}{V'_m}\left[\frac{C_o}{2} + Km\right] = \frac{K_m}{V'_m}(C_o) + \frac{1}{2V'_m}(C_o)^2 \quad (267)$$

If we let $K_m/V'_m = a_1$ and $1/2V'_m = a_2$ then Equation (267) may be written as Equation (268).

$$a_2C_o^2 + a_1C_o = \text{AUC} \quad (268)$$

If the drug is administered extravascularly then AUC in the above equation would be AUC $t(C_o)$–Inf in Figure 7.4. If two doses of drug are administered then one would obtain two equations as Equations (269) and (270).

$$a_2(C_o)_1^2 + a_1(C_o)_1 = (\mathrm{AUC})_1 \tag{269}$$

$$a_2(C_o)_2^2 + a_1(C_o)_2 = (\mathrm{AUC})_2 \tag{270}$$

The solutions for a_1 and a_2 in Equations (269) and (270) are obtained as a ratio of two determinants as follows.

$$a_2 = \frac{\begin{vmatrix} (\mathrm{AUC})_1 & (C_o)_1 \\ (\mathrm{AUC})_2 & (C_o)_2 \end{vmatrix}}{\begin{vmatrix} (C_o)_1^2 & (C_o)_1 \\ (C_o)_2^2 & (C_o)_2 \end{vmatrix}} = \frac{(C_o)_2(\mathrm{AUC})_1 - (C_o)_1(\mathrm{AUC})_2}{(C_o)_1^2(C_o)^2 - (C_o)_1(C_o)_2^2} \tag{271}$$

$$a_1 = \frac{\begin{vmatrix} (C_o)_1^2 & (\mathrm{AUC})_1 \\ (C_o)_2^2 & (\mathrm{AUC})_2 \end{vmatrix}}{\begin{vmatrix} (C_o)_1^2 & (C_o)_1 \\ (C_o)_2^2 & (C_o)_2 \end{vmatrix}} = \frac{(C_o)_1^2(\mathrm{AUC})_2 - (C_o)_2^2(\mathrm{AUC})_1}{(C_o)_1^2(C_o)_2 - (C_o)_1(C_o)_2^2} \tag{272}$$

To apply this method when there are three or four doses and C_o, AUC pairs, one applies Equations (271) and (272) to each possible pair and then obtains the median (not the average) V_m' and K_m values.

Method IIB. Parabolic Area Equation and Least Squares Fitting

If one has data from three or four doses one can fit C_o, AUC data to Equation (268) by nonlinear least squares regression and estimate the parameters a_1 and a_2, then from these coefficients one can obtain V_m' and K_m. Alternatively, one could fit the data to Equation (267) and directly estimate V_m' and K_m.

Method III. Fitting to Equation (263)

Individual sets of C, t data ranging from C_o at $t(C_o)$ (see Figure 7.4) to the last measured concentration may be computer-fitted to Equation (263) by nonlinear least squares. The program of Thomas et al. [66] has been used by the author with reasonable success. Figure 7.7 is an example of such a fit. It shows the fit of BAL (blood alcohol concentrations) of subject #2 after an oral dose of 0.160 g/kg alcohol, and only the data from C_o at $t(C_o)$ to C at 7

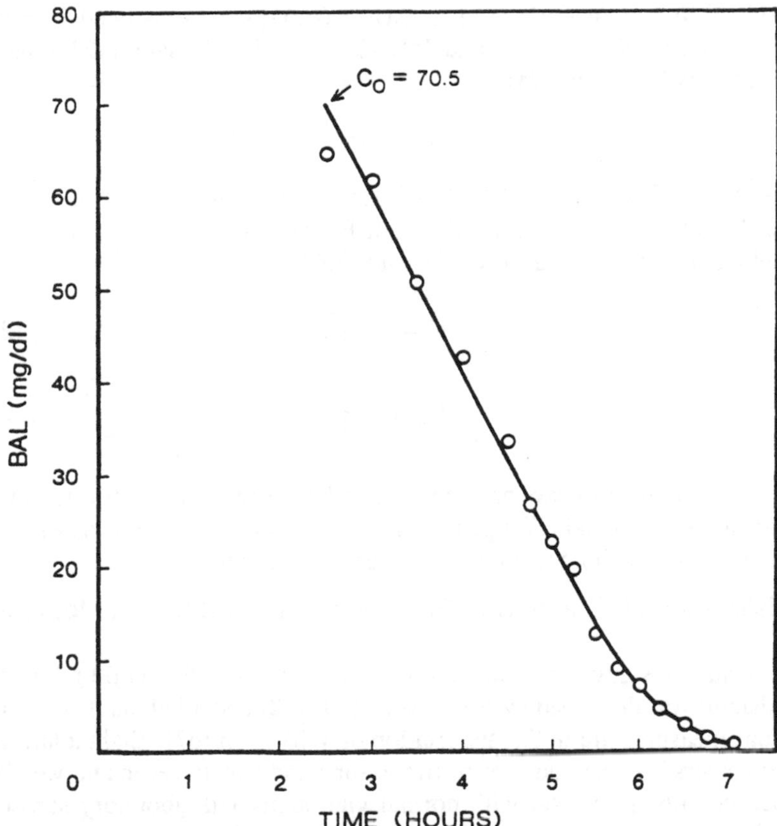

FIGURE 7.7. Least squares fit of alcohol blood concentrations of subject #2 to Equation (263) using the program of Thomas et al. [66]. This is an example of Method III.

hr were fitted. Weighting $1/Y_i^2$ caused poor model-predicted concentrations at early times, while equal weighting caused poor model-predicted concentrations at later times; weighting $1/Y_i$ provided good model-predicted concentrations at all times.

Method IV. Estimation of V_m', K_m and Volume of Distribution, V

If the numerator and denominator of the first term on the right-hand side of Equation (267) are multiplied by V and the numerator and denominator of the second term are multiplied by V^2 then Equation (267) becomes Equation (273), where

$$\text{AUC 0–Inf} = \left(\frac{K_m}{V_m}\right)D + \left(\frac{1}{2VV_m}\right)D^2 \tag{273}$$

AUC 0–Inf = AUC 0–$t(C_o)$ + AUC $t(C_o)$–Inf as illustrated in Figure 7.4. If we let $K_m/V_m = a_1'$ and $1/2VV_m = a_2'$ then Equation (273) may be written as Equation (274).

$$\text{AUC 0–Inf} = a_1'D + a_2'D^2 \tag{274}$$

With $X = D$ and $Y =$ AUC 0–Inf sets of data could be fitted to Equation (274) and a_1' and a_2' estimated. From Equations (267), (268), (273), and (274) one obtains Equations (275) and (276).

$$V = \sqrt{a_2/a_2'} \tag{275}$$

$$V_m = \frac{1}{2Va_2'} \tag{276}$$

Several simulations have been performed by the author and it was found that Equations (275) and (276) provided good estimates of V and V_m when parameters of alcohol metabolism in man were employed.

Estimation of V_m' and K_m of Ethanol in Man by Different Methods

A human study with four different doses (15, 30, 45, and 60 ml of 95% ethanol) was described by Wilkinson et al. [62] and all alcohol concentrations measured are in the dissertation of Wilkinson [67]. Eight adult male volunteers between the ages of twenty-one and twenty-seven and weighing between 66 and 89 kg, with normal vital signs and laboratory screening values, participated. The study was conducted using a Latin square design with four groups of two subjects each, arranged in order of increasing body weight. Each subject received 15, 30, 45, and 60 ml of 95% alcohol in orange juice (total 150 ml) orally, at one-week intervals. Just prior to dosing and at strategic times following each of the doses, 18, 25, 23, and 27 capillary blood samples were taken following the 15, 30, 45, and 60 ml doses, respectively. Each blood sample was collected from a fingertip, after lancing, in a 50 μl calibrated capillary tube. The samples were immediately mixed with an equal volume of internal standard solution, transferred to 6 ml amber glass serum vials, frozen, and kept in the frozen state until just prior to assay. The samples were assayed by a head-space GLC method.

V_m' and K_m were estimated by Methods IA, IB, IIA, IIB, and III described above. The estimated V_m' values are listed in Table 7.1 and the estimated K_m values are listed in Table 7.2 along with the intra- and intersubject means and coefficients of variation. No estimates of V_m' and K_m were made from the 15 ml data by Method III. For both parameters the intrasubject coefficients of variation are appreciably less than the intersubject coefficients of variation. This fact, coupled with the observation that the parameter values

TABLE 7.1. Summary of V'_m Values Estimated by Five Methods.

Subject	V'_m (mg/dl/hr)									
	Method IA	Method IB	Method IIA	Method IIB	Method III			Intrasubject		
					30 ml[a]	45 ml[a]	60 ml[a]	Average[b]	Mean[c]	C.V. (%)
1	16.6	16.1	17.9	16.1	17.8	20.6	19.0	19.1	17.2	7.62
2	18.8	15.9	17.0	16.7	15.6	15.5	22.2	17.8	17.2	6.43
3	12.9	13.2	12.9	13.4	17.1	16.8	15.5	16.5	13.8	9.30
4	12.2	11.7	14.5	14.4	18.2	15.3	16.3	16.6	13.9	14.2
5	16.7	16.4	16.5	17.2	18.0	18.8	19.1	18.6	17.1	5.29
6	15.9	15.9	17.9	18.0	19.9	17.6	17.1	18.2	17.2	6.82
7	14.6	14.4	16.4	16.1	16.9	14.6	16.8	16.1	15.5	5.85
8	28.0	26.3	26.9	24.4	24.4	22.0	28.4	24.9	26.1	5.62
Intersubj. mean	17.0	16.3	17.5	17.0				18.5	17.3	7.64
C.V. (%)	29.1	26.9	23.8	19.5				15.2		

[a]Dose in ml of 95% alcohol.
[b]Average of three doses.
[c]Mean of Methods IA, IB, IIA, IIB and mean of Method III.

TABLE 7.2. Summary of K_m Values Estimated by Five Methods.

	K_m									
					Method III			Intrasubject		
Subject	Method IA	Method IB	Method IIA	Method IIB	30 ml[a]	45 ml[a]	60 ml[a]	Average[b]	Mean[c]	C.V. (%)
1	3.19	3.11	3.48	2.81	3.08	8.29	2.48	4.62	3.44	20.4
2	8.66	5.52	6.02	4.72	4.01	3.10	6.40	4.50	5.88	28.4
3	1.89	2.38	3.22	1.99	5.63	5.29	4.02	4.98	2.89	44.3
4	2.49	2.93	2.98	3.28	4.90	3.64	3.34	3.96	3.13	17.4
5	2.86	3.10	2.28	3.10	4.22	3.96	3.47	3.88	3.04	18.9
6	3.82	4.23	5.46	5.49	7.09	3.15	3.93	4.72	4.74	15.6
7	5.05	5.33	5.68	5.79	6.58	2.51	0.44	3.18	5.01	21.2
8	13.2	13.0	9.58	7.34	15.0	8.25	9.87	11.0	10.8	26.2
Intersubj. mean	5.15	4.95	4.84	4.32				5.11	4.87	24.1
C.V. (%)	75.5	69.7	49.0	41.9				48.0	54.1	

[a]Dose in ml of 95% alcohol.
[b]Average of three doses.
[c]Mean of Methods IA, IB, IIA, IIB and mean of Method III.

estimated by Method III from data after three different doses agree reasonably well, indicate that there is small intrasubject variation of the V_m' and K_m of alcohol in man. Tables 7.1 and 7.2 also indicate that the V_m' and K_m values estimated by the five different methods agree quite well.

Advantages and Disadvantages of Methods of Estimating V_m' and K_m

The advantages are as follows:

(1) Methods IA and IB require C, t data only in the pseudolinear range and not below the K_m value.

(2) Method IB requires data after only two doses but also may be applied for any number of doses.

(3) Method IIA requires data after only two doses and does not require least squares fitting.

(4) Method III can be applied to data obtained after only one dose and provides standard deviations of the estimated values of V_m' and K_m.

The disadvantages are that Methods IA, IB, IIA, and IIB assume there is no intrasubject variation of V_m' and K_m, and require data collected after two or more doses.

Estimation of the Volume of Distribution of Ethanol in Man

Data from the above-described human alcohol study were fitted to Equations (268) and (274) and the coefficients a_2 and a_2' were estimated for each subject, then Equation (275) was employed to estimate the volume of distribution, V. Results are listed in Table 7.3.

TABLE 7.3. Volume of Distribution of Ethanol in Man.

	Volume of Distribution	
Subject	Liters	L/kg
1	42.9	0.651
2	50.0	0.709
3	49.5	0.702
4	57.7	0.808
5	45.9	0.612
6	51.8	0.691
7	44.7	0.562
8	60.3	0.681
Mean	50.4	0.677
C.V. (%)	12.2	10.8

The equations given in the next section were published by Wagner et al. [57].

MODEL XXVI. ONE-COMPARTMENT OPEN MODEL WITH BOLUS I.V. ADMINISTRATION AND MICHAELIS-MENTEN ELIMINATION KINETICS

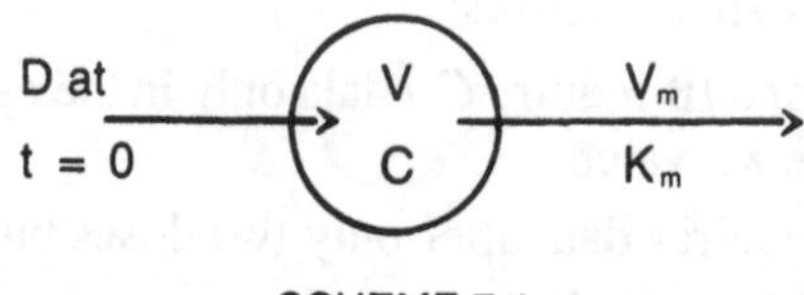

SCHEME 7.1.

This model has been considered under Michaelis-Menten elimination kinetics above. Equations (261) and (262) are the differential equations and Equation (263) is the integrated equation. Figures 7.2 and 7.3 show simulations with the model. The model may also be applied to extravascular data if one ignores the absorption phase as illustrated in Figures 7.4, 7.5, and 7.7. In such cases, absorption is assumed to cease at time $t(C_o)$ and data are treated as if a bolus I.V. dose is administered at $t(C_o)$, which gives a concentration equal to C_o. Equations (268) and (274) are the parabolic area equations which apply to the model. The clearance is not constant for this model since C is changing. Clearance is given by Equation (277).

$$\mathrm{CL} = \frac{V_m}{K_m + C} \tag{277}$$

The mean clearance, $\overline{\mathrm{CL}}$, is given by Equation (278).

$$\mathrm{CL} = \frac{V_m}{K_m + C_o/2} \tag{278}$$

MODEL XXVII. ONE-COMPARTMENT OPEN MODEL WITH INTERMITTENT BOLUS I.V. DOSING AND MICHAELIS-MENTEN ELIMINATION KINETICS

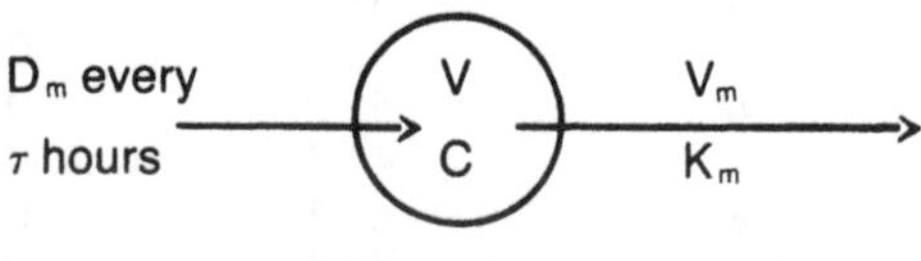

SCHEME 7.2.

The differential Equations (261) and (262) apply for this model also. In Scheme 7.2, D_m is the maintenance dose given every τ hours. Instead of using the single-dose limits C_o and 0 for C, one uses the steady-state limits of $(C_{ss})_{max}$ and $(C_{ss})_{min}$; also $(C_{ss})_{max} - (C_{ss})_{min} = C_o = D_m/V$; integration of the differential equation leads to Equation (279), which applies to the steady state. In Equation (279)

$$\text{AUC } 0\text{–}\tau = \left[\frac{(C_{ss})_{min} + K_m}{V_m}\right] D_m + \frac{D_m^2}{2VV_m} \tag{279}$$

AUC 0–τ is the area under the C, t curve during a dosage interval at steady state and $(C_{ss})_{min}$ is the value of C at τ hours after a dose. By comparing Equations (273) and (279) one can see that AUC 0–τ > AUC 0–Inf; i.e., the area in a dosage interval at steady state is greater than the area zero to infinity after a single dose, which is different than in linear kinetics with Model I single and multiple dose. The mean steady-state clearance is given by Equation (280) and is less than

$$\begin{aligned}\overline{\text{CL}}_{ss} &= V_m/[K_m + \{(C_{ss})_{max} + (C_{ss})_{min}\}/2] \\ &= V_m/[(C_{ss})_{min} + (C_o/2) + K_m] \end{aligned} \tag{280}$$

the mean clearance after a single dose given by Equation (278). This is so since Equation (280) has the extra term $(C_{ss})_{min}$ in the denominator.

MODEL XXVIII. ONE-COMPARTMENT OPEN MODEL WITH CONSTANT RATE (ZERO-ORDER) INPUT OVER A SHORT PERIOD OF TIME AND MICHAELIS-MENTEN ELIMINATION KINETICS

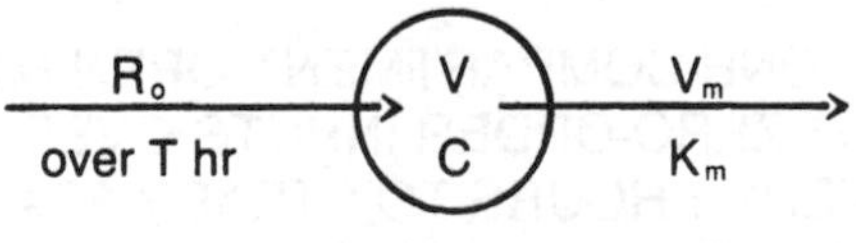

SCHEME 7.3.

The differential equation for the model is Equation (281) where R_o is the input rate in mass/time.

$$\frac{dC}{dt} = \frac{R_o}{V} - \frac{V'_m C}{K_m + C} \tag{281}$$

Integration of Equation (281) with limits $C = 0$ at $t = 0$ and $C = C_T$ at $t = T$ gives Equation (282).

$$\int_0^T C dt = \frac{1}{R_o - V_m}\left[VK_m C_T + \frac{VC_T^2}{2} - R_o K_m T\right] \tag{282}$$

After input ceases when $t > T$, integration yields Equation (283).

$$\int_T^{\text{Inf}} C dt = \frac{C_T}{V'_m}\left[\frac{C_T}{2} + K_m\right] = \frac{VC_T}{V_m}\left[\frac{C_T}{2} + K_m\right] \tag{283}$$

Addition of Equations (282) and (283), using the identity $R_oT = D_s$ and rearrangement, gives Equation (284).

$$\text{AUC 0–Inf} = \frac{K_m D_s}{V_m - R_o} - \left[\frac{1}{V_m - R_o} - \frac{1}{V_m}\right]\left[K_m - \frac{CT}{2}\right] VC_T$$

$$= \left[\frac{K_m C_o}{V'_m - (R_o/V)} - \frac{1}{V'_m - (R_o/V)} - \frac{1}{V'_m}\right]\left[K_m + \frac{C_T}{2}\right] C_T \tag{284}$$

As an example of application of this model, Wilkinson et al. [59] administered constant rate intravenous infusions of 720 ml of 8% v/v ethanol in physiological saline over two hours to six subjects and fitted the resulting capillary blood ethanol concentrations to the model by using Equation (281) during the infusion and Equation (262) after the infusion ceased. Fitting was done by numerical integration using the program NONLIN [68].

MODEL XXIX. ONE-COMPARTMENT OPEN MODEL WITH INTERMITTENT ZERO-ORDER INPUTS OVER T HOURS STARTING EVERY τ HOURS TO STEADY STATE WITH MICHAELIS-MENTEN ELIMINATION KINETICS

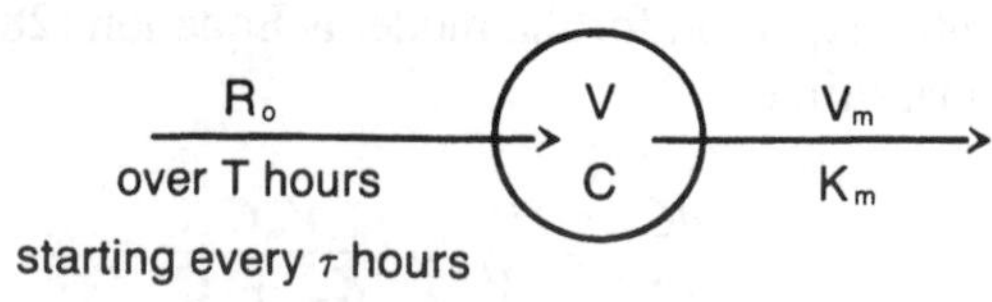

SCHEME 7.4.

Equation (281) is the same for this model. A similar integration is performed as that which led to Equation (282), except that $C_{ss}(T)$ replaces C_T to yield Equation (285). In Equation (285), $C_{ss}(T)$ is the steady-state concentration at time T at the end of zero-order input, and $C_{ss}(\tau)$ is the steady-state concentration at time τ at the end of the dosage interval. Thus, during zero-order input:

$$\int_0^T C dt = \frac{V}{R_o - V_m}\left[K_m\{C_{ss}(T) - C_{ss}(\tau)\} + \frac{1}{2}\{C_{ss}(T)^2 - C_{ss}(\tau)^2\} - \frac{R_o}{V}K_m T\right] \quad (285)$$

After zero-order input ceases when $t > T$:

$$\int_T^\tau C dt = \frac{K_m}{V'_m}[C_{ss}(T) - C_{ss}(t)] + \frac{1}{2V'_m}[C_{ss}(T)^2 - C_{ss}(t)^2] \quad (286)$$

Addition of Equations (285) and (286) gives Equation (287).

$$\int_0^\tau C dt = \frac{K_m R_o T}{V_m - R_o} - \left[\frac{1}{V'_m - (R_o/V)} - \frac{1}{V'_m}\right] \times \left[K_m\{C_{ss}(T) - C_{ss}(\tau)\} + \frac{1}{2}\{C_{ss}(T)^2 - C_{ss}(\tau)^2\}\right] \quad (287)$$

Equation (280) also holds for this model if Equation (287) is used for the area and $D_m = R_o T$.

MODEL XXX. ONE-COMPARTMENT OPEN MODEL WITH CONTINUOUS ZERO-ORDER INPUT TO STEADY STATE AND MICHAELIS-MENTEN ELIMINATION KINETICS

The model is the same as Model XXVIII except that zero-order input is continued until steady-state concentration, C_{ss}, is attained. Equation (288) gives this concentration.

$$C_{ss} = \frac{R_o}{\mathrm{CL}} = \frac{R_o K_m}{\mathrm{V}_m - R_o} = \frac{(R_o/V)K_m}{V'_m - (R_o/V)} = \frac{C_o K_m}{V'_m \tau - C_o} = \frac{C_o}{Q} \quad (288)$$

where

$$Q = \ln\left[\frac{(C_{ss})_{max}}{(C_{ss})_{min}}\right] = \frac{V'_m\tau - D_m}{VK_m} \tag{289}$$

It should be noted for the intermittent dosing models that $(C_{ss})_{min}$ is given by Equation (290) and the average steady-state concentration, $(C_{ss})_{av}$, by Equation (291). $(C_{ss})_{av}$ is analogous to C_{ss} for the continuous infusion model.

$$(C_{ss})_{min} = \frac{D_m}{V(e^Q - 1)} \tag{290}$$

$$(C_{ss})_{av} = \frac{\text{AUC } 0\text{–}\tau}{\tau} \tag{291}$$

From Equation (288) it can be seen the mean clearance for this model is given by Equation (292).

$$\overline{\text{CL}}_{ss} = D_m/\text{AUC } 0\text{–}\tau \tag{292}$$

MODEL XXXI. ONE-COMPARTMENT OPEN MODEL WITH FIRST-ORDER INPUT AND MICHAELIS-MENTEN ELIMINATION KINETICS

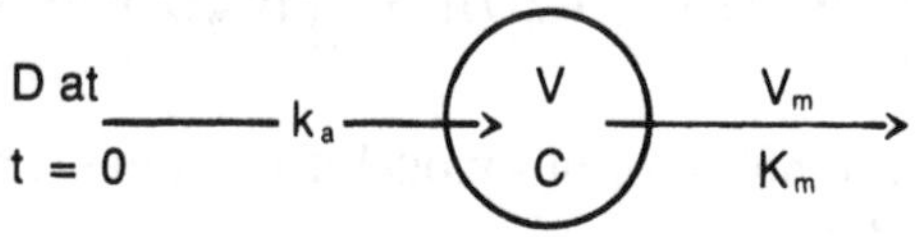

D at t = 0 —k_a→ (V, C) —V_m, K_m→

SCHEME 7.5.

The differential equation for the model is:

$$\frac{dC}{dt} = \left[\frac{k_a D}{V}\right] e^{-k_a t} - \frac{V'_m C}{K_m + C} \tag{293}$$

In Equation (293), C is the concentration in the volume V at time t, D is the absorbed dose, k_a is the first-order absorption rate constant, $V'_m = V_m/V$, where V_m is the maximal rate of metabolism, and K_m is the Michaelis constant and is equal to the concentration when the rate of metabolism is

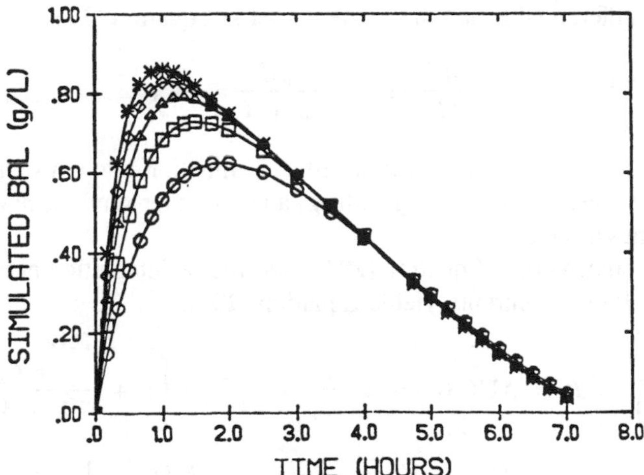

FIGURE 7.8. Simulations showing the effect of rate of availability on area under the *C,t* curve when Model XXXI applies.

equal to $V_m/2$. Equation (293) may only be integrated numerically, hence there is no simple expression for the area under the curve. The area depends upon k_a, hence the rate of absorption. Simulations showing this are given above.

Figure 7.8 shows results of simulations in which Equation (293) was numerically integrated using the program NONLIN [68]; average values of V_m' and K_m for alcohol in man were used and values of k_a of 3, 2.5, 2, 1.5, and 1 hr^{-1} were used for the curves from top to bottom. The results are due to the fact that when absorption is slowed C is smaller, and, since the clearance is $V_m/(K_m + C)$, the smaller the C, the larger the clearance, and the smaller the area during the absorption phase. It should be noted that after absorption has ceased the C,t profiles are essentially superimposable from four to seven hours.

MODEL XXXII. ONE-COMPARTMENT OPEN MODEL WITH BOLUS I.V. ADMINISTRATION AND PARALLEL FIRST-ORDER AND MICHAELIS-MENTEN ELIMINATION

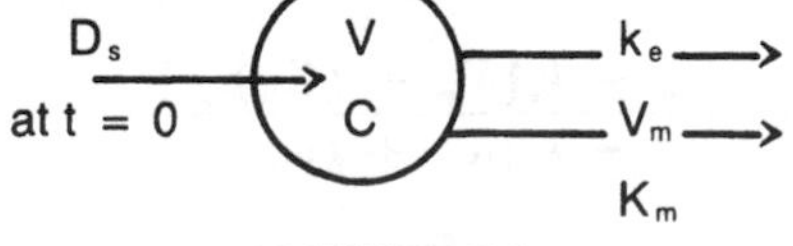

D_s at t = 0 → (V, C) → k_e →; V_m, K_m →

SCHEME 7.6.

The differential equation for the model is Equation (294).

$$\frac{dC}{dt} = -\frac{V'_m C}{K_m + C} - k_e C \tag{294}$$

In Equation (294) k_e is a first-order elimination rate constant (e.g., for urinary excretion of unchanged drug) and the other symbols have the same meanings as before.

Rearrangement of Equation (294) followed by integration between limits of $t = 0$ to $t =$ infinity yields Equation (295).

$$\int_0^{\mathrm{Inf}} C dt = \mathrm{AUC\ 0\text{–}Inf} = \frac{C_o}{k_e} - \frac{V'_m}{k_e^2} \ln\left[1 + \frac{k_e C_o}{V'_m + k_e K_m}\right]$$

$$= \frac{D_s}{V k_e} - \frac{V_m}{V k_e^2} \ln\left[1 + \frac{k_e D_s}{V_m + V k_e K_m}\right] \tag{295}$$

MODEL XXXIII. ONE-COMPARTMENT OPEN MODEL WITH INTERMITTENT BOLUS DOSING TO STEADY STATE AND PARALLEL FIRST-ORDER AND MICHAELIS-MENTEN ELIMINATION

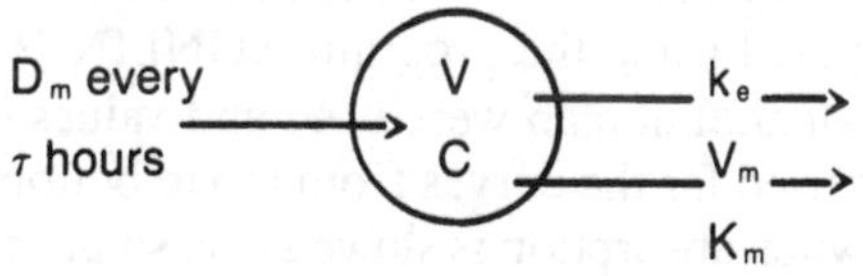

D_m every τ hours → V C → k_e →; V_m →, K_m

SCHEME 7.7.

The differential Equation (294) applies to this model also. In performing the integration, $(C_{ss})_{max}$ replaces C_o and $(C_{ss})_{min}$ replaces zero. Integration leads to Equation (296).

$$\mathrm{AUC\ 0\text{–}\tau} = \frac{1}{k_e}[(C_{ss})_{max} - (C_{ss})_{min}] + \frac{K_m}{k_e}$$

$$\times \ln\left[\frac{V'_m + k_e\{K_m + (C_{ss})_{max}\}}{V'_m + k_e\{V'_m + k_e\{K_m + (C_{ss})_{min}\}\}}\right] - \frac{V'_m + k_e K_m}{k_e^2}$$

$$\times \ln \frac{V'_m + k_e\{K_m + (C_{ss})_{max}\}}{V'_m + k_e\{K_m + (C_{ss})_{min}\}}$$

$$= \frac{D_m}{V k_e} - \frac{V_m}{V k_e^2} \ln\left[\frac{V_m + V k_e\{K_m + (C_{ss})_{max}\}}{V_m + V k_e\{K_m + (C_{ss})_{min}\}}\right] \tag{296}$$

Also we can define K as the limiting first-order rate constant for elimination of drug where

$$K = \frac{V'_m}{K_m} + k_e \tag{297}$$

and f as the fraction of drug reaching the circulation that is excreted unchanged in the urine, where

$$f = \frac{k_e}{K} \tag{298}$$

if the first-order path is assumed to be renal excretion of unchanged drug and V_m and K_m are assumed to refer to metabolism.

The instantaneous clearance for the model is given by Equation (299) and applies to Model XXXII also.

$$\text{CL} = \frac{VdC/dt}{C} = \frac{V_m}{K_m + C} + Vk_e \tag{299}$$

As an approximation, the mean clearance, $\overline{\text{CL}}_{\text{ss}}$, is given by Equation (300).

$$\overline{\text{CL}}_{\text{ss}} = \frac{V_m}{K_m + (C_{\text{ss}})_{\text{av}}} + Vk_e \tag{300}$$

MODEL XXXIV. ONE-COMPARTMENT OPEN MODEL WITH CONSTANT RATE (ZERO-ORDER) INPUT AND PARALLEL FIRST-ORDER AND MICHAELIS-MENTEN ELIMINATION

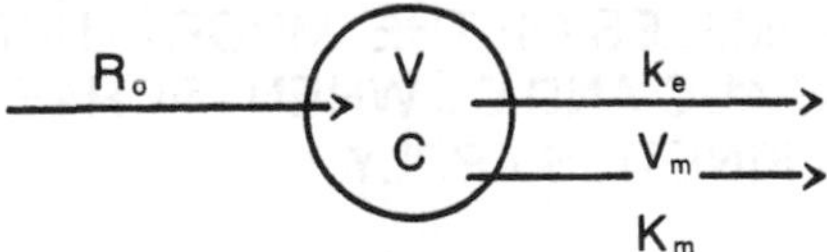

SCHEME 7.8.

The differential equation for the model is shown as Equation (301).

$$\frac{dC}{dt} = \frac{R_o}{V} - \frac{V'_m C}{K_m + C} - k_e C \tag{301}$$

The integration of Equation (301) was reported by Wagner et al. [57] under their Model VA and their Equation (52). For the corresponding model with zero-order inputs over T hours starting every τ hours to steady state, the integration of Equation (301) was also reported by Wagner et al. [57] under their model VB and their Equation (55).

MODEL XXXV. ONE-COMPARTMENT OPEN MODEL WITH CONTINUOUS ZERO-ORDER INPUT TO STEADY STATE AND PARALLEL FIRST-ORDER AND MICHAELIS-MENTEN ELIMINATION

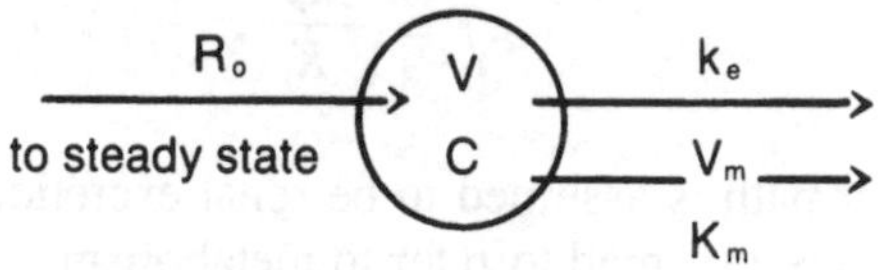

SCHEME 7.9.

The steady-state concentration, C_{ss}, is given by Equation (302).

$$C_{ss} = \frac{1}{2k_e}\left[\left\{\left(\frac{R_o}{V}\right) - V'_m - k_eK_m\right\} + \sqrt{\{(R_o/V) - V'_m - k_eK_m\}^2 + 4k_e(R_o/V)K_m}\right] \tag{302}$$

The steady-state clearance, CL_{ss}, for the model is given by Equation (303).

$$CL_{ss} = R_o/C_{ss} \tag{303}$$

FURTHER EXAMPLES OF THE IMPORTANCE OF INPUT RATE INFLUENCING AUC AND C_{ss} WHEN MICHAELIS-MENTEN ELIMINATION KINETICS APPLY

Example 1

Wagner [61] performed simulations with Model XXVIII, the results of which are reproduced here. Parameters $V'_m = 0.202$ mg/(ml × hr), $K_m = 0.082$ mg/ml, and $V = 40$ L for alcohol metabolism in man were employed. A dose of 10.5 g of alcohol was used and this dose was divided by the infusion time, T, and V to give the value of R_o/V in (mg/ml) × hr. When $R_o/V = 0$ the AUC 0–Inf was obtained with the bolus I.V. Equation

TABLE 7.4. AUCs Obtained in Input Rate Simulations with Models XXVI and XXVIII.

T (min)	R_o/V [(mg/ml) × hr]	AUC 0–Inf [(mg/ml) × hr]	AUC Ratio[a]
0	0	0.2769	1.000
3	5.25	0.2723	0.983
6	2.265	0.2702	0.976
9	1.75	0.2680	0.968
12	1.3125	0.2642	0.954
15	1.05	0.2618	0.945
18	0.875	0.2594	0.937
20	0.78758	0.2567	0.927
40	0.39379	0.2389	0.863
60	0.2625	0.2238	0.808
120	0.13125	0.1859	0.671

[a]Ratio = $(AUC)_{infusion}/(AUC)_{bolus}$.

(267), and when $R_o/V > 0$, AUC 0–Inf was obtained with Equation (284). Results are shown in Table 7.4.

As you can see both AUC and the $(AUC)_{inf}/(AUC)_{bolus}$ ratio decrease with decreasing input rate.

Example 2

Wagner [69] derived an equation to estimate the ratio $(AUC\ 0\text{–}\tau)_{zero\ order}/(AUC\ 0\text{–}\tau)_{bolus}$ where the numerator is the area in a dosage interval at steady state when input is zero order and the denominator is the area in a dosage interval at steady state when input is intermittent bolus. The derivation is reproduced here.

In Example 1 above, the dose was held constant and the infusion rate was altered. In this example the dose is a variable and the method of input also varies.

Under Model XXVII, Equation (279) is:

$$(AUC\ 0\text{–}\tau)_{bolus} = \frac{D_m}{V_m}\left[\frac{D_m}{2V} + (C_{ss})_{min} + K_m\right] \tag{279}$$

Under Model XXX, Equation (288) may be written as Equation (304).

$$(AUC\ 0\text{–}\tau)_{zero} = \frac{K_m D_m}{V_m - R_o} \tag{304}$$

Dividing Equation (304) by Equation (279), with some simplification, gives Equation (305).

$$(\text{AUC } 0\text{–}\tau)_{\text{zero}}/(\text{AUC } 0\text{–}\tau)_{\text{bolus}}$$

$$= \left[\frac{V_m}{V_m - R_o}\right]\left[\frac{K_m}{(D_m/2V) + (C_{\text{ss}})_{\min} + K_m}\right] \tag{305}$$

If we define $\beta = V_m/VK_m$ then Q is defined in Equation (289) as the first term on the right-hand side of Equation (306), and is also equal to the second term on the right-hand side of Equation (306) where $r = R_o/V_m$.

$$Q = \frac{V_m\tau - D_m}{VK_m} = (1 - r)\beta\tau \tag{306}$$

Under Model XXX, $(C_{\text{ss}})_{\min}$ is given by Equation (290).

$$(C_{\text{ss}})_{\min} = \frac{D_m}{V}\left[\frac{1}{e^Q - 1}\right] \tag{290}$$

Substituting from Equations (290) and (306) into Equation (305), followed by division of the numerator and denominator by K_m, gives Equation (307).

$$\frac{(\text{AUC } 0\text{–}\tau)_{\text{zero}}}{(\text{AUC } 0\text{–}\tau)_{\text{bolus}}} = \left[\frac{1}{1 - r}\right]\left[\frac{1}{1 + (D_m/VK_m)\{1/2 + 1/(e^Q - 1)\}}\right] \tag{307}$$

The terms D_m/VK_m and $1/(e^Q - 1)$ have opposite effects. As D_m/VK_m decreases, $1/(e^Q - 1)$ increases. Also, since $D_m = R_o\tau$, there is a minimum in the area ratio as R_o increases.

Wagner [69] also reported pooled Michaelis-Menten parameters for propranolol (mean V_m = 470 mg/day and mean K_m = 44.4 ng/ml) and verapamil (mean V_m = 575 mg/day and mean K_m = 133 ng/ml) which were derived from literature data. From these mean values and Equation (307), Figure 7.9 was constructed. This figure shows that for verapamil a zero-order sustained-release formulation would have equal or greater than 95% bioavailability at all dose rates, but for propranolol a zero-order sustained-release formulation would have only 65% bioavailability if tested at a dose rate of 160–180 mg/day. Literature data reported by Wagner [69] indicated that human studies on sustained-release propranolol had shown reduced bioavailability of the same magnitude as the above theory indicated.

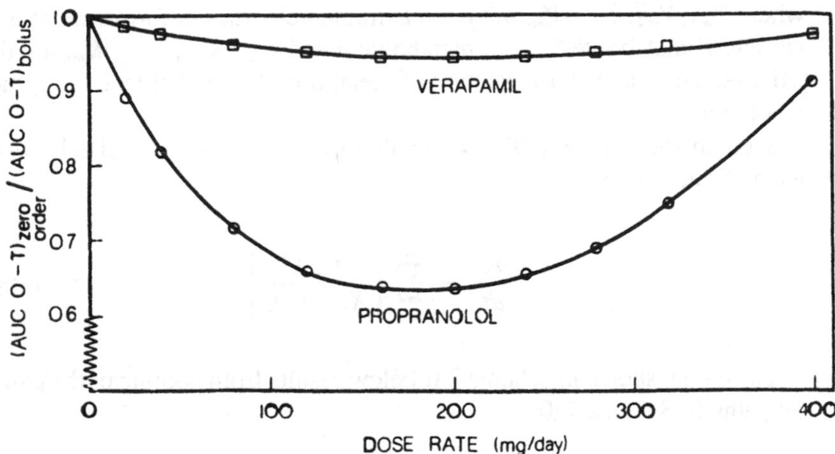

FIGURE 7.9. Ratio $(\text{AUC } 0-\tau)_{\text{zero order}}/(\text{AUC } 0-\tau)_{\text{bolus}}$ plotted *versus* dose rate, R_o, for propranolol and verapamil which are both subject to Michaelis-Menten elimination (from Reference [69], with permission from Mosby-Year Book, Inc.).

QUANTITATIVE POOLING OF MICHAELIS-MENTEN EQUATIONS IN MODELS WITH PARALLEL METABOLITE FORMATION PATHS

In classical first-order models, simplification is often achieved by the addition of first-order rate constants applicable to parallel paths for formation of two or more metabolites from the drug (i.e., $K = k_{m1} + k_{m2} + \ldots K_{mn}$ where K is the overall first-order elimination rate constant and K_{m1}, $k_{m2}, \ldots, k_{mn}$ are the first-order rate constants for formation of the first, second, . . ., and nth metabolite).

Scheme 7.10 for parallel metabolite paths involving Michaelis-Menten kinetics is shown below.

	$\underline{\quad V_{m1}, K_{m1} \quad}\!\!\rightarrow$	Metabolite 1
Concentration	$\underline{\quad V_{m2}, K_{m2} \quad}\!\!\rightarrow$	Metabolite 2
of drug in blood	$\underline{\quad V_{mn}, K_{mn} \quad}\!\!\rightarrow$	Metabolite n

SCHEME 7.10.

The differential equation for this model is:

$$-\frac{dC}{dt} = \frac{V_{m1}C}{K_{m1} + C} + \frac{V_{m2}C}{K_{m2} + C} + \ldots + \frac{V_{mn}C}{K_{mn} + C} \qquad (308)$$

where $V_{m1}, V_{m2}, \ldots, V_{mn}$ are the maximum velocities of formation of metabolite 1, metabolite 2, . . ., metabolite n and $K_{m1}, K_{m2}, \ldots, K_{mn}$ are the Michaelis constants for formation of metabolite 1, metabolite 2, . . ., metabolite n.

Sedman and Wagner [70] reported that Equation (308) may also be written as Equation (309).

$$-\frac{dC}{dt} = \sum_{i=1}^{n} \left\{ \frac{V_{mi} C}{K_{mi} + C} \right\} \tag{309}$$

The model shown in Scheme 7.11 below results from pooling of the parallel paths in Scheme 7.10.

Concentration of
drug in blood ——— V_p, K_p ———> Metabolites

SCHEME 7.11.

Equation (310) mathematically describes Scheme 7.11.

$$-\frac{dC}{dt} = \frac{V_p C}{K_p + C} \tag{310}$$

In Equation (310), V_p represents the pooled maximal velocity and K_p represents the pooled Michaelis constant.

If n Michaelis-Menten equations can be represented by a single pooled equation, then Equation (309) must be equivalent to Equation (310). Equating the right-hand sides of Equations (309) and (310) leads to Equation (311).

$$\frac{V_p C}{K_p + C} = \sum_{i=1}^{n} \left\{ \frac{V_{mi} C}{K_{mi} + C} \right\} \tag{311}$$

Division of both sides of Equation (311) by C gives Equation (312).

$$\frac{V_p}{K_p + C} = \sum_{i=1}^{n} \left\{ \frac{V_{mi}}{K_{mi} + C} \right\} \tag{312}$$

Imposing boundary conditions in Equation (312) leads to Equations (313) and (314). When $t = 0$, $C = C_o$, and substitution into Equation (312) yields Equation (313).

$$\frac{V_p}{K_p + C_o} = \sum_{i=1}^{n} \left\{ \frac{V_{mi}}{K_{mi} + C_o} \right\} \tag{313}$$

When $t = \text{Inf}$, $C = 0$, and substitution into Equation (312) yields Equation (314).

$$\frac{V_p}{K_p} = \sum_{i=1}^{n} \left\{ \frac{V_{mi}}{K_{mi}} \right\} \tag{314}$$

Solving Equations (313) and (314) for V_p and K_p, respectively, gives the dose-dependent Equations (315) and (316).

$$V_p = C_o \left[\sum_{i=1}^{n} \left\{ \frac{V_{mi}}{K_{mi} + C_o} \right\} \right] \left[\sum_{i=1}^{n} \left\{ \frac{V_{mi}}{K_{mi}} \right\} \right] \Bigg/ \left[\sum_{i=1}^{n} \left\{ \frac{V_{mi}}{K_{mi}} \right\} - \sum_{i=1}^{n} \left\{ \frac{V_{mi}}{K_{mi} + C_o} \right\} \right] \tag{315}$$

$$K_p = C_o \left[\sum_{i=1}^{n} \left\{ \frac{V_{mi}}{K_{mi} + C_o} \right\} \right] \Bigg/ \left[\sum_{i=1}^{n} \left\{ \frac{V_{mi}}{K_{mi}} \right\} - \sum_{i=1}^{n} \left\{ \frac{V_{mi}}{K_{mi} + C_o} \right\} \right] \tag{316}$$

In Equations (315) and (316), V_p and K_p are dependent upon the initial drug concentration, C_o, hence also upon the dose of drug.

However, when C_o is very much greater than K_p and all values of K_{mi}, then Equations (315) and (316) simplify to the dose-independent Equations (317) and (318), respectively.

$$V_p = \sum_{i=1}^{n} (V_{mi}) \tag{317}$$

$$K_p = \sum_{i=1}^{n} (V_{mi}) \Bigg/ \sum_{i=1}^{n} \left\{ \frac{V_{mi}}{K_{mi}} \right\} \tag{318}$$

Equations (317) and (318) relate the pooled constants, V_p and K_p, only to the microscopic constant, V_{mi} and K_{mi}, hence in this case V_p and K_p are independent of C_o and dose.

Sedman and Wagner [70] gave several examples with simulated data. Pooled maximal velocities of metabolism and Michaelis constants have been estimated for several drugs including phenytoin [71], verapamil [72], zonisamide (CI-912) [73], prednisone [74], theophylline [75], 5-fluorouracil [76–78], adinazolam [79], and nicardipine [80].

VENOUS EQUILIBRATION MODEL AND STEADY-STATE EQUATION INVOLVING ARTERIAL AND HEPATIC VENOUS DRUG CONCENTRATIONS

Model XXX above utilized the steady-state Equation (288), which is shown below.

$$C_{ss} = \frac{R_o K_m}{V_m - R_o} \tag{288}$$

The venous equilibration model of hepatic metabolism assumes that there is no concentration gradient in each hepatic sinusoid and that Equation (319) applies if the liver is the only site of metabolism.

$$v = R_o = Q\{(C_A)_{ss} - (C_V)_{ss}\} = \frac{V_m (C_V)_{ss}}{K_m + (C_V)_{ss}} \tag{319}$$

In Equation (319), R_o is the hepatic arterial infusion rate or the intravenous infusion rate if one assumes that all arterial concentrations are equal in the body, Q is the effective liver blood flow (volume/time), $(C_A)_{ss}$ is hepatic arterial drug concentration, $(C_V)_{ss}$ is the hepatic venous drug concentration, V_m is the maximal velocity of drug metabolism in the liver, and K_m is the Michaelis constant. Equating the third and fourth terms of Equation (319), and dividing through by Q, gives a quadratic equation whose positive root is Equation (320).

$$(C_V)_{ss} = 0.5\left[\left\{(C_A)_{ss} - \frac{V_m}{Q} - K_m\right\} + \sqrt{\{(C_A)_{ss} - V_m/Q - K_m\}^2 + 4K_m(C_A)_{ss}}\right] \tag{320}$$

In deriving Equation (320), one is not assuming that the infusion rate, R_o, is equal to the velocity of metabolism, v, as in Equation (319); for this reason Equation (320) applies to cases where there is both splanchnic and extrasplanchnic metabolism.

Wagner et al. [78] fitted steady-state arterial and hepatic venous plasma concentrations of 5-bromo-2′-deoxyuridine and 5-iodo-2′-deoxyuridine measured in cancer patients, dogs, and rabbits to Equation (320) by nonlinear least squares. These drugs are subject to both splanchnic and extrasplanchnic metabolism, and the fits were excellent. In such fittings the ordinate is $(C_v)_{ss}$ and the abscissa is $(C_A)_{ss}$. Figure 7.10 shows an example of such a fit [78].

It should be noted with respect to the above that V_m/Q is equivalent to the maximum arterial-venous concentration difference, i.e., $\{(C_A)_{ss} - (CV)_{ss}\}_{max}$.

The consequences of Michaelis-Menten elimination are:

(1) AUC and C_{ss} increase more than proportionately with increase in dose and dose rate, respectively.

(2) Drug clearance changes at every instant after single doses; and, in the elimination phase the clearance increases with decrease in the concentration.

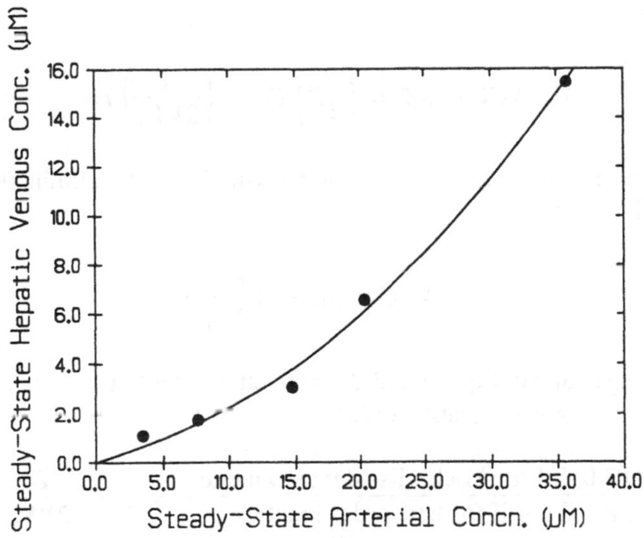

FIGURE 7.10. Plot of steady-state hepatic venous plasma concentration *versus* steady-state hepatic arterial plasma concentration with a fitted line based on Equation (320) for 5-bromo-2′-deoxyuridine in rabbit 10305D.

(3) The percentage of drug metabolized via a Michaelis-Menten pathway decreases with increase in the dose.
(4) The slower the rate of absorption, the smaller the AUC 0–Inf for a given dose; conversely, the greater the rate of absorption, the larger the AUC 0–Inf. These are true if the parameters V_m and K_m and the dose are in certain critical ranges.
(5) Rectilinear plots of blood concentration *versus* time are pseudolinear for approximately two-thirds of their length. This phase lasts after a single dose from some initial concentration, C_o, when input has ceased and Michaelis-Menten elimination kinetics alone are operative to $C_o/e = C_o/2.718 = 0.368\ C_o$.
(6) The time required to reach steady state increases with increase in the size of the maintenance dose given.

RELATIONSHIPS BETWEEN FIRST-ORDER AND MICHAELIS-MENTEN KINETICS [82]

Relationship No. 1

The one-compartment open model with Michaelis-Menten elimination Equation (273), rewritten below, relates the area under the curve to the single dose.

$$\text{AUC 0–Inf} = \left(\frac{K_m}{V_m}\right) D + \left(\frac{1}{2VV_m}\right) D^2 \tag{273}$$

For the one-compartment open model with first-order elimination, Equation (321) applies.

$$\text{AUC 0–Inf} = \left(\frac{K_m}{V_m}\right) D \tag{321}$$

Making a ratio of Equation (273) in the numerator to Equation (321) in the denominator gives Equation (322).

$$\frac{\text{AUC 0–Inf for Michaelis-Menten kinetics}}{\text{AUC 0–Inf for first-order kinetics}} = 1 + \frac{D}{2VK_m} \tag{322}$$

The last term of Equation (322) contains VK_m, which may be termed the "mass K_m."

Relationship No. 2

For the steady state,

$$(C_{ss})_{av} \text{ for Michaelis-Menten kinetics}$$

$$= \{(C_{ss})_{av} \text{ for first-order kinetics}\}/\{1 - v/V_m\} \qquad (323)$$

where v/V_m is the fraction of the enzyme system that is saturated and $1 - v/V_m$ is the fraction that is unsaturated.

Relationship No. 3

For the steady state,

$$\frac{\text{Dose if Michaelis-Menten elimination}}{\text{Dose if first-order elimination}} = 1 - v/V_m$$

$$= \text{Fraction of elimination system unsaturated} < 1 \qquad (324)$$

Michaelis-Menten elimination kinetics are cost-effective since the maintenance dose necessary to maintain a given steady-state concentration is less than it would be if elimination obeyed first-order kinetics, and the lower the dose, the lower the cost of the medication.

CHAPTER 8

More Complicated Nonlinear Models

MODEL XXXVI. FAMILY OF REVERSIBLE METABOLISM MODELS WITH MICHAELIS-MENTEN ELIMINATION

Model XVIII in Chapter 3 included three reversible metabolism models with first-order elimination kinetics. Model XXXVI includes the same models but with the conversion of the species in compartment #2 to the species in compartment #1 being Michaelis-Menten kinetics rather than first-order kinetics.

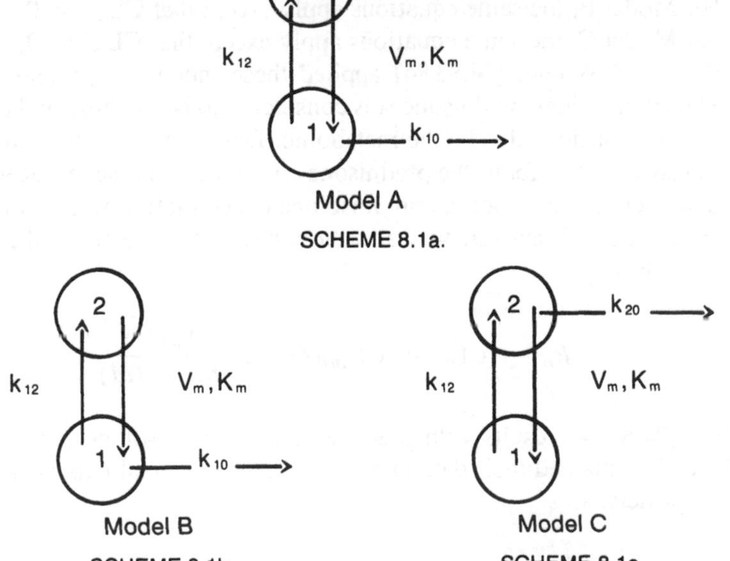

Model A

SCHEME 8.1a.

Model B

SCHEME 8.1b.

Model C

SCHEME 8.1c.

Symbolism

(D) = concentration of drug in compartment #1
(M) = concentration of metabolite in compartment #2
$(D)_{ss}$ = concentration of drug in compartment #1 at steady state
R_D = zero-order infusion rate of drug (mass/time)
CL_{12} = V_1k_{12} = conversion of drug to metabolite expressed as a clearance (volume/time)
CL_{10} = V_1k_{10} = clearance of drug from the body (volume/time)
CL_{20} = V_2k_{20} = clearance of metabolite from the body (volume/time)
V_1 = volume of distribution of drug (volume)
V_2 = volume of distribution of metabolite (volume)
V_m = maximum rate of conversion of metabolite to drug (mass/time)
K_m = concentration of metabolite at 0.5 V_m

The differential equations for Model A are:

$$\frac{d(D)}{dt} = R_D - CL_{12}(D) - CL_{10}(D) + \frac{V_m(M)}{K_m + (M)} \tag{325}$$

$$\frac{d(M)}{dt} = CL_{12}(D) - CL_{20}(M) - \frac{V_m(M)}{K_m + (M)} \tag{326}$$

For Model B, the same equations apply except that $CL_{20} = 0$.
For Model C, the same equations apply except that $CL_{10} = 0$.

Ferry and Wagner [74,83,84] applied these models to prednisone and prednisolone where prednisone was considered to be the drug and prednisolone was considered to be the metabolite. During intravenous infusion of prednisone in the rabbit, the prednisone concentration reached steady state in about ten minutes, but the prednisolone concentration was still rising at sixty minutes. When (D) was at steady state, equal to $(D)_{ss}$, then $d(D)/dt = 0$ thus,

$$R_D - (CL_{12} + CL_{10})(D)_{ss} + \frac{V_m(M)}{K_m + (M)} \tag{327}$$

Their [74,83,84] results with prednisone and prednisolone indicated that Model C explained their data in the rabbit. The original papers should be read for details.

MODEL XXXVII. THE FIRST-PASS TWO-COMPARTMENT OPEN MODEL WITH PERIPHERAL MICHAELIS-MENTEN ELIMINATION AND WITH AND WITHOUT FIRST-ORDER ELIMINATION OF DRUG BY URINARY EXCRETION FROM THE CENTRAL COMPARTMENT

Wagner et al. [85] published details of the derivations of the equations shown below. The model without urinary excretion of unchanged drug and for both intravenous infusion into the central compartment and zero-order input orally into the peripheral compartment (which includes the liver) is shown in Scheme 8.2.

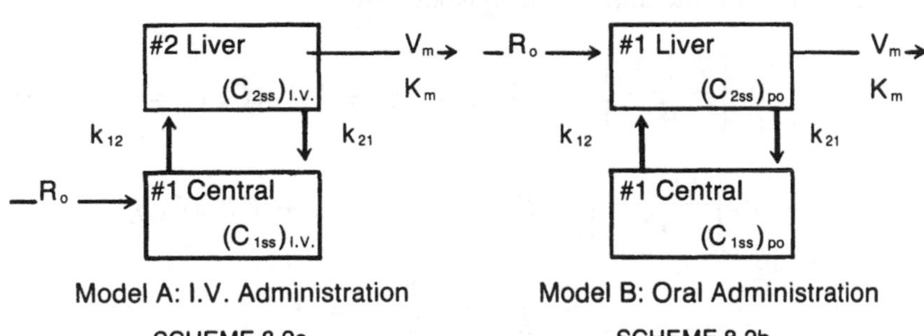

Model A: I.V. Administration

SCHEME 8.2a.

Model B: Oral Administration

SCHEME 8.2b.

Note that "1" refers to the central compartment #1, "ss" refers to the steady state, "I.V." refers to intravenous constant rate infusion into compartment #1, po refers to constant rate infusion into the peripheral compartment #2, C refers to whole blood concentration when Q is actually liver blood flow and to plasma concentration when Q is equal to rQ, where r = whole blood concentration/plasma concentration ratio.

For Model A the differential equations are:

$$V_1 \left\{ \frac{d(C_{1ss})_{\text{I.V.}}}{dt} \right\} = R_o - V_1 k_{12}(C_{1ss})_{\text{I.V.}} + V_2 k_{21}(C_{2ss})_{\text{I.V.}} = 0 \quad (328)$$

$$V_2 \left\{ \frac{d(C_{2ss})_{\text{I.V.}}}{dt} \right\} = V_1 k_{12}(C_{1ss})_{\text{I.V.}} + V_2 k_{21}(C_{2ss})_{\text{I.V.}}$$

$$- \frac{V_m(C_{2ss})_{\text{I.V.}}}{K_m + (C_{2ss})_{\text{I.V.}}} = 0 \quad (329)$$

Also,

$$Q = V_1 k_{12} = V_2 k_{21} \tag{330}$$

The following can be derived.

$$(C_{1ss})_{\text{I.V.}} = \left(\frac{1}{Q}\right) R_o + \left\{\frac{K_m}{V_m - R_o}\right\} R_o \tag{331}$$

$$(C_{2ss})_{\text{I.V.}} = \left\{\frac{K_m R_o}{V_m - R_o}\right\} \tag{332}$$

For Model B the differential equations are:

$$V_1 \left\{\frac{d(C_{1ss})_{\text{po}}}{dt}\right\} = -V_1 k_{12}(C_{1ss})_{\text{po}} + V_2 k_{21}(C_{2ss})_{\text{po}} = 0 \tag{333}$$

$$V_2 \left\{\frac{d(C_{2ss})_{\text{I.V.}}}{dt}\right\} = R_o - V_2 k_{21}(C_{2ss})_{\text{po}} + V_1 k_{12}(C_{1ss})_{\text{po}} - \frac{V_m (C_{2ss})_{\text{po}}}{K_m + (C_{2ss})_{\text{po}}} = 0 \tag{334}$$

The following can be derived.

$$(C_{2ss})_{\text{po}} = (C_{1ss})_{\text{po}} = \frac{K_m R_o}{V_m - R_o} \tag{335}$$

The steady-state bioavailability, F_{ss}, is given by Equation (336).

$$F_{ss} = \frac{1}{1 + (V_m - R_o)/QK_m} \tag{336}$$

The steady-state extraction ratio, E_{ss}, is given by Equation (337).

$$E_{ss} = 1 - F_{ss} = \frac{(C_{1ss})_{\text{I.V.}} - (C_{1ss})_{\text{po}}}{(C_{1ss})_{\text{I.V.}}} \tag{337}$$

Figure 8.1 shows a simulation with Equations (331) and (332), where the parameters were $Q = 0.02$ L/kg/min, $V_m = 1.5$ μmoles/kg/min, and $K_m = 25$ μM.

Figure 8.2 shows a simulation with Equation (336) using the same pa-

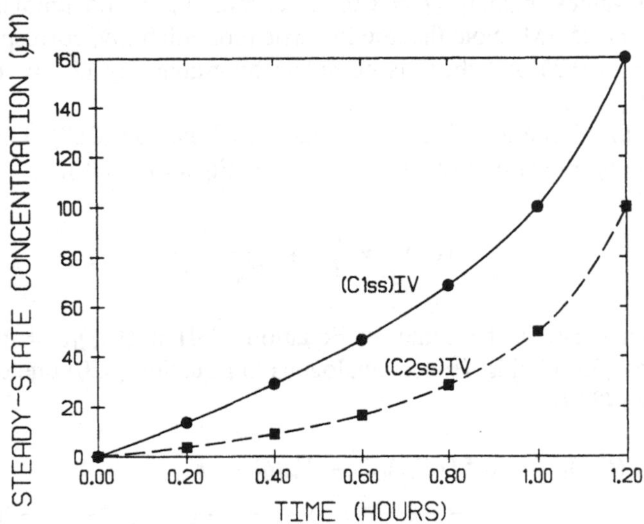

FIGURE 8.1. Simulation using Equations (331) and (332) with Q = 0.02 L/kg/min, V_m = 1.5 μmoles/kg/min, and K_m = 25 μM.

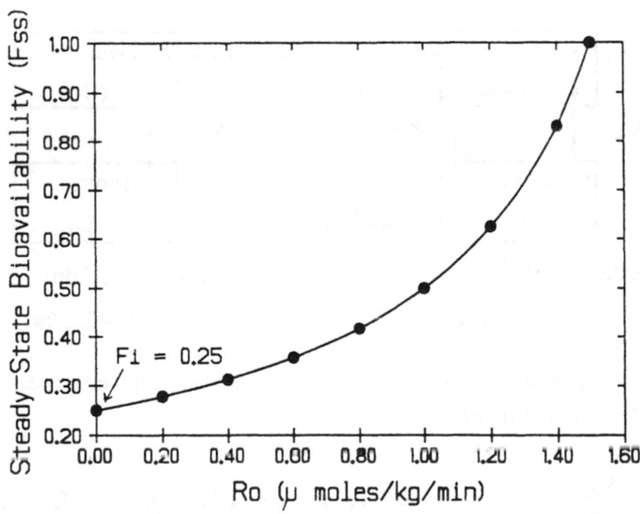

FIGURE 8.2. Simulation using Equation (336) with Q = 0.02 L/kg/min, V_m = 1.5 μmoles/kg/min, and K_m = 25 μM.

rameter values, namely $Q = 0.02$ L/kg/min, $V_m = 1.5$ μmoles/kg/min, and $K_m = 25$ μM. Note that the intrinsic bioavailability, corresponding to $R_o = 0$, is 0.25 and there is complete bioavailability ($F_{ss} = 1.0$) when $R_o = V_m = 1.5$.

Note that Equation (332) is the same as Equation (288) if $(C_{2ss})_{\text{I.V.}} = C_{ss}$. Solving Equation (319) for $(C_A)_{ss}$ gives Equation (338).

$$(C_A)_{ss} = \frac{R_o}{Q} + \frac{K_m R_o}{V_m - R_o} \tag{338}$$

Equation (338) is the same as Equation (331) if $(C_{1ss})_{\text{I.V.}} = (C_A)_{ss}$ and Equation (332) holds. Hence, analogous to Equation (320) one can obtain Equation (339).

$$(C_{2ss})_{\text{I.V.}} = 0.5\Big[(C_{1ss})_{\text{I.V.}} - V_m/Q - K_m + \sqrt{\{(C_{1ss})_{\text{I.V.}} - V_m/Q - K_m\}^2 + 4K_m(C_{1ss})_{\text{I.V.}}}\Big] \tag{339}$$

A simulation with Equation (339) using the same parameter values as for Figures 8.1 and 8.2, namely $V_m/Q = 75$ μm and $K_m = 25$ μM, is shown in Figure 8.3.

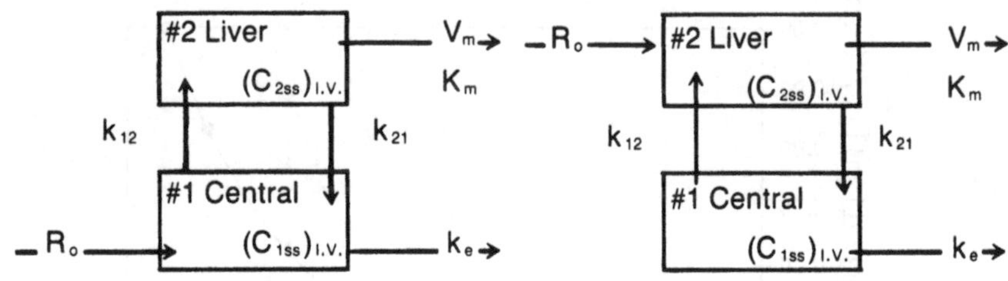

Model C: I.V. Administration

SCHEME 8.2c.

Model B: I.V. Administration

SCHEME 8.2d.

Utilizing Equation (330) and $\text{CL}_r = V_1 k_e$, the differential equations for Model C are as follows.

$$V_1\left(\frac{d(C_{1ss})_{\text{I.V.}}}{dt}\right) = R_o - \text{CL}_r(C_{1ss})_{\text{I.V.}} - Q\{(C_{1ss})_{\text{I.V.}} - (C_{2ss})_{\text{I.V.}}\} = 0 \tag{340}$$

$$V_2\left(\frac{d(C_{2ss})_{\text{I.V.}}}{dt}\right) = Q\{(C_{1ss})_{\text{I.V.}} - (C_{2ss})_{\text{I.V.}}\} - \frac{V_m(C_{2ss})_{\text{I.V.}}}{K_m + (C_{2ss})_{\text{I.V.}}} = 0 \tag{341}$$

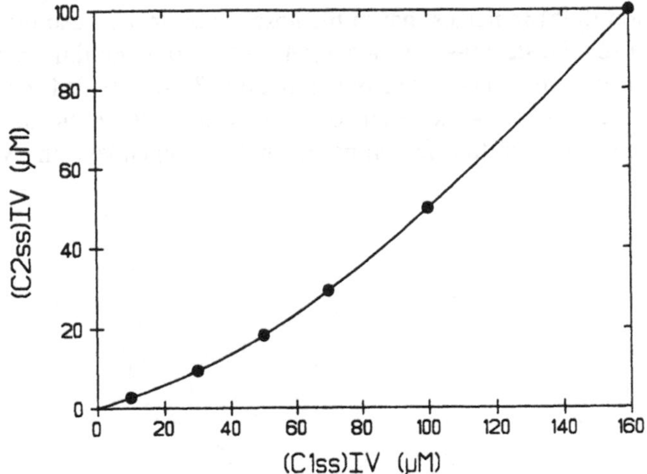

FIGURE 8.3. Simulation using Equation (339) with V_m/Q = 75 μM and K_m = 25 μM.

Equations (340) and (341) lead to Equations (342) and (343).

$$(C_{1ss})_{I.V.} = \left(\frac{1}{Q}\right)\{R_o - CL_r(C_{1ss})_{I.V.}\} + \frac{K_m\{R_o - CL_r(C_{1ss})_{I.V.}\}}{V_m - \{R_o - CL_r(C_{1ss})_{I.V.}\}} \tag{342}$$

$$(C_{2ss})_{I.V.} = \frac{K_m\{R_o - CL_r(C_{1ss})_{I.V.}\}}{V_m - \{R_o - CL_r(C_{1ss})_{I.V.}\}} \tag{343}$$

The differential equations for Model B are as follows.

$$V_1\left(\frac{d(C_{1ss})_{po}}{dt}\right) = -Q\{(C_{1ss})_{I.V.} - (C_{2ss})_{po}\} - CL_r(C_{1ss})_{po} = 0 \tag{344}$$

$$V_2\left(\frac{d(C_{2ss})_{po}}{dt}\right) = R_o + Q\{(C_{1ss})_{po} - (C_{2ss})_{po}\} - \frac{V_m(C_{2ss})_{po}}{K_m + (C_{2ss})_{po}} = 0 \tag{345}$$

From Equations (344) and (345), one obtains Equations (346) and (347).

$$(C_{1ss})_{po} = \left(\frac{Q}{Q + CL_r}\right)\frac{K_m\{R_o - CL_r(C_{1ss})_{po}\}}{V_m - \{R_o - CL_r(C_{1ss})_{po}\}} \tag{346}$$

$$(C_{2ss})_{po} = \frac{K_m\{R_o - CL_r(C_{1ss})_{po}\}}{V_m - \{R_o - CL_r(C_{1ss})_{po}\}} \tag{347}$$

Equation (335) indicates that in the absence of urinary excretion of unchanged drug the steady-state concentration after oral administration is independent of liver blood flow; but Equation (346), derived for the case where there is urinary excretion of unchanged drug, indicates that the steady-state concentration in compartment #1 is dependent on liver blood flow.

CHAPTER 9

Urinary Excretion

RENAL CLEARANCE

Renal clearance, CL_r, is the ratio of the urinary excretion rate, dA_e/dt, to the drug concentration, C, in whole blood or plasma as indicated in Equation (348).

$$\mathrm{CL}_r = \frac{dA_e/dt}{C} \tag{348}$$

Renal clearance may be estimated in several ways:

(1) By dividing the average urinary excretion rate in some period of time, $\Delta A_e/\Delta t$, by the drug concentration in plasma at the time corresponding to the midpoint of the urine collection period as indicated by Equation (349).

$$\mathrm{CL}_r = \frac{\Delta A_e}{\Delta t}\bigg/ C \tag{349}$$

(2) By plotting the average urinary excretion rate *versus* the drug concentration at the midpoints of the excretion intervals and fitting the least squares straight line forced through the origin (i.e., 0,0 point).

$$\frac{\Delta A_e}{\Delta t} = \mathrm{CL}_r C \tag{350}$$

Figure 9.1 shows an example of this method using some of the peni-

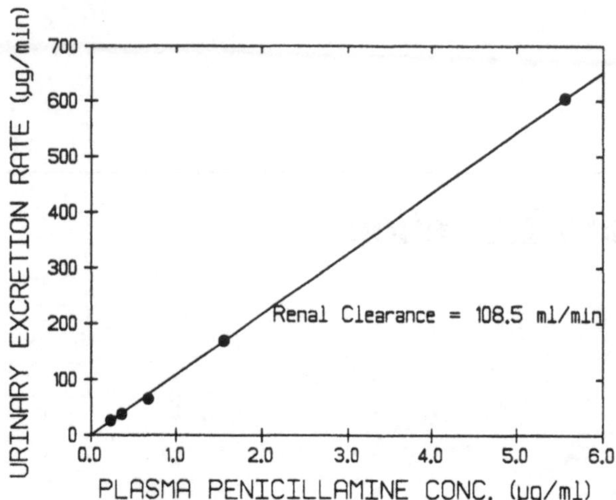

FIGURE 9.1. Example of estimation of renal clearance using Equation (350). Subject RB was administered 1 g of penicillamine orally. The slope of the line is the renal clearance.

cillamine data of Bergstrom [86]. The slope of the line is the renal clearance, in this example, 108.5 ml/min. In Figure 9.1 only data in the range 120 to 720 minutes was used since penicillamine exhibits double peaks in the plasma concentration-time plot below 120 minutes, but data from 120 minutes on were monoexponential as shown in Figure 9.2.

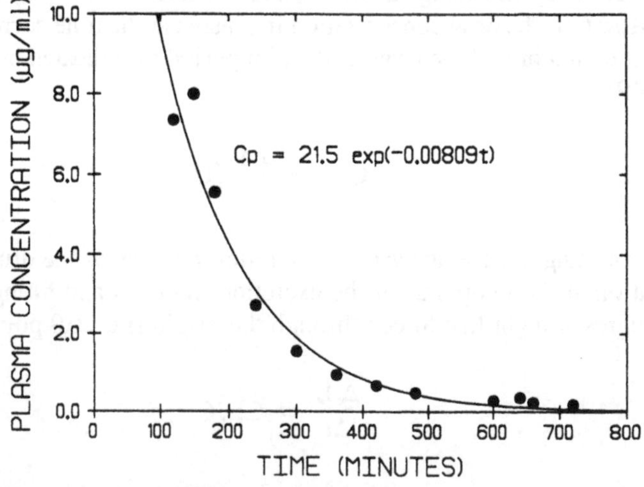

FIGURE 9.2. Monoexponential decline of plasma concentrations of penicillamine in subject RB after a 1 g oral dose of penicillamine.

(3) Integration of Equation (348) between limits of t_1 and t_2 gives Equation (351), where A_e is the amount excreted in urine between t_1 and t_2. The penicillamine data used in

$$\mathrm{CL}_r = (A_e)\int_{t_1}^{t_2} C dt \tag{351}$$

Figures 9.1 and 9.2 indicated that 108.2 mg of penicillamine was excreted in the period 120 to 720 minutes and that the area under the plasma concentration-time curve as 1027 (μg $\times$ hr)/ml in the same time interval, hence the renal clearance was (108.2 $\times$ 103)/1027 = 105.4 ml/min, which agrees with the value of 108.5 ml/min shown in Figure 9.1.

In Equation (351), t_1 could be equal to zero and t_2 equal to infinity; thus, the whole time course of urinary excretion and plasma concentration could be utilized and the overall mean renal clearance estimated.

RENAL EXCRETION RATE CONSTANT, k_e

Equation (352) indicates that renal clearance is the product of the renal excretion rate constant, k_e, and a volume of distribution, V.

$$\mathrm{CL}_r = k_e V \tag{352}$$

There are two simple ways of estimating k_e, as follows:

(1) By obtaining the slope of the least squares line when the natural logarithm of the urinary excretion rate is plotted *versus* time, as indicated by Equation (353).

$$\ln\left\{\frac{\Delta A_e}{\Delta t}\right\} = \text{Intercept} - k_e t \tag{353}$$

An example, using the penicillamine data used for Figures 9.1 and 9.2, is shown in Figure 9.3. A k_e value of 0.00649 min^{-1} was estimated.

(2) By obtaining the slope of the least squares line when the natural logarithm of the amount of drug not yet excreted in the urine is plotted *versus* time as indicated by Equation (354), where $(A_e)_{\mathrm{Inf}}$ is the amount excreted in

$$\ln\{(A_e)_{\mathrm{Inf}} - (A_e)\} = \text{Intercept} - k_e t \tag{354}$$

the urine in infinite time and A_e is the amount excreted in time t.

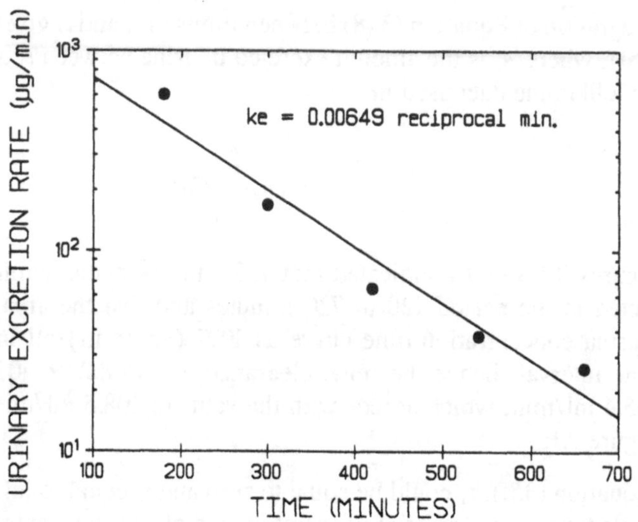

FIGURE 9.3. Example of estimation of renal excretion rate constant, k_e, based on Equation (353) using the penicillamine data of subject RB [86].

An example is shown in Figure 9.4 which utilizes the same penicillamine data as shown in Figures 9.1 through 9.3. A k_e value of 0.00690 min^{-1} was estimated. Note that the scatter of points about the line is much less in Figure 9.4 than in Figure 9.3.

Using the average renal clearance of 106.95 ml/min and the average k_e of 0.006695 min^{-1} from the above, a volume of distribution of 16 L was esti-

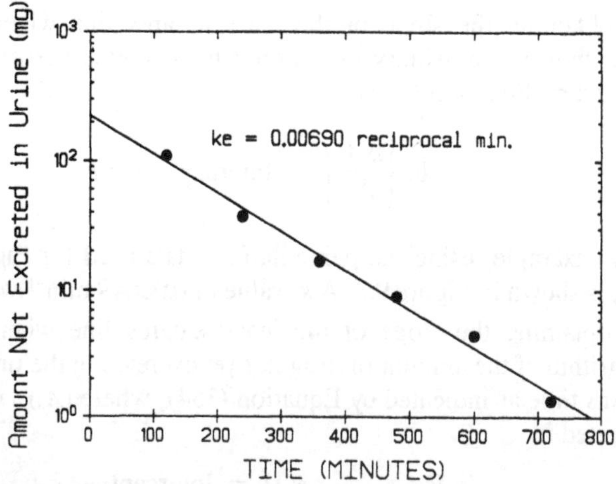

FIGURE 9.4. Example of estimation of renal excretion rate constant, k_e, based on Equation (354) using the penicillamine data of subject RB [86].

mated using Equation (352); this is the same as that accepted for extracellular fluid.

MECHANISMS OF RENAL EXCRETION OF DRUGS

The mechanisms involved in renal excretion are [87]:

(1) *Glomerular filtration.* Part of the blood flowing through the kidney is ultrafiltered in the glomeruli and this produces about 125 ml of filtrate per minute. Only free (unbound) drug is filtered.

(2) *Passive back diffusion.* Nearly all of the water in the glomerular filtrate is reabsorbed in the renal tubuli. As a result, the concentration of the drug in the tubular fluid increases markedly and a concentration gradient is established between the tubular fluid and plasma. This gradient is the driving force for diffusion of the drug from tubular fluid to plasma.

(3) *Tubular secretion.* Some drugs are excreted from the capillaries surrounding the renal tubuli into the tubular fluid by a carrier-mediated and active process. This process is saturable and there is a transport maximum value for each drug which is actively secreted.

In most cases the clearance as a result of tubular secretion may be adequately described by Michaelis-Menten kinetics as in Equation (355), where CL_{TS} is that part of the renal clearance that proceeds by tubular secretion, T_m is the transport maximum (i.e., the maximum amount of drug that can be secreted per unit time), K_T is the apparent Michaelis constant for the drug with respect to its secretory carrier, with units of concentration, and C_p is the free (unbound) drug concentration.

$$CL_{TS} = \frac{T_m}{K_T + C_p} \tag{355}$$

(4) *Active tubular reabsorption.* Some drugs are actively reabsorbed from the tubular fluid into the capillaries around the renal tubuli. This process is analogous to that for endogenous substances such as glucose. This process is saturable.

When $K_T \gg C_p$, kinetics are linear and excretion rate is directly proportional to the free (unbound) drug concentration. If the protein binding of drug in plasma is linear (i.e., independent of concentration) then urinary excretion rate will be directly proportional to total (free + bound) drug concentration.

The renal excretion rate of a drug is the net result of filtration, secretion, and reabsorption, as indicated by Equation (356).

$$\text{Rate of excretion} = \text{Rate of filtration} + \text{Rate of secretion} - \text{Rate of reabsorption} \quad (356)$$

Since only free (unbound) drug in plasma is filtered at the glomerulus, the rate of filtration is given by Equation (357), where C_p is the total (bound + free) plasma concentration and f_p is the free fraction of the drug in plasma.

$$\text{Rate of filtration} = f_p \text{GFR}\, C_p \quad (357)$$

FLOW AND pH DEPENDENCE OF RENAL CLEARANCE

Passive reabsorption depends upon: (1) the degree of ionization and the lipophilicity of the drug; (2) the urine flow rate; and (3) the pH of the luminal fluid in the renal tubule. Forced diuresis hastens the renal elimination of drugs and shortens the time required to detoxify patients overdosed with certain drugs. Renal clearances of small molecules, such as ethanol, barbiturates, and theophylline, are flow-dependent as a result of convective reabsorption, but the phenomenon is not important for molecules with molecular weights greater than about 200. Basic drugs are renally eliminated faster under acidic urinary conditions because the drugs are more ionized; basic drugs are renally eliminated slower under basic urinary conditions because the concentration of unionized molecules is greater and these are back-extracted to a greater degree than ions. Acidic drugs are renally excreted faster under basic urinary conditions since the drugs are more ionized and less back-extracted; acidic drugs are renally excreted slower under acidic urinary conditions since the concentration of unionized molecules is greater and these are back-extracted to a greater degree. In therapeutics, ammonium chloride and ascorbic acid are used to acidify the urine and sodium bicarbonate is used to alkalinize the urine.

The following generalizations apply to most drugs:

(1) Polar weak acids do not get reabsorbed regardless of their degree of ionization.

(2) Very weak acids ($pK_a > 7.5$) are extensively reabsorbed at all pH values of urine and are insensitive to changes in urine pH since they have a large percent unionized.

(3) Strongly acidic drugs ($pK_a < 3$) are reabsorbed to a very small extent as a result of extensive ionization at all urine pH values.
(4) Nonpolar acids (pK_a from 3 to 7.5) have renal clearances which are very sensitive to urinary pH.
(5) Nonpolar bases (pK_a from 6 to 12) have renal clearances which are pH-sensitive.

(1) Strongly acidic drugs (pK_a < 3) are absorbed to a very small extent as a result of extensive ionization at all usual pH values.

(2) Weak acidic drugs (pK_a from 3 to 7.5) have gradual variations which are very sensitive to a change in pH.

(3) Strong bases (pK_a from 5 to 11) have considerable variations which are pH-sensitive.

CHAPTER 10

Absorption Analysis and Bioavailability

ABSORPTION ANALYSIS

As indicated in the Introduction chapter, drug delivery scientists and pharmacokineticists in academia and industry should be able to estimate *in vivo* drug absorption rates following administration of different dosage forms of a drug. This chapter will give details on how to do this.

Peak Time and Peak Concentration

The peak concentration and time of the peak concentration are used in bioavailability testing as parameters to make comparisons. These reflect rate of absorption but only indirectly.

Simulations were carried out with Model II of Chapter 1. Simulations of both single-dose and steady-state data were made. The elimination rate constant (K) used was 0.231 hr^{-1}, corresponding to an elimination half-life of three hours. For steady state, a dosage interval, τ, of three hours was used. Absorption rate constants, k_a, of 0.5, 0.75, 1, 2, 3, and 4 hr^{-1} were employed. The single-dose equation giving the concentration, C, as a function of time is Equation (358). The $t_{\max}$ was estimated with Equation (359) and $C_{\max}$ with Equation (360). An FD/V value of 100 was employed.

$$C = 100\left\{\frac{k_a}{k_a - 0.231}\right\}\{e^{-0.231t} - e^{-k_a t}\} \tag{358}$$

$$t_{\max} = \frac{1}{k_a - K}\ln\left\{\frac{k_a}{0.231}\right\} \tag{359}$$

$$C_{\max} = 100\left\{\frac{k_a}{k_a - 0.231}\right\}\{e^{-0.231t_{\max}} - e^{-k_a t_{\max}}\} \tag{360}$$

For steady state, the concentration, C, as a function of time was obtained with Equation (361), t_{max} with Equation (362), and C_{max} with Equation (363).

$$C = 100\left\{\frac{k_a}{k_a - 0.231}\right\}\left[(2e^{-0.231t}) - \frac{e^{-k_a t}}{1 - e^{-3k_a}}\right] \quad (361)$$

$$t_{max} = \left\{\frac{1}{k_a - 0.231}\right\} \ln\left[\frac{0.5k_a}{0.231(1 - e^{-3k_a})}\right] \quad (362)$$

$$C_{max} = 100\left\{\frac{k_a}{k_a - 0.231}\right\}\left\{\frac{e^{-0.231t_{max}}}{1 - e^{-(0.231)(3)}} - \frac{e^{-k_a t_{max}}}{1 - e^{-3k_a}}\right\} \quad (363)$$

Table 10.1 gives values obtained with Equations (359), (360), (362), and (363).

The single-dose time courses are shown in Figure 10.1. Note that the faster the rate of absorption (the greater the k_a) the shorter the time of the peak concentration and the higher the maximum concentration. There is also considerable separation of the curves for different absorption rate constants.

The steady-state time courses are shown in Figure 10.2. The faster the rate of absorption, the shorter the time of the peak concentration and the higher the maximum concentration as for the single doses. But, at steady state, the rate of absorption does not affect the shape of the curves nearly as much as it does after single doses. In Figure 10.2 the curves are much closer together than they are in Figure 10.1.

In Figures 10.3 and 10.4 the data shown in Table 10.1 are plotted. In Figure 10.3 t_{max} is plotted *versus* k_a and in Figure 10.4, C_{max} is plotted *versus* k_a. These plots emphasize the observations made above.

TABLE 10.1. t_{max} and C_{max} after Single Doses and at Steady State for the Simulations Outlined in Equations (358) through (363).

k_a (hr^{-1})	t_{max} (hr) Single Dose	t_{max} (hr) Steady State	C_{max} (μg/ml) Single Dose	C_{max} (μg/ml) Steady State
0.5	2.8706	1.2324	51.525	150.45
0.75	2.2697	1.1481	59.205	153.41
1	1.9055	1.0706	64.393	156.18
2	1.2202	0.8297	75.438	165.11
3	0.9259	0.6757	80.744	171.10
4	0.7566	0.5727	83.965	175.22

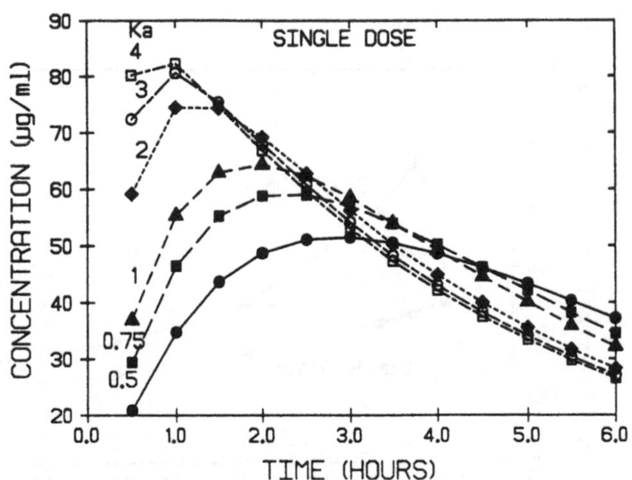

FIGURE 10.1. Showing how the time of the peak concentration increases and the peak concentration decreases as the absorption rate constant decreases after single doses. The simulation used Model II of Chapter 1.

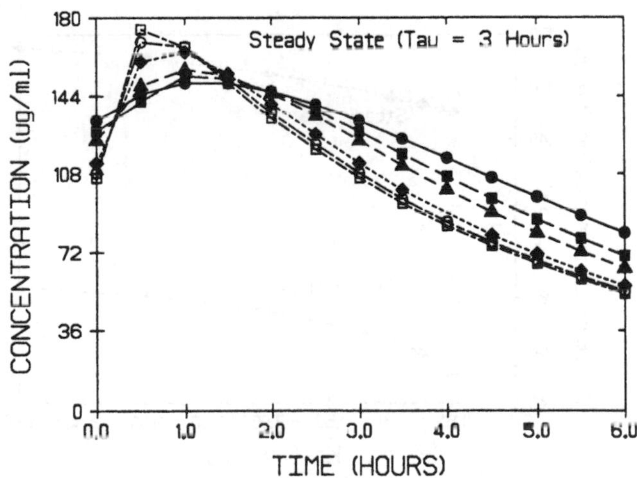

FIGURE 10.2. Qualitatively similar results to those in Figure 10.1 but quantitatively much different when steady-state conditions hold.

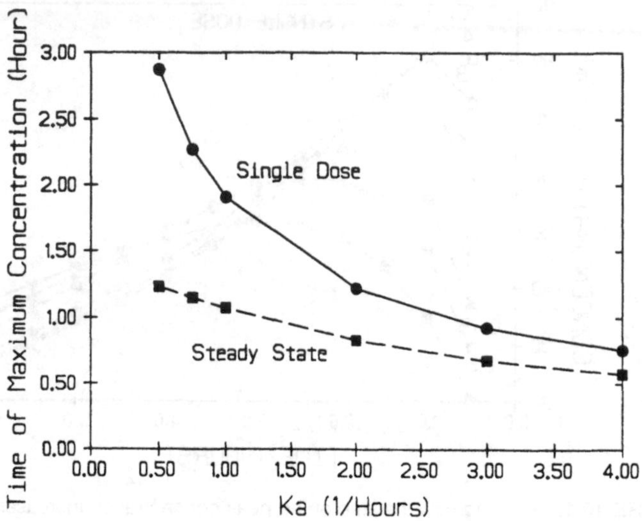

FIGURE 10.3. Plots of t_{max} *versus* k_a after single doses and at steady state when Model II of Chapter 1 and data from Table 10.1 are used.

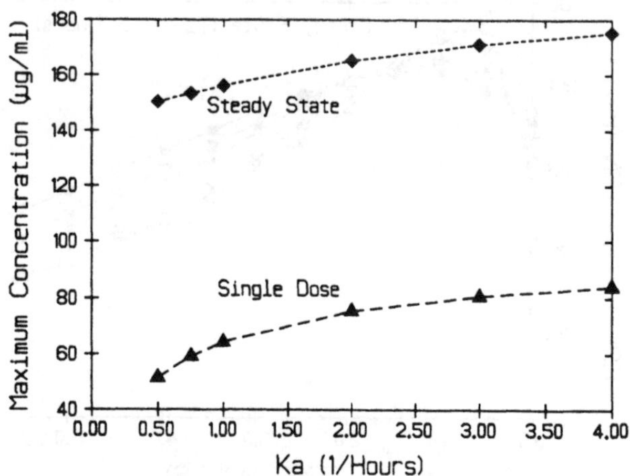

FIGURE 10.4. Plots of C_{max} *versus* k_a after single doses and at steady state when Model II of Chapter 1 and data from Table 10.1 are used.

It should be noted that the above observations with respect to the effect of rate of absorption on t_{max} and C_{max} were made when FD/V and K were held constant. The situation becomes much more complicated when these variables also vary along with the rate of absorption.

(A) WAGNER-NELSON SINGLE-DOSE LINEAR EQUATION

The model is shown in Scheme 10.1.

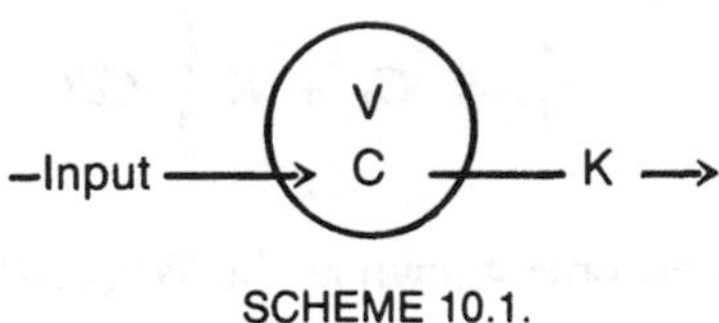

SCHEME 10.1.

Symbolism

A_T = amount of drug absorbed or which reaches the central compartment in time T
A_b = amount of drug in the "body" or central compartment at time T
A_e = amount of drug eliminated in time T
V = volume of central compartment
C = concentration of drug in the central compartment
C_T = concentration of drug in the central compartment at time T
K = first-order elimination rate constant

$$A_{max}/V = \text{asymptotic value of } C_T + K\int_0^T C dt$$

$$\text{FA} = \text{fraction absorbed} = (A_T/V)/(A_{max})/V = A_T/A_{max}$$

Derivation

$$A_T = A_b + A_e \tag{364}$$

$$A_b = VC \tag{365}$$

$$A_e = \text{CL}\int_0^T C dt = VK\int_0^T C dt \tag{366}$$

Substituting for A_b and A_e in Equation (364) from Equations (365) and (366) gives:

$$A_T = VC_T + VK \int_0^T Cdt \tag{367}$$

Dividing Equation (367) through by V gives:

$$\frac{A_T}{V} = C_T + K \int_0^T Cdt \tag{368}$$

Equation (368) has become known as the Wagner-Nelson equation [6].

$$\frac{A_{max}}{V} = \frac{1}{n} \sum_{i=1}^{n} \left\{ C_T + K \int_0^T Cdt \right\} \tag{369}$$

where n is the number of C, t points used to estimate K at the tail end of the C, t curve, and the right-hand side of Equation (369) is the average value of $C_T + K \int_0^T Cdt$ for those points used to estimate K. Often

$$\frac{A_{max}}{V} = K \int_0^{Inf} Cdt \tag{370}$$

is a good estimate of A_{max}/V but sometimes it is better to use Equation (369).

$$FA = \frac{A_T/V}{A_{max}/V} = \frac{A_T}{A_{max}} \tag{371}$$

(B) WAGNER-NELSON SINGLE-DOSE EQUATION MODIFIED FOR MICHAELIS-MENTEN ELIMINATION KINETICS

The model is the same as Scheme 10.1 except elimination obeys Michaelis-Menten kinetics instead of first-order kinetics. The equation is:

$$\frac{A_T}{V} = C_T + \int_0^T \frac{V_m' C}{K_m + C} dt \tag{372}$$

To apply Equation (372) one estimates V'_m and K_m by one of the methods described in Chapter 7, then obtains the values of $V'_m C/(K_m + C)$, then uses the combined linear and logarithmic trapezoidal rules to estimate the integrals, then adds the concentrations to obtain values of A_T/V as a function of time.

(C) WAGNER-NELSON LINEAR MULTIPLE DOSE EQUATION [88]

The model is the same as Scheme 10.1 except that multiple doses are administered at uniform time intervals.

Equation (373) is the appropriate equation.

$$\frac{A_T}{V} = C_n + K \int_0^T C dt - (C_n)_o \tag{373}$$

In Equation (373), C_n is the drug concentration at time T in the dosage interval after the nth dose and $(C_n)_o$ is the drug concentration at the beginning of the particular dosage interval. It should be noted that one performs the Wagner-Nelson calculations on the multiple dose C_n, t data to obtain values of $C_n + K \int_0^T C dt$ and then subtracts the initial concentration, $(C_n)_o$. The wrong answers would be obtained if one first subtracted $(C_n)_o$ from the C_n values and then performed the Wagner-Nelson calculations.

Performing Wagner-Nelson Calculations Using Method (A)—The Single-Dose Linear Equations (368) to (371)

One plots *terminal* C, t data on semilogarithmic graph paper then chooses those points that appear to be linear. One then obtains the rate constant, K, by least squares regression using those points chosen and $\ln C = \ln C_o - Kt$.

Using a combination of the linear and logarithmic trapezoidal rules (the linear rule when C is increasing or remaining constant and the logarithmic rule when C is decreasing) one obtains the cumulative AUC values for each time. One then multiplies each AUC by K and adds the C_T values to the products to obtain the A_T/V values. One then averages those A_T/V values for the points used to estimate K in order to estimate A_{max}/V. One obtains the FA values by dividing each A_T/V value by the A_{max}/V according to Equation (371). One can then plot FA *versus* t and fit the data to the appropriate equation to determine the kinetics of absorption.

EXAMPLE OF APPLICATION OF METHOD (A)

The data were theophylline serum concentrations after Theo-Dur, 300 mg, Lot 6A222 in subject #1. Data and the calculations are shown in Table 10.2.

The FA, t data from one to eight hours was linear on rectilinear graph paper, indicating zero-order absorption of the theophylline from the sustained-release product Theo-Dur. The plot of these data is shown in Figure 10.5.

EXAMPLE OF APPLICATION OF METHOD (B)

The ethanol whole capillary blood concentrations of subject #5 during and following a two-hour intravenous infusion of 720 ml of 8% v/v ethanol in physiological saline [59,67] were used and are shown in columns 1 and 2 of Table 10.3. The C, t data from 2 to 6.25 hours were used with the program of Thomas et al. [66] and Equation (263) to give least squares estimates of V'_m = 0.1727 (mg × hr)/ml and K_m = 0.0133 mg/ml. These values were used to produce the values shown in columns 3 to 5 of Table 10.3.

The fraction available (FA) values shown in the last column of Table 10.3 were plotted *versus* time and are shown in Figure 10.6. The least squares line free to pass through any intercept was FA = $0.520t - 0.00055$, but

TABLE 10.2. Example of the Wagner-Nelson Method (A).

Time (Hours)	C (μg/ml)	AUC (μg × hr/ml)	A_T/V (μg/ml)	FA = $(A_T/V)/20.057$ (Dimensionless)
0	0	0	0	0
0.5	2.2	0.55	2.248	0.112
1	3.3	1.925	3.470	0.173[c]
2	4.6	5.875	5.118	0.255[c]
3	7.7	12.025	8.759	0.437[c]
4	8.5	20.1250	10.273	0.512[c]
5	10.9	29.825	13.528	0.674[c]
6	11.8	41.175	15.428	0.769[c]
8	12.9	65.875	18.704	0.933[c]
10	10.9	89.619	18.795	0.937
12	10.7[a]	111.218	20.498[b]	1.00
14	8.2[a]	130.117	19.763[b]	
16	7.1[a]	145.486	19.917[b]	
20	5.1[a]	169.665	20.048[b]	

[a] ln C *versus* t gave $K = 0.0881\ \text{hr}^{-1}$.
[b] A_{max}/V = average value = 20.057 μg/ml.
[c] Least squares regression gave: FA = $0.0676 + 0.113t$ (see Figure 10.5 for plot of these data).

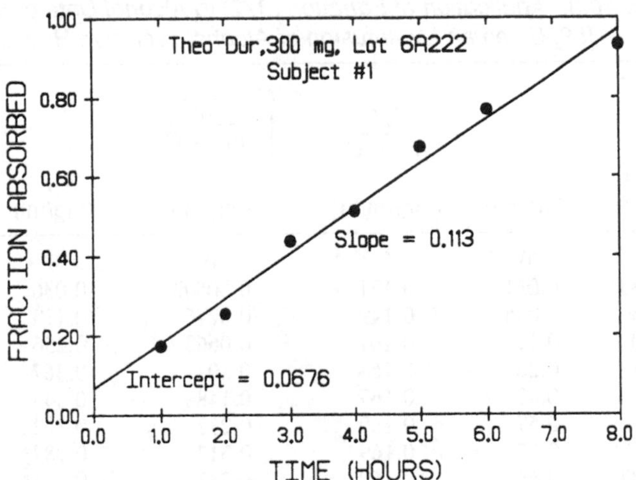

FIGURE 10.5. Example of use of the Wagner-Nelson method after a single dose [Equations (368) to (371)]. Data in the first and fifth columns of Table 10.2 in the 1 to 8 hr time range are plotted. Linearity of the plot indicates zero-order absorption.

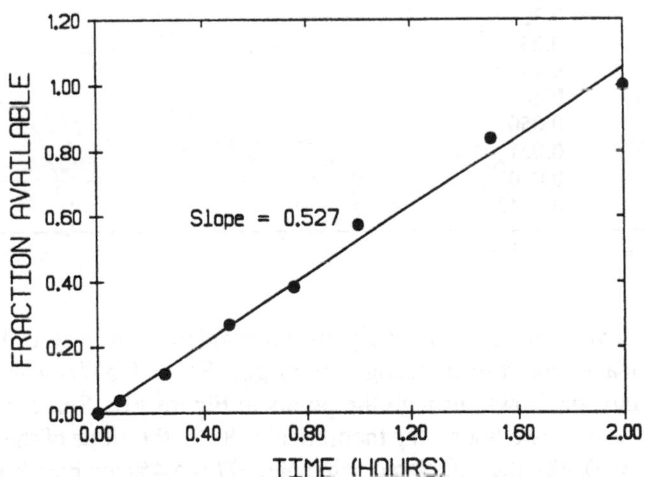

FIGURE 10.6. Example of use of the Wagner-Nelson method during a 2-hr infusion using the alcohol data in columns one and six of Table 10.3 in the 0 to 2 hr time period. The linearity of the plot shows that the method correctly indicated that input of alcohol was at a constant rate and was zero order.

TABLE 10.3. Application of Equation (372) to Alcohol Data of Subject #5 [59] Given an I.V. Infusion of Alcohol over Two Hours.

Time (Hours)	C (mg/ml)	$\frac{V'_m C}{K_m + C}$ (mg/ml/hr)	$\int_0^T \frac{V'_m C}{K_m + C} dt$ (mg/ml)	A_T/V (mg/ml)	FA = $\frac{A_T/V}{0.956}$
0	0	0	0	0	0
0.083	0.031	0.121	0.00502	0.036	0.038
0.25	0.085	0.149	0.0276	0.113	0.118
0.50	0.19	0.161	0.0663	0.256	0.268
0.75	0.26	0.164	0.107	0.367	0.384
1.0	0.40	0.167	0.148	0.548	0.573
1.5	0.57	0.169	0.232	0.802	0.839
2.0	0.67	0.169	0.317	0.987[a]	1.00
2.083	0.64	0.169	0.331	0.971[a]	
2.167	0.58	0.169	0.345	0.925[a]	
2.25	0.58	0.169	0.359	0.939[a]	
2.5	0.55				
2.75	0.50				
3.0	0.42				
3.5	0.37				
4.0	0.32				
4.5	0.23				
5.0	0.14				
5.25	0.10				
5.5	0.050				
5.75	0.021				
6.0	0.010				
6.25	0.0042				

[a] A_{max}/V = mean = 0.956.

the negative intercept was not significantly different from zero. Hence, the least squares line forced through the origin (FA = $0.527t$) was calculated and is the one drawn through the points in Figure 10.6. Since the infusion was given over two hours, the theoretical value of the slope of the FA *versus* t is 1/2 = 0.500; the calculated value of 0.527 is 5.4% too high but is a good estimate.

The known dose of alcohol was 47,800 mg, hence the actual k_o = 47,800/2 = 23,900 mg/hr. The estimated k_o = 0.527 × 47,800 = 25,191 mg/hr. The estimated $A_{max}/V = D/V$ = 0.956 mg/ml. Hence, the estimated $V = D/(D/V)$ = (47,800/0.956) × 10^{-3} = 50.0 L.

Why use I.V. infusion data to study the kinetics of absorption? Since the kinetics (zero-order), the dose, and input rate are known, infusion data constitute a good way of checking the accuracy of absorption methods.

EXAMPLE OF APPLICATION OF METHOD (C)

The superposition method (see Table 5.2 in Chapter 5) was applied to the single-dose data in Table 10.2 so that simulated data after the third dose at twelve-hour intervals was produced. Equation (373) was applied to the multiple dose data with results shown in Table 10.4. The FA, t data in the interval one to eight hours were plotted and the least squares line was found to be: FA = 0.0657 + 0.113t; these data and this line are shown in Figure 10.7. Hence, the simulated multiple dose data gave the same FA, t plot as the original single-dose data. This may be seen by comparing Figures 10.5 and 10.7 and Tables 10.2 and 10.4.

The Wagner-Nelson Method Applied to the Two-Compartment Open Model with Zero-Order Input (Model XII) [89]

If one performs the operations indicated by Equations (368), (370), and (371) on Equation (74), one obtains Equation (374).

$$FA = \frac{T}{\tau} + \frac{1}{\tau}\left[\frac{1}{k_{21}} - \frac{1}{\lambda_2}\right][1 - e^{-\lambda_2 T}] \qquad (374)$$

In Equation (374), T is some specific time t, τ is the duration of the zero-order input, k_{21} is the first-order rate constant for transfer of drug from the peripheral to the central compartment of the two-compartment open model,

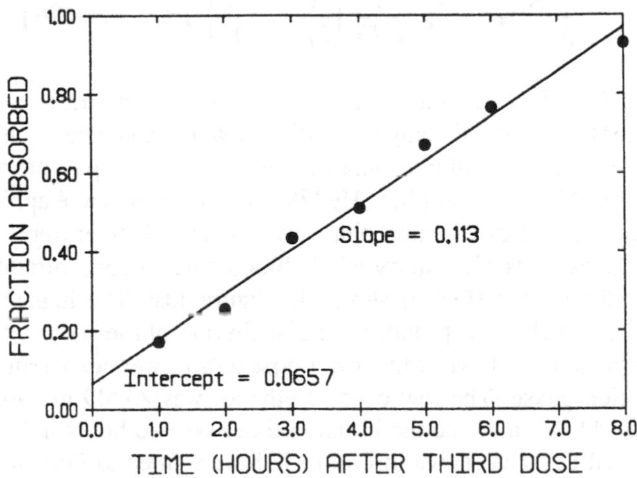

FIGURE 10.7. Example of application of the Wagner-Nelson multiple dose Equation (373) using simulated data. Data in columns one and four of Table 10.4 are plotted in the 1 to 8 hr time range.

TABLE 10.4. Simulated Third-Dose Data and Results Obtained by Applying Equation (373).

Time (Hours)	C (μg/ml)	A_T/V (μg/ml)	FA = $(A_T/V)/20.219$ (Dimensionless)
0	13.80	0	0
0.5	15.38	2.234	0.111
1	15.90	3.456	0.171[a]
2	16.12	5.112	0.253[a]
3	18.23	8.763	0.433[a]
4	18.13	10.293	0.509[a]
5	19.70	13.560	0.671[a]
6	19.84	15.473	0.765[a]
8	19.62	18.793	0.929[a]
10	16.51	18.916	0.936
12	14.99	20.219	1.00

[a]Values plotted in Figure 10.7.

and λ_2 is the hybrid rate constant defined by Equation (25). The term T/τ in Equation (374) is the actual cumulative fraction absorbed or available to the central compartment to time T. The second term of Equation (374) is an error term which provides an intercept on the plot of FA *versus t*. In most cases this error term approaches zero quite fast so that the slope of the FA *versus* T or t plot is nearly equal to the correct $1/\tau$ as indicated by Equation (375).

$$\frac{\Delta \text{FA}}{\Delta T} = \frac{1}{\tau} + \frac{1}{\Delta T \tau}\left[\frac{1}{k_{21}} - \frac{1}{\lambda_2}\right][e^{-\lambda_2 T_1} - e^{-\lambda_2 T_2}] \tag{375}$$

where $\Delta T = T_2 - T_1$ and the latter are two sampling times.

Wagner [89] reported many simulations that indicated the accuracy of the Wagner-Nelson method in estimating the zero-order input rate constant in cases when Model XII applies. He [89] also reported some applications to real data where a distribution phase was evident. One example of this involved the two sets of capillary whole blood ethanol concentrations of subject #7 (the author J.G.W.), shown in Figure 10.8. The higher dose data (upper curve) shows a pronounced distribution phase indicating a multicompartment model, while the lower dose data (lower curve) shows a slight distribution phase. The low dose of ethanol was 23,519 mg and the high dose was 44,327 mg, and the infusion time was two hours in both cases.

Terminal C, t data shown in Figure 10.8 were fitted to Equation (263) by numerical integration of Equation (262) using the program NONLIN [68]; V_m' values of 0.161 and 0.192 mg $\times$ ml^{-1} $\times$ hr^{-1} and K_m values of 0.0208

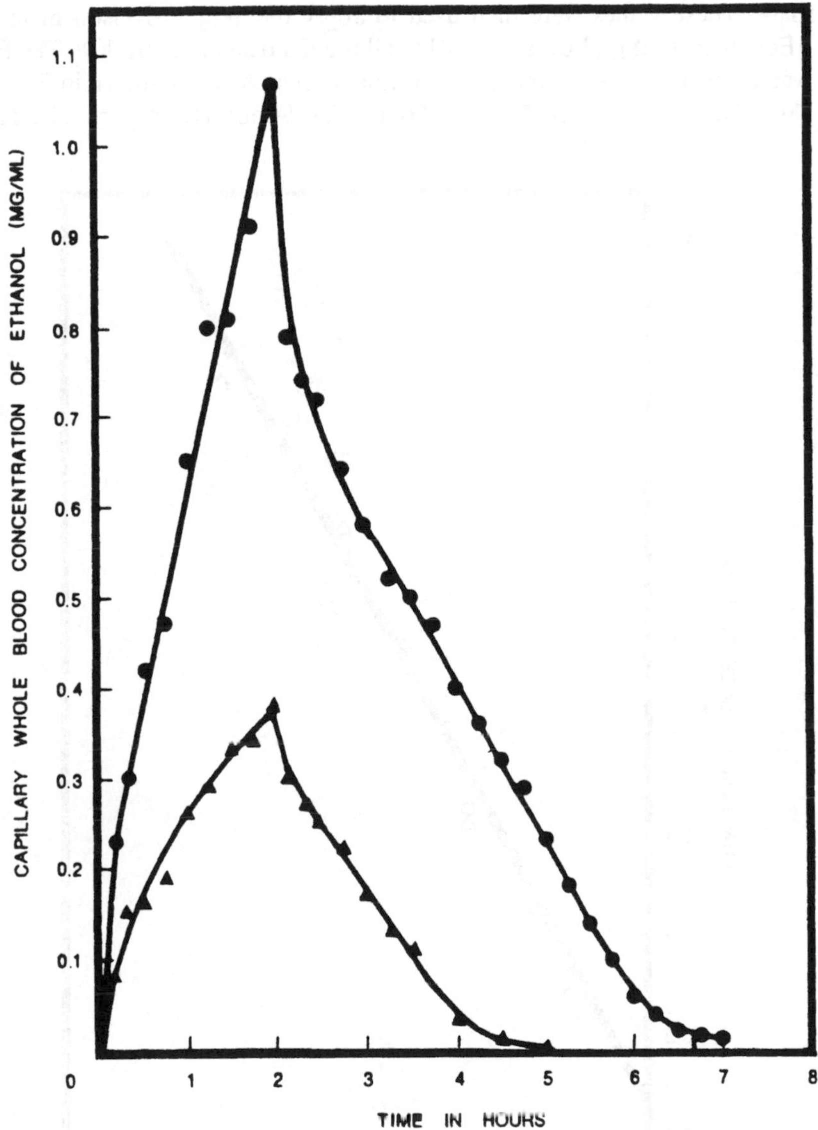

FIGURE 10.8. Capillary whole blood ethanol concentrations of subject #7 during and subsequent to two constant rate infusions of ethanol. V_m' and K_m values were estimated as described in the text.

and 0.0335 mg × ml^{-1}, respectively, were estimated for the two sets of data. These values were then used to apply the Wagner-Nelson method [Equations (368), (370), and (371)] to all the *C, t* data in Figure 10.8. The FA values obtained were plotted *versus* time to give the plots shown in Figure 10.9. The slopes are 0.486 and 0.556 hr^{-1} for 7A and 7B, respectively. The

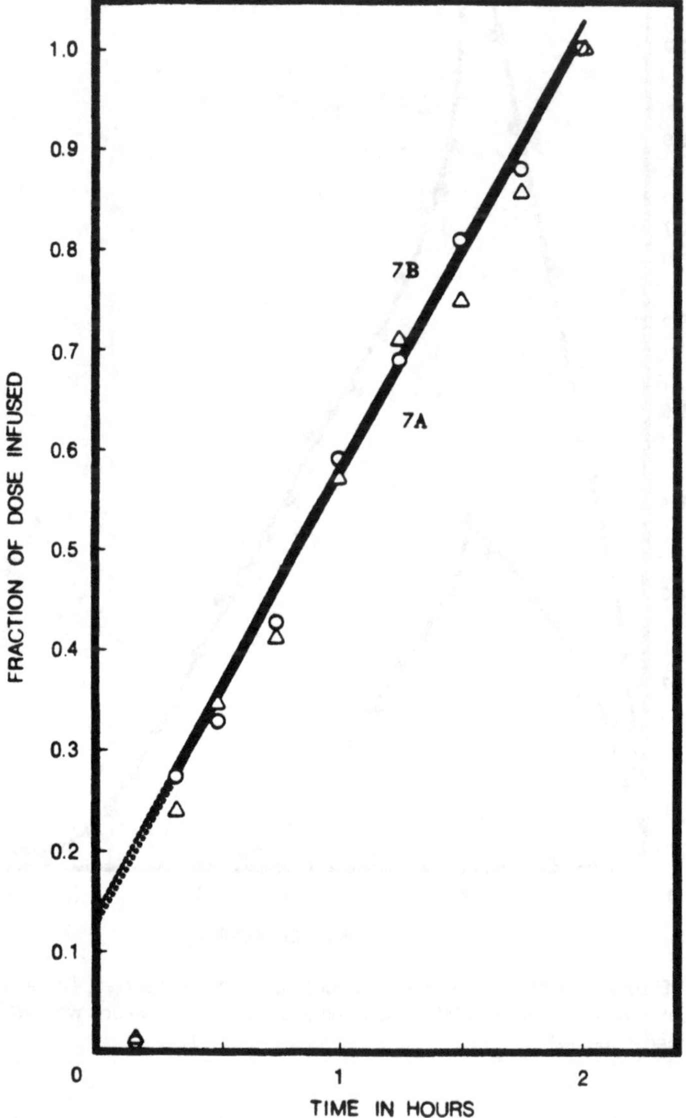

FIGURE 10.9. Results after Wagner-Nelson method [Equations (368), (370), and (371)] were applied to the data shown in Figure 10.8. See text for slopes of lines.

theoretical slopes are both $1/\tau = 1/2 = 0.500$ hr^{-1}. These slopes are equivalent to k_o/D, where k_o is the infusion rate in mg/hr and D is the dose $= k_o\tau$ in mg. The average slope is 0.521 hr^{-1}, which is 4.2% higher than the theoretical 0.5. Hence with these data the accuracy was quite good.

The Wagner-Nelson Method Applied to the Two-Compartment Open Model with First-Order Input (Model X) [90]

If one performs the operations indicated by Equations (368) and (370) [under Method (A) above] on Equation (45) (Chapter 2, under Model X) one obtains Equation (376).

$$\frac{A_T}{V} = C_1(T) + \lambda_1 \int_0^T C_1(t)dt$$

$$= C_o \left[\frac{k_{21}}{\lambda_2} + \frac{1}{\lambda_2(k_a - \lambda_2)} \{k_a(\lambda_2 - k_{21})e^{-\lambda_2 T} - \lambda_2(k_a - k_{21})e^{-k_a T}\}\right] \tag{376}$$

Equation (376) gives the time course of the Wagner-Nelson function if it is applied to the equation, giving C as a function of time for the classical two-compartment open model with first-order absorption. Note that the time course has two exponential terms – one involving k_a and the other involving λ_2. The shape of the A_T/V *versus* time curve depends upon whether the relative magnitudes of the rate constants are $k_a > \lambda_2 > \lambda_1$ or $\lambda_2 > k_a > \lambda_1$.

Figure 10.10 shows the Wagner-Nelson plot for the oral flurbiprofen C,t data of Model X. Note that there is a peak on the A_T/V *versus* time plot then the curve drops down and reaches an asymptote. In this case the relative magnitudes are $k_a > \lambda_2 > \lambda_1$.

Figure 10.11 is the Wagner-Nelson plot for the spectinomycin I.M. data of Model X. In this case $\lambda_2 > k_a > \lambda_1$ and the A_T/V *versus* time plot rises and reaches its asymptote in the same manner as one-compartment open model data and without a peak.

Figure 10.12 is a semilogarithmic plot of the spectinomycin I.M. C,t data. Note that the early C,t points come up beneath the extrapolated terminal log-linear line and do not go above it.

Figure 10.13 is a semilogarithmic plot of the oral flurbiprofen C,t data. Note that the early C,t points come up beneath the extrapolated terminal log-linear line then go above this line and reach a peak before falling.

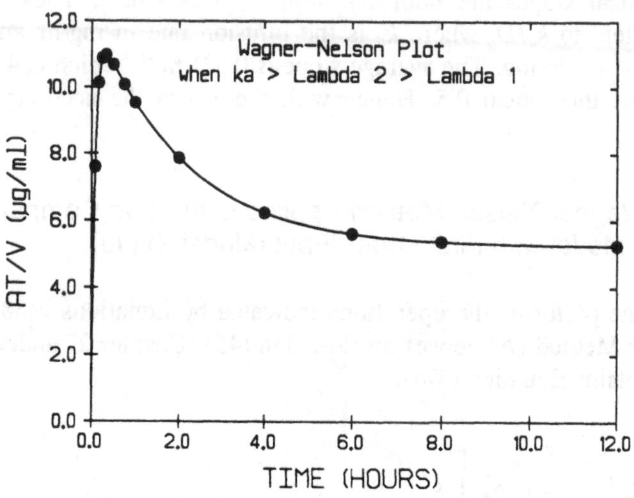

FIGURE 10.10. Example of use of the Wagner-Nelson method applied to the two-compartment open model with first-order input (Model X). Data were the oral flurbiprofen example of Model X.

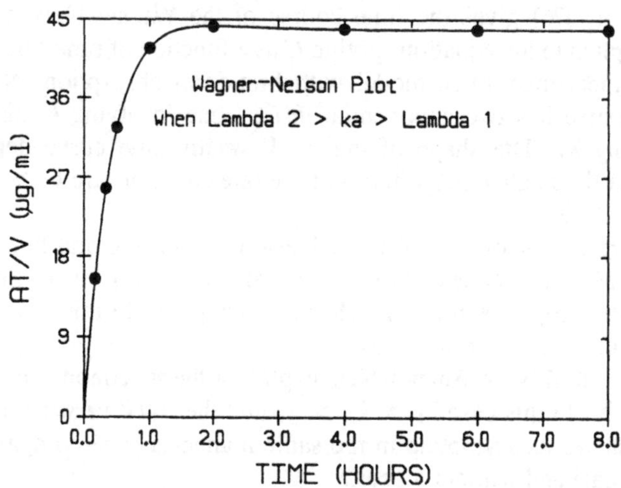

FIGURE 10.11. Wagner-Nelson plot for the spectinomycin I.M. data of Model X. See text for details.

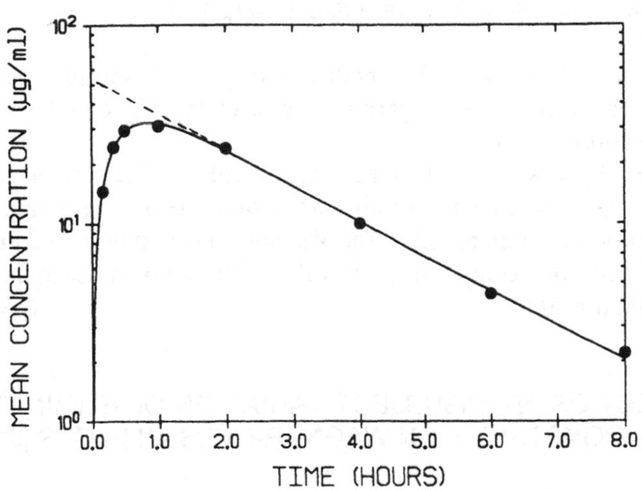

FIGURE 10.12. Semilogarithmic plot of spectinomycin C,t data. Note that there is no "nose" in the plot.

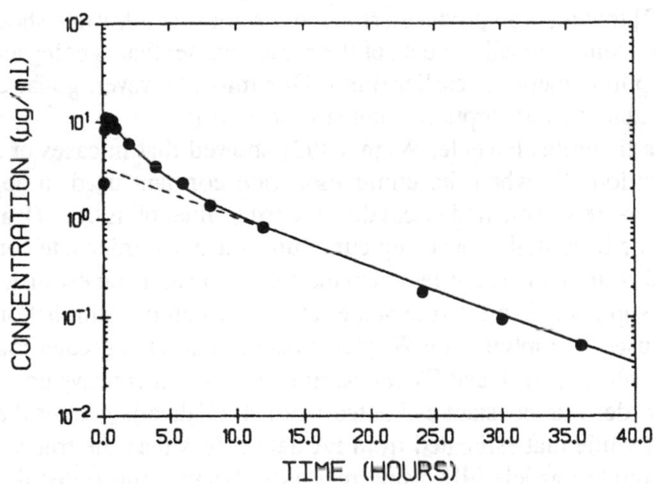

FIGURE 10.13. Semilogarithmic plot of C,t data after flurbiprofen solution given orally and where $k_a > \lambda_1 > \lambda_2$. Note that there is a "nose" in the plot.

Diagnostic Procedure to Determine If $k_a > \lambda_2 > \lambda_1$ or $\lambda_2 > k_a > \lambda_1$ When C, t Data Are Triexponential

When $k_a > \lambda_2 > \lambda_1$ then both the Wagner-Nelson plot on rectilinear graph paper and a semilogarithmic plot of the C, t data have peaks like Figures 10.10 and 10.13.

When $\lambda_2 > k_a > \lambda_1$ then the semilogarithmic C, t plot has a shape like one-compartment data with all early points below the extrapolated log-linear line as in Figure 10.12. The Wagner-Nelson plot on rectilinear graph paper is like one-compartment data also with a smooth rise to the asymptote as in Figure 10.11.

EFFECT OF INTRASUBJECT VARIATION OF ELIMINATION RATE CONSTANT *K* ON WAGNER-NELSON PLOTS [91,92]

In their proposed protocol for bioavailability of sustained-release theophylline formulations the Food and Drug Administration recommends using the elimination rate constant, K, derived from another treatment (e.g., a solution), to perform a Wagner-Nelson calculation on C, t data derived from a solid oral sustained-release dosage form. The author has shown [91,92] that this can produce erroneous results. The K value should be estimated from terminal C, t data of the same data set that is being analyzed for absorption kinetics at earlier times. One must, however, go far enough out on the end that absorption is not still occurring.

In a theoretical article, Wagner [92] showed that in cases of zero-order absorption: (1) when the elimination rate constant used in applying the Wagner-Nelson method is less than the true value, plots of A_T/V and FA *versus* time have slight convex up curvature; the zero-order rate constant estimated from A_T/V data is less than the true value and that estimated from FA data is greater than the true value; and (2) when the elimination rate constant used in applying the Wagner-Nelson method is greater than the true value, plots of A_T/V and FA *versus* time have slight concave up curvature; a zero-order rate constant estimated from A_T/V data is greater than the true value, while that estimated from FA data is less than the true value.

In another article [91], using real data, Wagner illustrated the problem. Figures 10.14 and 10.15 are from that chapter. The raw data were C, t data following administration of Theo-Dur Sprinkle to two different asthmatic children. K values were estimated both from terminal oral data being analyzed for absorption and from C, t data obtained after intravenous administration of theophylline to the same children.

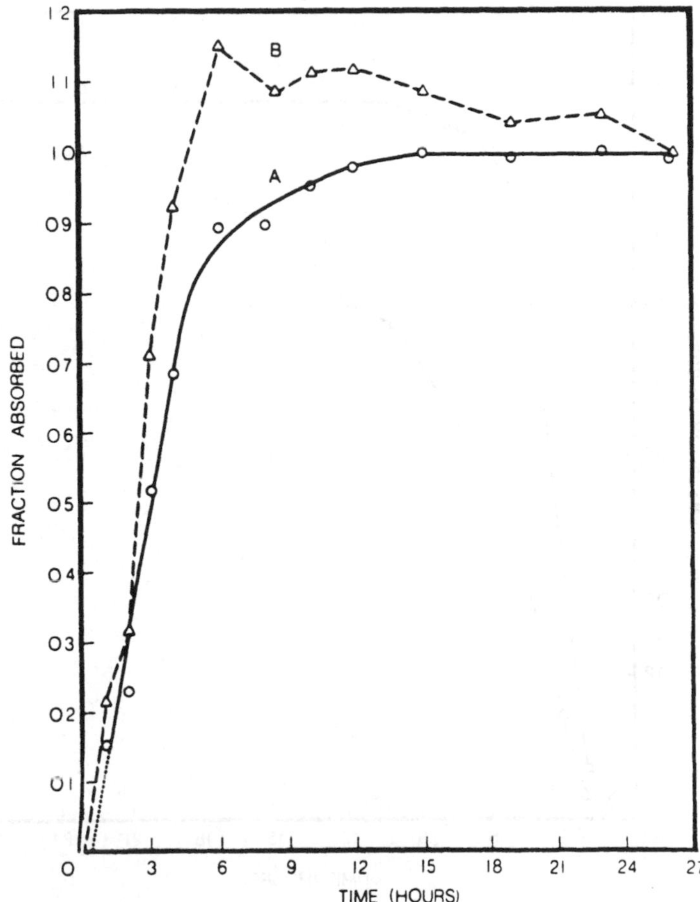

FIGURE 10.14. Wagner-Nelson plots after Theo-Dur Sprinkle orally. Curve A: K = 0.147 hr^{-1} from oral data. Curve B: K = 0.102 hr^{-1} from I.V. data (from Reference [91], with permission from Excerpta Medica).

In Figure 10.14, curve A is the Wagner-Nelson FA *versus t* plot when $K = 0.147\ hr^{-1}$ from the oral data was used – here there is a linear part in the FA *versus t* plot; and curve B is when $K = 0.102\ hr^{-1}$ from the I.V. data was used – here there is no linear region and there is a peak in the FA *versus t* plot. Note in this case the po K value is greater than the I.V. K value.

In Figure 10.15, curve A is when $K = 0.108\ hr^{-1}$ from the oral data was used; there is a linear region in the FA *versus t* plot and an asymptote is reached in about nine hours. Curve B was obtained when $K = 0.288\ hr^{-1}$

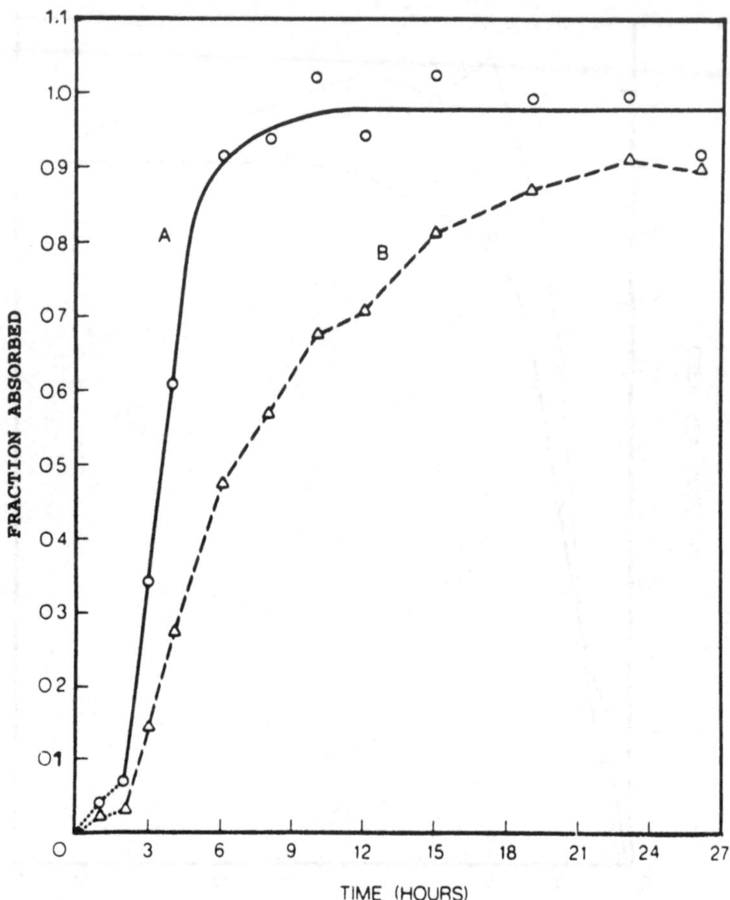

FIGURE 10.15. Wagner-Nelson plots after Theo-Dur Sprinkle orally. Curve A: K = 0.108 hr^{-1} from oral data. Curve B: K = 0.288 hr^{-1} from I.V. data (from Reference [91], with permission from Excerpta Medica).

from I.V. data was used; there is no linear region and the asymptote was not reached until about twenty-four hours, which indicates absorption proceeded over twenty-four hours – which is most unlikely with this product.

These results obtained with real data [91] were predicted by the theoretical article [92] since part of the theophylline in Theo-Dur Sprinkle is absorbed at a zero-order rate. Hence the author believes that curve A in Figure 10.14 and curve A in Figure 10.15 are the correct ones and that the B curves are erroneous and misleading.

Drug Delivery Example with Wagner-Nelson Equation

A drug (nameless for proprietary reasons) was tested in four different formulations in four Cynomolgus monkeys. The monkeys were acclimated to laboratory conditions for seven days prior to the study. They were fasted for sixteen hours prior to dosing. Dosing was at weekly intervals and was according to a crossover design. The suspension [treatment (A)] was administered by nasal intubation. The tablets were inserted in a size 0 gelatin capsule and administered via a catheter cut at one end to accept the capsule, while the other end was attached to an air-filled syringe; the end containing the tablet was placed in the esophagus, and the capsule was expelled by depressing the syringe. Treatment B was a rapid-release tablet containing micronized drug. Treatment C was a medium-release rate tablet. Treatment D was a slow-release rate tablet with nonmicronized drug.

One milliliter blood samples were taken at 0, 0.25, 0.5, 0.75, 1, 1.5, 2, 3, 4, 6, 8, 12, 24, 32, and 48 hours. Serum was harvested from each blood sample and assayed for drug by a sensitive and specific method.

The Wagner-Nelson method [Equations (368), (369), and (371)] was applied to each set of C, t data. The FA values at each sampling time were averaged and fitted to Equation (16) but with an added lag time, t_o (see Model VI, Chapter 1). The concentrations were also averaged. FA values were obtained also from mean concentrations. Estimated parameters from fitting mean FA values are shown in Table 10.5.

Figure 10.16 shows plots of mean FA *versus* t for the four treatments with the theoretical lines according to Equation (377). Figure 10.17 shows plots of the mean concentrations following the four treatments; inset are the mean AUCs (not the area under the plots in this figure but the average of areas for the four monkeys). Analysis of variance indicated that the differences in areas was highly significant ($p < 0.001$). Parameters of Equation

TABLE 10.5. Estimated Parameters from Fitting FA, t Data to:

$$FA = 1 - [k_1 e^{-k_2(t - t_o)} - k_2 e^{-k_1(t - t_o)}]/(k_1 - k_2) \quad (377)$$

Treatment	k_1 (hr^{-1})	k_2 (hr^{-1})	t_o (hr)
A	5.31	0.707	0
B	1.52	0.769	0.237
C	1.17	0.816	0.921
D	1.50	0.511	0.316

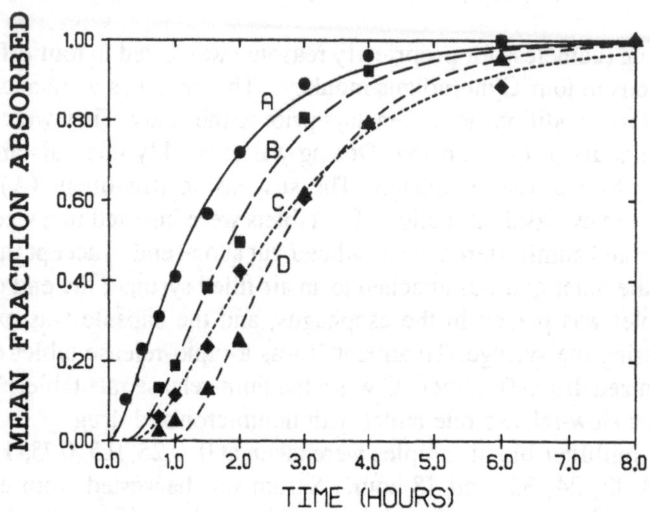

FIGURE 10.16. Drug delivery example showing mean FA,*t* plots for treatments A, B, C, and D.

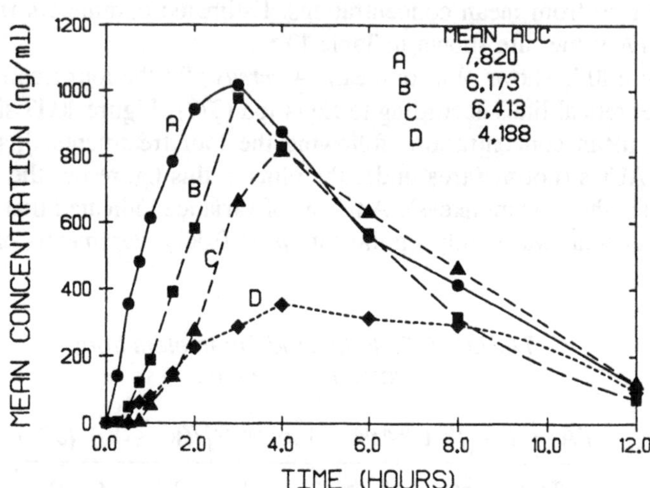

FIGURE 10.17. Drug delivery example showing mean concentration profiles for treatments A, B, C, and D with mean AUCs inset.

(377) derived from fitting mean concentrations showed differences similar to those derived from mean FA values and shown in Table 10.5.

Rates of dissolution measured *in vitro* correlated with the FA, t data.

The model that explains the data in the above drug delivery example is shown in Scheme 10.2.

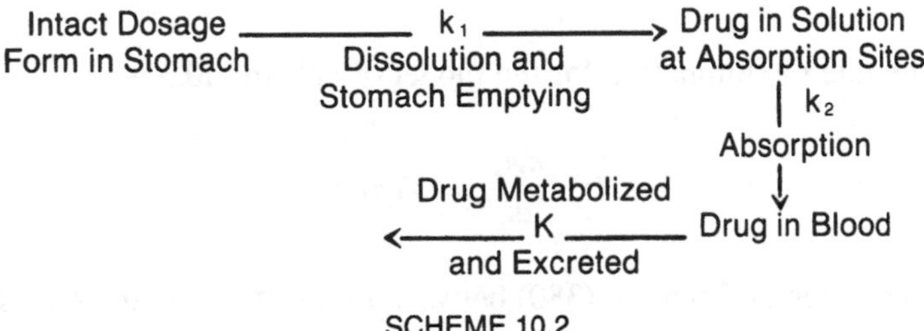

SCHEME 10.2.

EXACT LOO-RIEGELMAN EQUATION [10,11] WHEN DISPOSITION IS BIEXPONENTIAL

Loo and Riegelman [27] published a method to estimate A_T/V_1 as a function of time based on the two-compartment open model with central compartment elimination. The method assumes a linear segment between each pair of C, t data points. Wagner [10] published his Exact Loo-Riegelman method, which does not assume such linear segments and has just one term to account for drug in the peripheral compartment and which replaces four terms in the Loo-Riegelman equation [27]. The model is shown in Scheme 10.3.

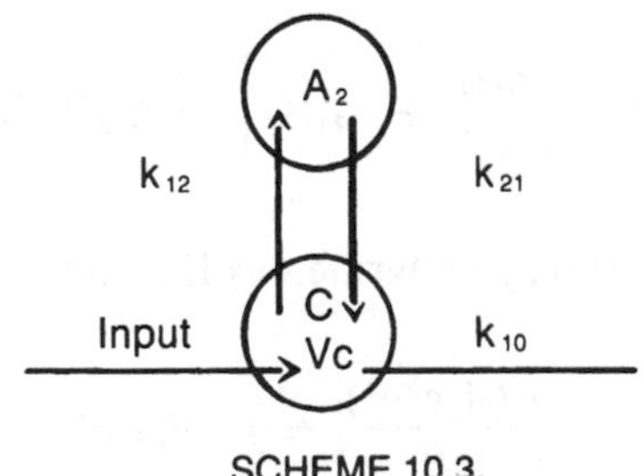

SCHEME 10.3.

Derivation

Let D = the dose of drug, A_r = the amount of drug remaining at the absorption site, A_T = amount of drug absorbed to time T, $A_1 = VcC$ = amount of drug in the central compartment at time t, A_2 = amount of drug in the peripheral compartment at time t, and A_e = amount of drug which has been eliminated by metabolism and excretion in time T. Then mass balance gives:

$$D = A_r + A_1 + A_2 + A_e \tag{378}$$

$$A_T = D - A_r = A_1 + A_2 + A_e \tag{379}$$

The rate of elimination for the model of Scheme 10.3 is:

$$\frac{dA_e}{dt} = Vck_{10}C \tag{380}$$

Integration of Equation (380) between limits of $t = 0$ and $t = T$ yields:

$$A_e = Vck_{10}\int_0^T Cdt \tag{381}$$

The differential equation for the peripheral compartment of the model of Scheme 10.3 is:

$$\frac{dA_2}{dt} = k_{12}VcC - k_{21}A_2 \tag{382}$$

Rearrangement of Equation (382) and multiplication of both sides by $e^{k_{21}t}$ gives:

$$e^{k_{21}t}\left[\frac{dA_2}{dt} + k_{21}A_2\right] = k_{12}VcCe^{k_{21}t} \tag{383}$$

But Equation (383) may be written as Equation (384).

$$\frac{d(A_2e^{k_{21}t})}{dt} = k_{12}VcCe^{k_{21}t} \tag{384}$$

Integrating Equation (384) between the limits $t = 0$ and $t = T$ gives:

$$A_2e^{k_{21}T} = k_{12}Vc\int_0^T Ce^{k_{21}t}dt \tag{385}$$

hence

$$A_2 = k_{12}e^{-k_{21}T}Vc\int_0^T Ce^{k_{21}t}dt \tag{386}$$

Substituting $A_1 = VcC$ and for A_2 and A_e from Equations (381) and (386) into Equation (379), followed by division of both sides by Vc, gives Equation (387).

$$\frac{A_T}{Vc} = C_T + k_{10}\int_0^T Cdt + k_{12}e^{-k_{21}T}\int_0^T Ce^{k_{21}t}dt \tag{387}$$

Equation (387) is the Exact Loo-Riegelman equation when disposition is biexponential.

A simulation example is shown in Table 10.6. Model X was used with $k_{12} = 1.5\ \text{hr}^{-1}$, $k_{21} = 0.5\ \text{hr}^{-1}$, $k_{10} = 0.5\ \text{hr}^{-1}$, $k_a = 4\ \text{hr}^{-1}$, $Vc = 10$ L and $FD_{po} = 1000$ mg. These values gave $\lambda_1 = 0.10436\ \text{hr}^{-1}$ and $\lambda_2 = 2.3956\ \text{hr}^{-1}$ and Equation (388).

$$C = 17.7301e^{-0.10436t} + 206.2640e^{-2.3956t} - 223.9941e^{-4t} \tag{388}$$

Equation (388) was used to generate the C, t data shown in columns 1 and 2 of Table 10.6. In Table 10.6: A_e/Vc was obtained from Equation (381) by dividing through by Vc. A_2/Vc was obtained from Equation (386) by dividing through by Vc. A_T/Vc was obtained with Equation (387). The actual $\text{FA} = 1 - e^{-4t}$. The experimental FA values in the second to last column of Table 10.6 were fitted by the method of least squares to the equation: $\text{FA} = 1 - e^{-k_a t}$ and the estimated k_a was 3.93 hr^{-1}, which is a -1.75% error, since the actual value was 4 hr^{-1}. Results in Table 10.6 were obtained with the program ABSPLOTS [93] which makes such calculations easy to perform.

It should be noted that in the calculation of AUCs in Equations (381) and (386) the Proost [11] method involving the second derivative should be used; this method is used in the program ABSPLOTS [93].

EXACT LOO-RIEGELMAN [10,11] METHOD WHEN DISPOSITION IS TRIEXPONENTIAL

The disposition parameters of Model XIX must be estimated to apply the method. The extravascular model is shown in Scheme 10.4.

TABLE 10.6. Exact Loo-Riegelman Example.

T (hrs)	C (μg/ml)	$\int_0^T Cdt$	A_e/Vc	A_2/Vc	A_T/Vc	$FA = \frac{A_T/Vc}{100.295}$	Actual FA
0	0	0	0	0	0	0	0
0.05	17.23	0.43	0.22	0.65	18.09	0.180	0.181
0.1	29.72	1.60	0.80	2.37	32.90	0.328	0.330
0.2	44.46	5.31	2.66	7.71	54.83	0.549	0.551
0.3	50.25	10.05	5.02	14.28	69.55	0.693	0.699
0.4	50.9	15.11	7.55	20.98	79.44	0.792	0.798
0.5	48.78	20.09	10.05	27.25	86.08	0.858	0.865
0.6	45.33	24.80	12.40	32.80	90.53	0.903	0.909
0.75	39.45	31.15	15.58	39.60	94.63	0.944	0.950
1.00	30.67	39.87	19.94	47.23	97.83	0.975	0.982
1.50	20.28	52.43	26.22	53.30	99.80	0.995	0.9975
2.00	16.03	61.47	30.73	53.45	100.21*	1.00	0.9997
3.00	13.12	75.99	38.00	49.42	100.54*	–	1.00
4.00	11.69	88.39	44.19	44.53	100.42*		
6.00	9.48	109.48	54.74	36.04	100.26*		
12	5.07	151.76	75.88	19.22	100.17*		
18	2.71	174.36	87.18	10.28	100.17*		

*Mean = A_{max}/Vc = 100.295.

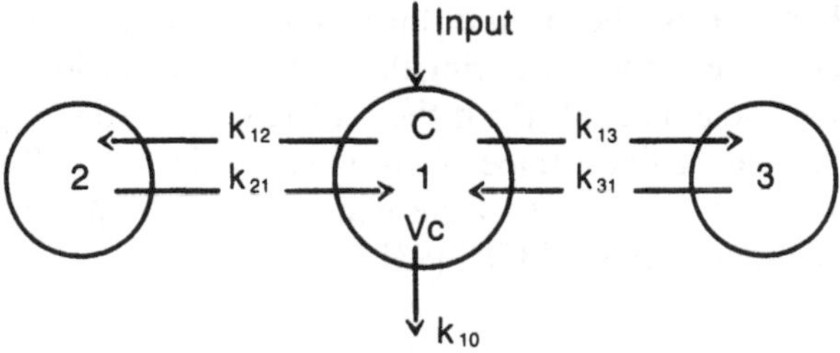

SCHEME 10.4.

The equations are similar to those when disposition is biexponential except that there are two peripheral compartments (#2 and #3), hence there are terms for A_2/Vc and A_3/Vc. A_T/Vc is given by Equation (389).

$$A_T/Vc = C_T + k_{10}\int_0^T Cdt + k_{12}e^{-k_{21}T}\int_0^T Ce^{k_{21}t}dt + k_{13}e^{-k_{31}T}\int_0^T Ce^{k_{31}t}dt \quad (389)$$

The program ABSPLOTS [93] is capable of performing all calculations using Equation (389) as well as those using Equation (387) and the Wagner-Nelson method with Equations (368) and (370). This author does not like this program, estimating A_{max}/Vc with Equation (370) rather than Equation (369) and calculating rates of absorption from the A_T/Vc values. This author thinks it is better, if possible, to fit the $A_T/Vc, T$ data with a function to estimate absorption rate constants. However, in some cases absorption is irregular and such fitting is not possible, but the A_T/Vc *versus* t or FA *versus* t plots are still useful, particularly for comparing results from different formulations of the same drug, as illustrated in the section on Drug Delivery Example with Wagner-Nelson Equation.

In Chapter 4, examples of estimating triexponential disposition parameters from extravascular downslope C, t data and applying Equation (389) are illustrated with diazepam data in three subjects.

Criterion of Proost [10] to Decide Whether to Use the Linear or Logarithmic Trapezoidal Rule

In applying the Exact Loo-Riegelman Equation (387) or (389) the criterion of Proost [10] should be used. He showed that the correct criterion to decide whether to use the linear or logarithmic trapezoidal rule is: (a) if the second derivative is positive during the interval then the logarithmic trapezoidal rule should be used; (b) if the second derivative is negative during the interval then the linear trapezoidal rule should be used. These rules are most important in evaluation of the integral $\int_0^T Ce^{k_{21}t}dt$. This criterion is used in the program ABSPLOTS [93].

Relationship of Exact Loo-Riegelman Rate Equation to a Deconvolution Equation

The rate of absorption equation corresponding to Equation (387) is Equation (390).

$$\left[\frac{d(A/V)}{dt}\right]_T = (k_{12} + k_{10})C + \left[\frac{dC}{dt}\right] - k_{12}k_{21}e^{-k_{21}T}\int_0^T Ce^{k_{21}t}dt \tag{390}$$

The deconvolution Equation (67) of Veng-Pedersen [42] is the same as Equation (390) and the conversion of his Equation (67) to Equation (390) requires the following:

Veng-Pedersen		Wagner	
$-\left[\dfrac{a_1\lambda_1 + a_2\lambda_2}{(a_1 + a_2)^2}\right]$	$=$	$\dfrac{Vc}{D_{\mathrm{I.V.}}}(k_{12} + k_{10})$	(391)
$\dfrac{a_1a_2(\lambda_1 - \lambda_2)^2}{(a_1 + a_2)^3}$	$=$	$\dfrac{Vc}{D_{\mathrm{I.V.}}}(k_{12}k_{21})$	(392)
$\dfrac{a_1\lambda_2 + a_2\lambda_1}{a_1 + a_2}$	$=$	$\dfrac{Vc}{D_{\mathrm{I.V.}}}(k_{21})$	(393)
$-\dfrac{\lambda_1\lambda_2}{a_1\lambda_2 + a_2\lambda_1}$	$=$	$\dfrac{Vc}{D_{\mathrm{I.V.}}}(k_{10})$	(394)

Use of Equation (390) or Equation (67) of Veng-Pedersen [42] requires values of the derivatives, dC/dt, from extravascular data. If C,t data have been fitted with polyexponential or polynomial equations then such derivatives can be obtained readily. But if only raw C,t data are available then such derivatives are obtainable only with great errors by usual methods.

This author believes it is better to obtain values of A_T/Vc or FA as a function of time and fit these data to an appropriate equation to obtain the kinetics of absorption.

Kinetics of Absorption via Deconvolution

Veng-Pedersen [28,42–46] has published extensively on deconvolution methods to estimate absorption rates of drugs. The basic deconvolution Equation (214) was given in Chapter 5. An illustration of the Point-Area Deconvolution Method was given in Table 5.4 in the same chapter. Rescigno and Segre [40], Benet and Chiang [38], Kiwada et al. [39], Cutler [41], and others have all made significant contributions to deconvolution.

BIOAVAILABILITY

Bioavailability is a term used to indicate the measurement of both the absolute and relative amount of an administered drug that reaches the general circulation and the rate at which this occurs.

Areas of application of bioavailability knowledge are:

(1) Generic equivalence and inequivalence [bioequivalence or the comparison of drug products containing the same active ingredient(s), but made by different manufacturers]
(2) Effect of food on the absorption of drugs from products
(3) Effect of one drug on the absorption of another drug
(4) Effect of age on the absorption of drugs
(5) Effect of disease states on the absorption of drugs
(6) Assessment of the "first-pass" effect to explain the difference in potency when a drug is administered extravascularly (particularly orally) compared with intravenously
(7) Drug-drug interaction studies

Absolute bioavailability is when one is comparing an extravascular route to the intravenous route. *Relative bioavailability* is when one is comparing two different dosage forms by the same extravascular route or when two different dosage forms are given by different extravascular routes, such as oral and rectal.

The Food and Drug Administration published "Bioavailability and Bioequivalence Requirements" in the *Federal Register,* Vol. 42, No. 5, on Friday, January 7, 1977, Book I, pages 1624–1653 and other subsequent publications. These regulations made the collecting and filing of bioavailability data on specific drug products a requirement.

Equations Used to Estimate Relative or Absolute Efficiency of Absorption

Bioavailability equations are based on the following mass balance equations for I.V. and oral administrations.

$$D_{\text{I.V.}} = \text{CL}(\text{AUC 0–Inf}) = (\text{CL}_r + \text{CL}_{nr})(\text{AUC 0–Inf}) \quad (395)$$

$$FD_{\text{po}} = \text{CL}(\text{AUC 0–Inf}) \quad (396)$$

In Equations (395) and (396), $D_{\text{I.V.}}$ is the intravenous dose, D_{po} is the oral dose, CL is the total clearance, CL_r is the renal clearance, CL_{nr} is the non-renal clearance $= \text{CL} - \text{CL}_r$, F is the fraction of the oral dose that is absorbed or, in the case of a "first-pass" drug, the amount that is bioavailable, and AUC 0–Inf is the area under the whole blood (serum or plasma) concentration-time curve. Note that from Equation (396) one obtains the oral clearance $= \text{CL}/F = D_{\text{po}}/(\text{AUC 0–Inf})_{\text{po}}$.

When urinary data are available another mass balance equation is Equation (397).

$$(A_e\ 0\text{–Inf}) = \text{CL}_r \text{AUC 0–Inf} \quad (397)$$

where (A_e 0–Inf) is the amount of drug excreted unchanged in infinite time after a single dose. If one is doing a study at steady state then AUC 0–τ replaces AUC 0–Inf in Equations (395) to (397) and (A_e 0–τ) replaces (A_e 0–Inf) in Equation (397).

Different bioavailability equations are based on different assumptions—which are not always stated by authors.

Assumption #1: blood or plasma clearance (CL) is constant intrasubject, i.e., $CL_A = CL_B$*, where these are total clearances after two treatments, A and B, in the same subject.*

$$\text{Bioavailability of B relative to A} = \frac{F_{\text{B}}}{F_{\text{A}}}$$

$$= \frac{(\text{AUC 0–Inf})_{\text{B}}/D_{\text{B}}}{(\text{AUC 0–Inf})_{\text{A}}/D_{\text{A}}} = \frac{D_{\text{A}}(\text{AUC 0–Inf})_{\text{B}}}{D_{\text{B}}(\text{AUC 0–Inf})_{\text{A}}} \quad (398)$$

To obtain the doses for use in Equation (398) at least twenty aliquots of the two dosage forms should be assayed and the mean potencies used as the doses. However, the Food and Drug Administration does not usually allow such a dose correction, but the two dosage forms must have potencies not more than 5% different.

Assumption #2: the nonrenal component of total clearance remains constant, but the renal clearance can vary with the treatment or time of administration of the two treatments, A and B. There are two forms of the equation for this assumption.

- Form #1:

$$\text{Bioavailability of B relative to A} = \frac{F_B}{F_A}$$

$$= \frac{D_A(\text{AUC 0–Inf})_B}{D_B(\text{AUC 0–Inf})_A} - \frac{\{(\text{CL}_r)_A - (\text{CL}_r)_B\}(\text{AUC 0–Inf})_B}{D_B} \tag{399}$$

- Form #2:

$$\text{Bioavailability of B relative to A} = \frac{F_B}{F_A}$$

$$= \frac{[D_A - (A_e\,\text{0–Inf})_A]\left[\dfrac{(\text{AUC 0–Inf})_B}{(\text{AUC 0–Inf})_A}\right] + (A_e\,\text{0–Inf})_B}{D_B} \tag{400}$$

Equations (399) and (400) have been used for two oral treatments but actually are only exact when treatment A is intravenous – i.e., when absolute bioavailability is being measured.

Assumption #3: a change in renal clearance is accompanied by a proportional change in nonrenal clearance, i.e.,

$$\frac{(\text{CL}_r)_B}{(\text{CL}_r)_A} = \frac{(\text{CL}_{nr})_B}{(\text{CL}_{nr})_A} \tag{401}$$

or, if the renal clearance changes X%, then total clearance changes X% also. Then,

$$\text{Bioavailability of B relative to A} = \frac{F_B}{F_A}$$

$$= \frac{D_A(\text{CL}_r)_B(\text{AUC 0–Inf})_B}{D_B(\text{CL}_r)_A(\text{AUC 0–Inf})_A} = \frac{D_A(A_e\,\text{0–Inf})_B}{D_B(A_e\,\text{0–Inf})_A} \tag{402}$$

Bioequivalence-Bioinequivalence Testing

The Food and Drug Administration does not require use of any of the above bioavailability equations. Instead they require a two-way crossover study with two treatments—test and reference—and analysis of C_{max} and AUC data; they formerly required t_{max} to meet certain requirements, but this parameter proved to be too variable. The type of analyses of C_{max} and AUC and criteria to be met are described below.

Schuirmann [94], of the FDA, discussed the advantages of a Two One-Sided t-Test Procedure over the previously accepted analysis of variance and power of the ANOVA. Let s be the square root of the "error" mean square from the analysis of variance for crossover design of the bioavailability parameter of interest (e.g., AUCs or peak concentrations). Let X_T be the mean parameter after the test formulation, X_R be the mean parameter after the reference formulation, and ϕ_1, ϕ_2 be the "equivalence interval." Since it has become accepted that the test formulation should be within 80 to 120% of the reference formulation, then $\phi_1 = -0.2\ X_R$ and $\phi_2 = 0.2\ X_R$. Then the two one-sided ts are:

$$t_1 = \frac{(X_T - X_R) - \phi_1}{s\sqrt{2/n}} \tag{403}$$

and

$$t_2 = \frac{\phi_2 - (X_T - X_R)}{s\sqrt{2/n}} \tag{404}$$

If μT and μR are the true means, corresponding to the estimates X_T and X_R, then if t_1 and t_2 are both greater than $t_1 - a(\nu)$ [the point that isolates probability α in the upper tail of the Student's t distribution with n degrees of freedom (the degrees of freedom associated with the "error" mean square)], then one can conclude bioequivalence, i.e.,

$$\phi_1 < \mu T - \mu R < \phi_2$$

An example is shown in detail below. This is a prednisone study carried out by the author in twelve normal volunteers. Treatment A was 10 mg of prednisone in a new tasteless formulation and treatment B was 10 mg of prednisone as Deltasone, 5 mg tablets (The Upjohn Company). Both prednisone and prednisolone were measured in plasma but only the results with prednisone are reported here. Table 10.7 lists the AUC 0–12 hr and C_{max} values.

The analysis of variance for crossover design of the AUCs is given in Table 10.8.

TABLE 10.7. AUCs and C_{max} Values in Prednisone Study.

Subject	AUC (ng × hr/ml) A	AUC (ng × hr/ml) B	C_{max} (ng/ml) A	C_{max} (ng/ml) B
1	281.1	227.7	51.0	37.3
2	200.8	266.8	35.8	53.8
3	379.7	366.5	52.1	52.4
4	192.8	183.6	38.4	32.2
5	325.3	313.6	46.8	54.7
6	351.3	284.3	39.9	39.4
7	295.4	327.8	46.2	43.7
8	293.8	301.1	34.1	56.8
9	316.3	281.7	50.6	43.5
10	296.9	288.0	40.2	43.3
11	252.3	406.2	37.9	39.4
12	259.5	262.9	32.6	37.0
Mean	287.9	292.5	42.1	44.5
C.V. (%)	19.1	20.1	16.4	18.2

Now

$$\phi_1 = (-0.2)(292.5) = -58.5 \quad \text{and} \quad \phi_2 = (0.2)(292.5) = 58.5$$

Substituting into Equation (403) gives:

$$t_1 = [(287.9 - 292.5) - (-58.5)]/(43.2)(2/12)1/2 = 3.06 \quad (405)$$

Substituting into Equation (404) gives:

$$t_2 = 58.5 - (287.9 - 292.5)/(43.2)(2/12)1/2 = 3.58 \quad (406)$$

The tabled $t_{n-2,0.95} = 1.8125$, hence both t_1 and t_2 are greater than the

TABLE 10.8. ANOVA for Crossover Design of AUCs.

Source of Variation	d.f.	Sum of Squares	Mean Square	*F*
Subject sequences	1	14,084.41	14,084.41	3.70
Subjects/sequence	10	38,044.82	3804.48	–
Treatments	1	123.31	123.31	0.07
Time periods	1	346.56	346.56	0.19
"Error"	10	18,662.61	1866.26	–
Total	23	71,2621.71		

$s = (1866.26)^{1/2} = 43.2$

tabled t and the AUCs indicate bioequivalence. The same result was obtained with the C_{max}s for prednisone and the AUCs and C_{max}s for prednisolone. Hence bioequivalence was shown.

Bioinequivalence as a Result of Variability of the Reference Formulation, Not the Test Formulation

The author performed a three-way crossover bioavailability study in eighteen normal volunteers with three digoxin formulations. The test formulation was a Collett digoxin tablet. The reference formulations were Lanoxin tablets (B & W) and an aqueous solution of digoxin. AUCs 0–Inf were estimated in the usual manner and are listed in Table 10.9.

ANOVA for crossover design of the areas in Table 10.9 gave $s = 5.366$ and $s(2/n)^{1/2} = (5.366)(2/15)^{1/2} = 1.9595$.

Using the aqueous solution as reference:

$$\phi_1 = (-0.2)(26.18) = -5.236 \quad (407)$$

$$\phi_2 = (0.2)(26.18) = 5.236 \quad (408)$$

$$t_1 = [(27.22 - 26.18) - (-5.231)]/1.9595 = 3.203 \quad (409)$$

$$t_2 = [5.236 - (27.22 - 26.18)]/1.9595 = 2.141 \quad (410)$$

The tabled $t = 2.056$.

Since both t_1 and t_2 are greater than the tabled t, the formulations are bioequivalent according to areas.

Using Lanoxin tablets as the reference:

$$\phi_1 = (-0.2)(22.42) = -4.484 \quad (411)$$

$$\phi_2 = (0.2)(22.42) = 4.484 \quad (412)$$

$$t_1 = [(27.22 - 22.42) - (-4.484)]/1.9595 = 4.738 \quad (413)$$

$$t_2 = [4.484 - (27.22 - 22.42)]/1.9595 = -0.16 \quad (414)$$

The tabled t is 2.056 as above.

Since the tabled t is greater than t_2 then the Collett tablet is *bioinequivalent* when Lanoxin tablets are the reference. However, the Collett tablet is bioequivalent when the aqueous solution of digoxin is the reference. The

TABLE 10.9. Areas 0–Inf in the Digoxin Study.

	AUC 0–Inf [(ng × hr)/ml]		
Subject	Lanoxin Tab.	Collett Tab.	Aq. Solution
1	14.09	20.86	14.06
2	18.80	28.09	23.18
3	22.87	31.95	39.55
4	17.53	14.55	20.69
5	16.79	17.24	23.03
6	27.69	37.17	28.03
7	19.17	24.65	22.16
8	33.09	46.14	35.37
9	19.36	19.33	24.95
10	17.91	28.21	30.28
11	16.89	27.40	19.95
12	48.69	27.52	35.17
13	26.13	21.42	24.10
14	13.71	26.20	20.69
15	23.52	37.57	31.44
Mean	22.42	27.22	26.18
C.V. (%)	40.1	30.9	26.6

reason for these results is that the variation of areas was considerably greater with Lanoxin tablets than with the aqueous solution. This may be seen by comparing the coefficients of variation at the bottom of Table 10.9–40.1% *versus* 26.6%.

Bioavailabilities of Formulations in Individual Subjects

In the above analyses bioequivalence or inequivalence was based only on bioavailability estimated from mean areas–i.e., Equation (398) without a dose correction. However, one can estimate bioavailability for each subject and obtain an array of F values from ratios of individual subjects' areas. If the mean ratio is designated as R then the expectation is that the mean ratio will be equal to unity if there is bioequivalence. This may be tested statistically with a Student t-test as follows:

$$t = |1 - R|/(SE_R) \tag{415}$$

where the bars around $1 - R$ indicate absolute value and SE_R is the standard error of the ratio R. The above t value is compared with the tabled $t_{.05,n-1}$ where n is the number of ratios. SE_R is estimated as indicated below with Equations (416) and (417) providing $R = \Sigma_1^n X_1 / \Sigma_1^n X_2$, where X_1 refers to the test formulation and X_2 refers to the reference formulation.

$$\text{Var. of } R = \frac{\Sigma X_1^2 - (2)\left[\frac{\Sigma X_1}{\Sigma X_2}\right][\Sigma X_1 X_2] + \left[\frac{\Sigma X_1}{\Sigma X_2}\right]^2 (\Sigma X_2)^2}{NR(NR - 1)[\Sigma X_2/NR]^2} \tag{416}$$

$$\text{SE}_R = \sqrt{\text{Var. of } R} \tag{417}$$

An example is shown in Table 10.10 based on the AUCs in Table 10.9. From column 2 of Table 10.10 – i.e., when the aqueous solution was the reference – substituting into Equation (415) gives:

$$t = \frac{[1 - 1.04]}{0.0660} = 0.606$$

The tabled $t_{.05,14} = 2.145$. In this case the experimental t is less than the tabled t and bioequivalence is concluded as before.

TABLE 10.10. Individual Subject Bioavailabilities of Digoxin.

Subject	$\frac{(\text{AUC})_{\text{Collett}}}{(\text{AUC})_{\text{solution}}}$	$\frac{(\text{AUC})_{\text{Collett}}}{(\text{AUC})_{\text{Lanoxin}}}$
1	1.48	1.48
2	1.21	1.49
3	0.808	1.40
4	0.703	0.803
5	0.749	1.03
6	1.33	1.34
7	1.11	1.29
8	1.30	1.39
9	0.775	0.998
10	0.932	1.58
11	1.37	1.62
12	0.782	0.565
13	0.889	0.820
14	1.27	1.91
15	1.19	1.60
R	1.04	1.21
SE_R	0.066	0.121

From column 3 of Table 10.10 – i.e., when the Lanoxin tablet was the reference – substituting into Equation (415) gives:

$$t = \frac{[1 - 1.214]}{0.121} = 1.736$$

In this case also the experimental t is less than the tabled t and one would conclude bioequivalence – which is a different conclusion than is made above. This method is not FDA approved, whereas the two one-sided t-test method is FDA approved. However, the data in Table 10.10 does indicate intersubject variation in bioavailability, and shows that the variance of the bioavailability is greater when Lanoxin tablets were the reference compared with the aqueous solution as reference.

It is instructive to make plots of $(\mathrm{AUC})_{\mathrm{test}}$ *versus* $(\mathrm{AUC})_{\mathrm{reference}}$. Figure 10.18 shows plots of AUC for the Collett digoxin tablet *versus* AUC for the aqueous solution of digoxin. The top part of the figure has the least squares line free to pass through any intercept with slope = 0.779; the intercept of 6.83 is not significantly different from zero. The bottom part of the figure has the least squares line forced through the origin with slope equal to 1.02. This figure strongly supports the bioequivalence decision above.

Figure 10.19 is a plot of AUC for the Collett digoxin tablet *versus* AUC for the Lanoxin tablet. The least squares line free to pass through any intercept is drawn in the figure with slope = 0.462 and intercept = 9.85. Here r = only 0.432 *versus* 0.609 in the case when the aqueous solution was the reference; the greater scatter in this case is the result of greater variation of AUC with Lanoxin than with the aqueous solution as reference.

Figure 10.20 is a plot of $(\mathrm{AUC})_{\mathrm{test}}$ *versus* $(\mathrm{AUC})_{\mathrm{ref}}$ using the data of L. Yuh [95]. These data indicate no significant correlation between the two areas since $r = 0.110$ and the critical $r_{.05,16} = 0.468$. This indicates to this author that one would have to be very careful how these data should be analyzed for bioequivalence or bioinequivalence. Yuh [95] described two new procedures – both based on the individual subject ratios. One method is based on the median ratio of areas. The other method is based on the assignment of weights to the area ratios. Both methods provide 90% confidence intervals for the average ratio of areas. The Yuh method using weights gave an average ratio of 1.09 with a 90% C.I. of 0.79 to 1.13. The R based on sums above was 0.962 and the 90% C.I. was 0.771 to 1.15. Here the 90% C.I. = 0.961 ± (1.740)(0.1098) where $t_{.10,17} = 1.740$ and $\mathrm{SE}_R = 0.1098$ with Equations (416) and (417). Thus the new Yuh method and the ratio method described by Equations (415) through (417) give almost exactly the same 90% C.I. from Yuh's data [95].

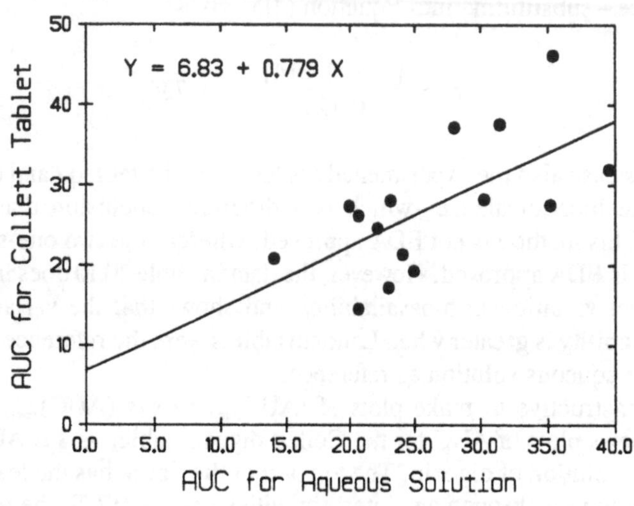

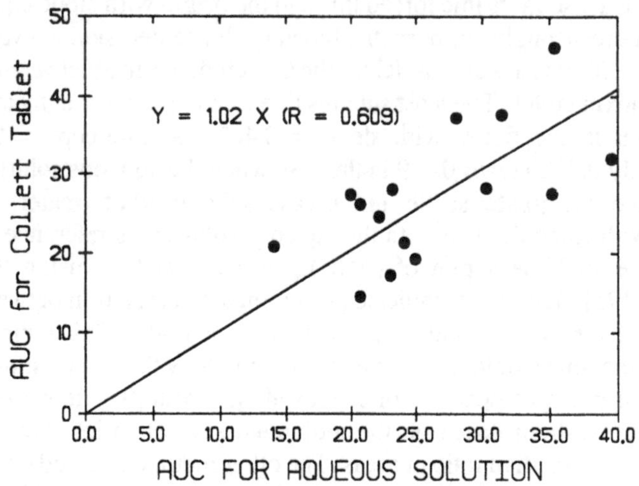

FIGURE 10.18. Example of plots of $(AUC)_{test}$ *versus* $(AUC)_{reference}$ using Collett digoxin tablet and digoxin aqueous solution results. The upper part has the least squares line free to pass through any intercept with an intercept of 6.83 which was not significantly different from zero. The lower part has the least squares line forced through the origin with slope = 1.02. The figure supports bioequivalence.

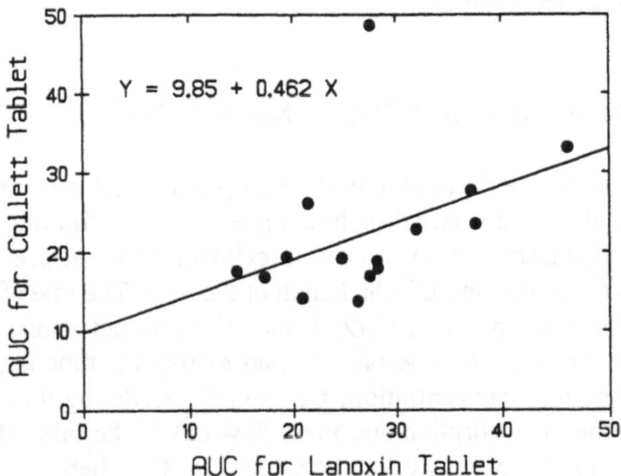

FIGURE 10.19. Example of plot of $(AUC)_{test}$ *versus* $(AUC)_{reference}$ using the Collett tablet as test and Lanoxin tablets as reference. The greater scatter and lower *r* value compared with Figure 10.18 is the result of greater variation of the reference Lanoxin tablets than the digoxin aqueous solution.

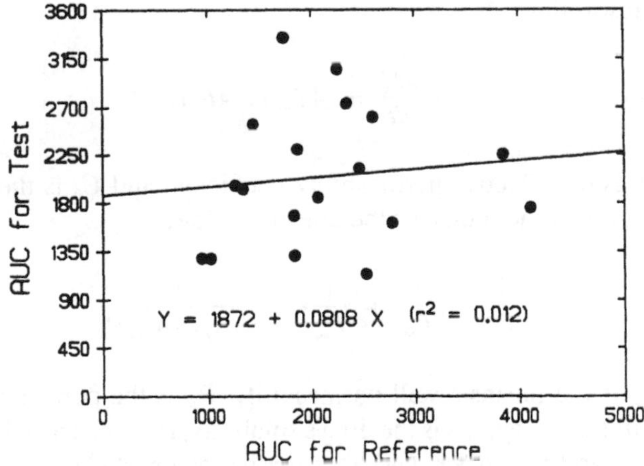

FIGURE 10.20. Example of plot of $(AUC)_{test}$ *versus* $(AUC)_{reference}$ using data of Yuh [95]. There is no significant correlation of the areas suggesting that these data would have to be analyzed very carefully.

THEORETICAL APPROACHES FOR ESTIMATING FRACTION OF DOSE ABSORBED

Cases Where Dissolution Rate Is Not Rate Limiting

Sinko, Leesman and Amidon [96] developed a model to estimate fraction of the dose absorbed where dissolution rate is not rate limiting. The model of the small intestine is taken to be a cylinder with surface area $2\pi RL$, where R is the radius and L is the length of the tube. The rate of mass entering the tube is the product, $C_o Q$, of the inlet concentration, C_o, and the volumetric flow rate, Q. The rate of mass exiting the tube is the product, $C_m Q$, of the outlet concentration, C_m, and Q. Assuming that mass is lost from the tube by absorption and mass flow out of the tube, the mass absorbed per unit time is the difference, $Q(C_o - C_m)$, hence,

$$-\frac{dM}{dt} = Q(C_o - C_m) = \iint_s J_w dA \tag{418}$$

where J_w is the wall flux and A is the absorptive surface area. If there is only passive absorption,

$$-\frac{dM}{dt} = AJ_w = AP_e C_b \tag{419}$$

where P_e is the effective permeability coefficient and C_b is the bulk drug concentration in the lumen of the intestine. Now,

$$\frac{1}{P_e} = \frac{1}{P_w} + \frac{1}{P_{aq}} + \frac{1}{(\mathrm{SA}_p/\mathrm{SA}_m)P_s} \tag{420}$$

where P_w is the intrinsic wall permeability, P_{aq} is the permeability of the aqueous diffusion layer, S is the drug's solubility, P_s is the particle boundary layer permeability and SA_p and SA_m are the total particle and wall surface areas, respectively.

Since the two ideal flow models, namely the Mixing Tank Model (MT) and the Complete Radial Mixing Model (CRM), used do not account for radial concentration variation, there is no distinction made between P_e and P_w. The P_w used in the correlations is the experimentally determined P_w from the *in situ* single-pass perfusion experiment in rats.

From Equation (419) and assuming cylindrical tube geometry and constant permeability this equation becomes Equation (421).

$$-\frac{dM}{dt} = Q(C_o - C_m) = 2\pi RP_e \int_0^L C_b dz \tag{421}$$

Introducing the dimensionless variables, S^*, z^*, and C_b^*, where

$$S^* = S/C_o \tag{422}$$

$$z^* = z/L \tag{423}$$

$$C_b^* = C_b/C_o \tag{424}$$

and S/C_o is the dimensionless solubility, z/L is the fractional length and C_b/C_o is the dimensionless concentration.

Equation (421) simplifies to Equation (425).

$$F = 1 - \frac{C_m}{C_o} = \left[\frac{2\pi RL}{Q}\right] P_e \int_0^1 C_b^* dz^* \tag{425}$$

where F is the fraction of the dose absorbed. Now,

$$Q = \pi R2\nu_z \tag{426}$$

where ν_z is the mean axial fluid velocity. Substituting for Q in Equation (425) from Equation (426) gives Equation (427).

$$F = 2\frac{L}{R}\frac{P_e}{\nu_z}\int_0^T C_b^* dz^* \tag{427}$$

If we let,

$$A_n = \frac{L}{R}\frac{P_e}{\nu_z} \tag{428}$$

where A_n is the absorption number and substitute from Equation (428) into Equation (427) we get Equation (429).

$$F = 2A_n \int_0^1 C_b^* dz^* \tag{429}$$

Case I Where $C_o < S$, $C_m < S$: Mixing Tank Model

Here,

$$F = 1 - C_b^* \tag{430}$$

whence

$$C_b^* = 1 - F \tag{431}$$

Substituting for C_b^* in Equation (429) from Equation (431) gives Equation (432).

$$F = 2A_n \int_0^1 (1 - F) dz^* \tag{432}$$

Integrating and simplifying Equation (432) gives Equation (433).

$$F_1 = \frac{2A_n}{1 + 2A_n} \tag{433}$$

Case II. Complete Radial Mixing Model

$$C_b^* = e^{-2A_n z^*} \tag{434}$$

Substituting for C_b^* in Equation (429) from Equation (434) gives Equation (435).

$$F = 2A_n \int_0^1 e^{-2A_n z^* dz^*} \tag{435}$$

Integrating Equation (435) gives Equation (436).

$$F_{\mathrm{I}} = 1 - e^{-2A_n} \tag{436}$$

Case III Where $C_o > S$ and $C_m > S$

The drug concentration at the wall is equal to the drug solubility, S. Since C_b^* is constant and equal to the solubility, the CRM model can be treated as the MT model. Substituting $C_b = S$ into Equation (429) and performing the integration leads to Equation (437).

$$F_{\mathrm{III}} = 2A_n S^* \tag{437}$$

Case II Where $C_b > S$, $C_m > S$

Equation (438) may be derived for the CMA model.

$$\begin{aligned} F_{\mathrm{II}} &= F_{\mathrm{III}} + F_{\mathrm{I}}(1 - F_{\mathrm{III}}) \\ &= 1 - S^* e^{(1/S^* - 1 - 2A_n)} \end{aligned} \tag{438}$$

The solution for the MT model for Case II is the same as for Case I.

Nonlinear Wall Permeability

For compounds that are absorbed by a carrier-mediated process, a concentration-dependent permeability requires use of a mean permeability, $(P_w^*)_{\mathrm{av}}$, defined by the following equations.

$$(P_w^*)_{\mathrm{av}} = \int_{C_o}^{0} P_w^* dC \Big/ \int_{C_o}^{0} dC \tag{439}$$

where $C_o = D_o/\mathrm{VL}$, D_o = dose, and VL = volume of intestinal lumen.

For Case I,

$$(P_w^*)_{\mathrm{av}} = P_m^* + P_c^* \frac{K_m}{C_o} \ln\left(1 + \frac{C_o}{K_m}\right) \tag{440}$$

For Case III,

$$(P_w^*)_{av} = P_c^* \bigg/ \left(1 + \frac{S}{K_m}\right) + P_m^* \tag{441}$$

For Case II,

$$(P_w^*)_{av} = X_p/L(P_w^*)_{I} + (1 - X_p/L)(P_w^*)_{III} \tag{442}$$

Application to Amoxicillin Data

Rat intestinal perfusion experiments with amoxicillin gave: $K_m = 9.1$ mM at pH 4.0 (approx. 3.33 mg/ml), $P_c^* = 0.65$ (carrier permeability), and P_m^* (passive membrane permeability) = not different from zero.

A scaling parameter = 1.27, a mean amoxicillin solubility of 6.0 mg/ml and an intestinal volume equal to the volume of water given with the drug dose were used in calculating the predicted fraction of dose absorbed. All three cases above were used since the amoxicillin dose was varied over a wide range.

Thirteen sets of amoxicillin human data were used in the correlations. The initial concentration (dose/volume of fluid taken with the dose) varied from 1.25 to 20 mg/ml. The relationship between predicted and observed fraction of the dose absorbed is shown in Figure 10.21; the correlation

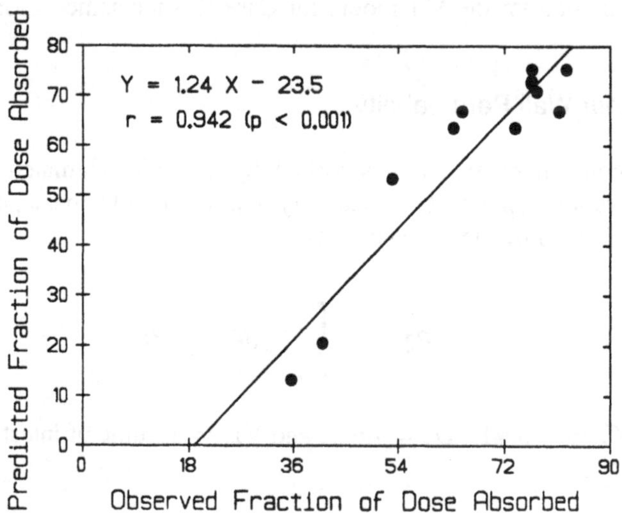

FIGURE 10.21. Plot of predicted *versus* observed fraction of dose absorbed using the amoxicillin human data of Sinko et al. [96]. The large negative intercept indicates a significant bias of unknown source.

coefficient was highly significant ($p < 0.001$), but the large negative intercept of -23.5 indicates a significant bias of unknown source.

Cases Where Dissolution Rate Is Rate Limiting

Dressman and Fleisher [97] used a simple mixing tank model (Scheme 10.5) to predict amount of drug absorbed, and hence fraction of dose absorbed, in cases where absorption was dissolution rate controlled.

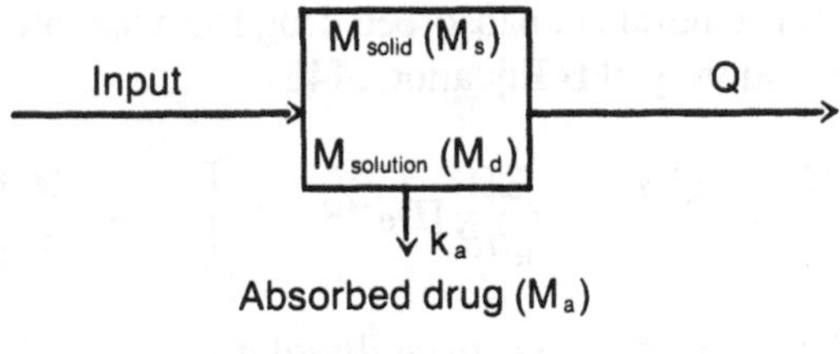

SCHEME 10.5.

Spherical particles of drug were assumed to dissolve according to the Noyes-Whitney Equation (446) below. Some of the symbolism is the same as in Equations (418) through (442) above. The initial number of particles, N_o, is given by Equation (443).

$$N_o = \frac{M_o/\varrho}{4/3\ \pi R^3} \tag{443}$$

where M_o is the dose of drug, ϱ is the density, and R is the radius of the particles.

Since drug leaves the "tank" by a first-order process, the number of particles at time t, N, is given by Equation (444).

$$N = N_o e^{-(Q/V)t} \tag{444}$$

where Q is the flow rate and V is the volume of intestinal fluid.

The surface area, A, of the N particles at time t is given by Equation (445).

$$A = 4\pi R^2 N = 4\pi N \left\{\frac{3M_d}{4\pi\varrho N}\right\}^{2/3} \tag{445}$$

where M_d is the mass of dissolved drug.

The Noyes-Whitney dissolution Equation (446) is:

$$-\frac{dM_s}{dt} = \frac{DA}{h}\left(C_s - \frac{M_d}{V}\right) \tag{446}$$

where M_s is the mass of solid drug, D is drug diffusivity, C_s is the aqueous solubility and h is the boundary layer thickness.

Substituting Equation (444) into Equation (445) then the resultant equation into Equation (446) gives Equation (447).

$$-\frac{dM_s}{dt} = \frac{3D}{\varrho hR}[De^{-(Q/V)t}]\left[C_s - \frac{M_d}{V}\right]M_s^{2/3} \tag{447}$$

Accounting for removal of unabsorbed drug from the intestine by outflow as well as dissolution, yields Equation (448).

$$-\frac{dM_s}{dt} = \frac{QM_s}{V} + \frac{3D}{\varrho hR}[De^{-(Q/V)t}]\left[C_s - \frac{M_d}{V}\right]M_s^{2/3} \tag{448}$$

Drug in solution is generated from dissolution of solid drug and is removed by absorption and transit, i.e.,

$$-\frac{dM_d}{dt} = k_a M_d + \frac{QM_d}{V} + \frac{3D}{\varrho hR}[De^{-(Q/V)t}]\left[C_s - \frac{M_d}{V}\right]M_s^{2/3} \tag{449}$$

where k_a is the first-order absorption rate constant.

Absorbed drug is generated from drug in solution,

$$\frac{dM_a}{dt} = k_a M_d \tag{450}$$

Equations (448) to (450) form a complete set of differential equations describing absorption for dissolution rate–limited systems. Initial conditions at $t = 0$ are M_s = dose, $M_d = 0$, and $M_a = 0$. These equations may be integrated by the Runge-Kutta numerical integration method to yield M_d and M_a and hence $F = M_a$/dose, where F is fraction of dose absorbed.

The authors [97] made quite accurate predictions of digoxin and griseofulvin absorption using parameter values listed in Table 10.11.

Case Where There Is a Polydisperse Powder and Dissolution Is Rate Limiting

Johnson and co-workers [98,99] extended the work of Dressman and Fleisher [97] to cover the case of a polydisperse powder. The powder was assumed to have a log-normal distribution and the diffusion layer thickness

TABLE 10.11. Parameters Used to Simulate Absorption of Digoxin and Griseofulvin [97].

Drug and Parameter	Parameter Value	Range Studied
Griseofulvin		
Dose, mg	500	100–600
Solubility, mg/ml	0.015	0.01–1
Diffusivity, cm^2/min	4.9×10^{-4}	
Initial particle diameter, μm		
Micronized	4	4–100
Nonmicronized	30–50	
h, mm	50	10–100
k_a, min^{-1}	0.125	0.0125–0.125
Digoxin		
Dose, mg	0.25	0.125–1
Solubility, mg/ml	0.024	
Diffusivity, cm^2/min	2.65×10^{-4}	
Particle size, μm	20–29	Solution–102
h, mm	50	10–30
k_a, min^{-1}	0.08	0.005–0.1

assumed to vary with the initial particle radius, r_o, according to Equation (451), where M_o is the dose of the drug.

$$h = r_o \left[\frac{M_s}{M_o} \right] \tag{451}$$

where r_o is the initial particle radius and M_o is the dose of drug.

CHAPTER 11

Physiological Pharmacokinetics

Physiological pharmacokinetics was initiated by Dedrick and Bischoff [100–102]. A typical model is shown in Figure 11.1.

BLOOD FLOW RATE–LIMITED MODELS

In Figure 11.1, the V_is are volumes, the C_is are concentrations, and the Q_is are blood flow rates. A through D indicate sites of administration: A–intravenous into an arm; B–hepatic arterial; C–intramuscular; and D–oral. Blood from the hepatic artery and portal vein is usually considered to mix before entering the sinusoids of the liver. The arrow out of the liver indicates loss of drug as a result of metabolism. The arrow out of the kidney indicates loss of drug as a result of urinary excretion of unchanged drug. Each organ has its own characteristic blood flow rate hence the Q_is differ from one another. There is an assumption of flow limitation, which means that the drug concentrations in arterial blood rapidly reach equilibrium with the drug concentrations in the tissue space of each compartment. For this assumption to be valid, the diffusion time for the drug to cross the tissue cell membrane and enter the tissue space has to be relatively fast in comparison with the transport of the drug by the blood. As a consequence, the tissue concentration is proportional to the outgoing venous drug concentration, and the total concentration of the drug in any particular compartment can be expressed as its venous plasma concentration multiplied by a constant, R, which is the partition coefficient or the distribution ratio of the drug between the tissue and blood, i.e., R = tissue concentration/whole blood concentration at steady state.

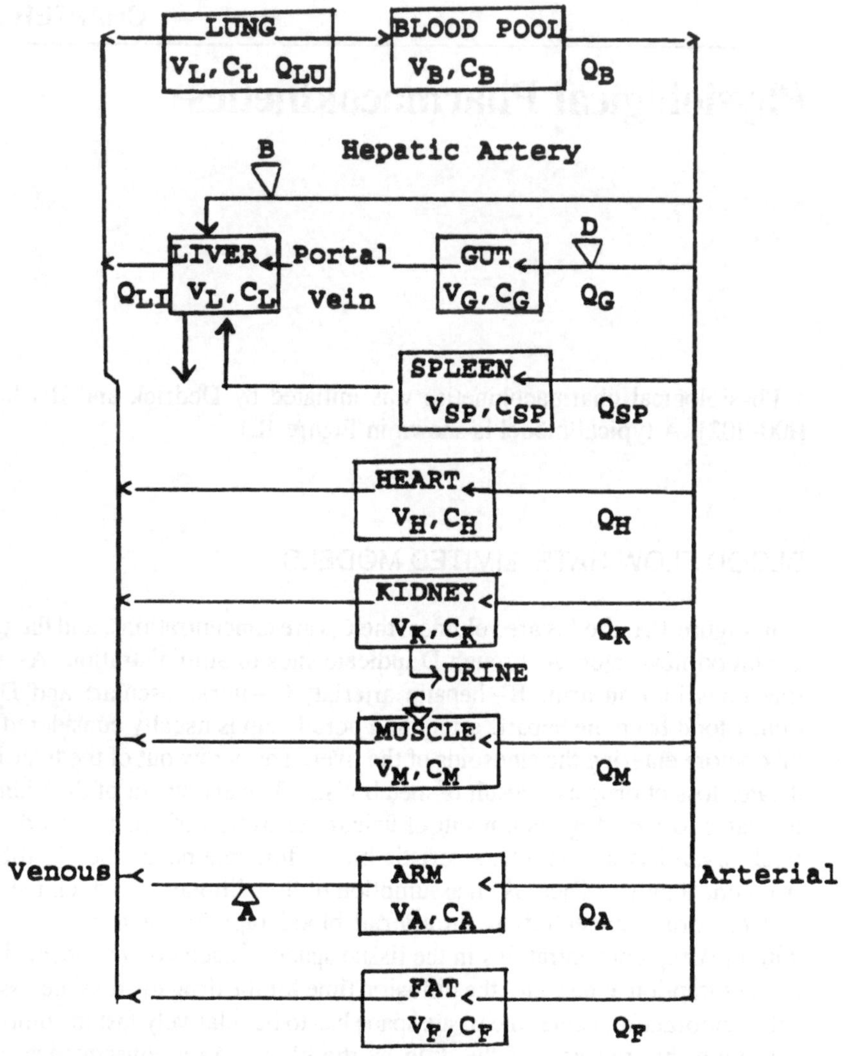

FIGURE 11.1. Model XXXVIII. Physiological model for drug disposition and elimination.

The differential equations for the model are mass balance equations. The mass balance rate equation for total drug concentration in a noneliminating compartment such as the heart, muscle, fat, and the arm in Figure 11.1 is given by equations such as Equations (452) and (453).

$$V_M \left\{ \frac{dC_M}{dt} \right\} = Q_M \left\{ C_B - \frac{C_M}{R_M} \right\} \tag{452}$$

$$V_F \left\{ \frac{dC_F}{dt} \right\} = Q_F \left\{ C_B - \frac{C_F}{R_F} \right\} \tag{453}$$

The mass balance rate equation for a compartment, such as the liver, which eliminates the drug, must also contain a term to account for the elimination, such as Equation (454).

$$V_L \left\{ \frac{dC_L}{dt} \right\} = \{Q_L - Q_G\}C_B + Q_G \left\{ \frac{C_G}{R_G} \right\} - Q_L \left\{ \frac{C_L}{R_L} \right\} - \frac{f_B \mathrm{CL}_L' C_L}{R_L} \tag{454}$$

In Equation (454), f_B is the free fraction in blood and CL_L' is the intrinsic clearance with respect to free drug concentration in the liver. For other equations see References [100–104].

Tissue volumes and blood flow rates for such models have been published by several authors [100–107].

An advantage of such models is that predictions may be made for one species from data collected on a different species, i.e., predictions can be made for man based on animal data. A major disadvantage is that the models are based on average tissue volumes and blood flow rates and hence tell one nothing about the individual animal or man.

To simulate data, equations for each compartment, such as Equations [452–454], are numerically integrated via computer to obtain the model-predicted concentrations, then these are graphically compared with the observed concentrations.

Harris and Gross [108] used a flow-limited model, similar to that shown in Figure 11.1, to simulate data collected after intravenous injection of adriamycin in the rabbit. Figure 11.2 compares predicted concentrations (solid lines) to the observed concentrations of adriamycin.

MEMBRANE-LIMITED MODELS

Physiologic pharmacokinetic modelling of some drugs has indicated that tissue uptake of these drugs is not consistent with a blood flow rate limita-

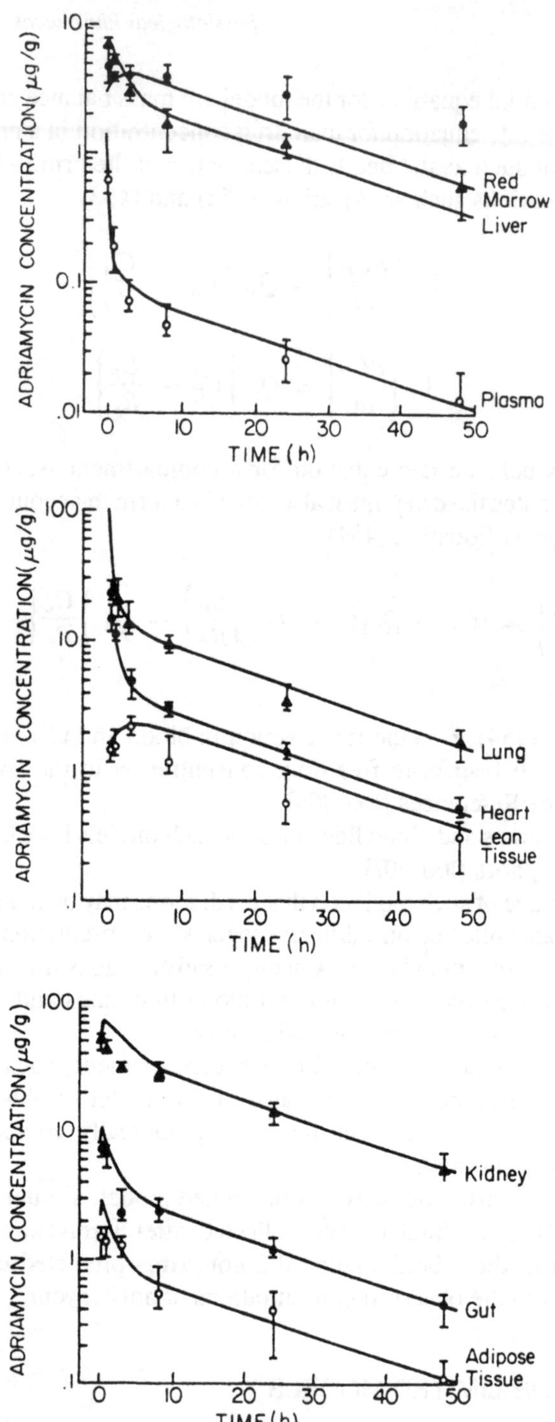

FIGURE 11.2. Tissue and plasma concentrations of adriamycin in the rabbit [108] after I.V. injection of 3 mg/kg of adriamycin.

tion. In such cases decline in tissue drug concentration does not parallel decline in drug concentration in blood. In such cases there may be membrane limitation rather than blood flow rate limitation. Chen and Gross [103] discussed physiologic pharmacokinetic modelling of actinomycin D in the beagle dog. Figure 11.3(a) shows a comparison of simulated data

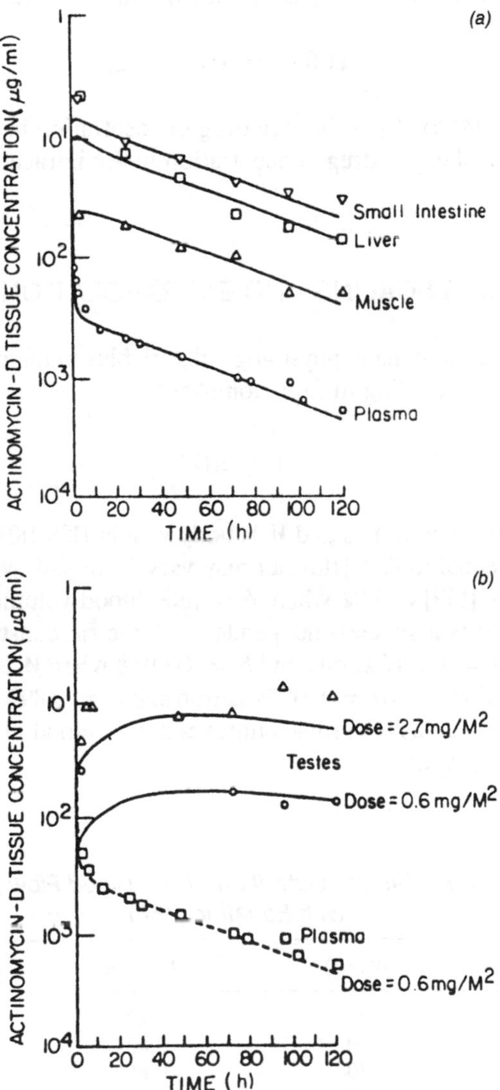

FIGURE 11.3. Tissue concentration profiles of actinomycin D in the dog after 0.03 mg/kg I.V.: (a) flow rate-limited model; (b) membrane-limited model (from Reference [103], with permission from Springer-Verlag).

(solid lines) with observed concentrations of actinomycin D after a single 0.6 mg/m² (0.03 mg/kg) dose when a flow rate–limited model similar to that shown in Figure 11.1 was used. However, Figure 11.3(b) shows results obtained in the testes following two different doses and those in plasma following the lower dose. Note that the testes profile differs from the plasma profile. The solid lines for the testes in Figure 11.3 are based on a membrane-limited model [Equation (455)] with $k = 0.2\ \text{hr}^{-1}$.

$$\text{Net flux} = k(C_E - C_I) \tag{455}$$

In Equation (455), C_E is the free drug concentration in the extracellular space and C_I is the free drug concentration in the intracellular fluid.

INTERSPECIES SCALING AND EXTRAPOLATION

The magnitudes of many physiological variables in mammals are related to body weight according to Equation (456).

$$Y = aW^b \tag{456}$$

where a and b are constants and W is body weight [109,110]. The constant b is frequently equal to 0.75 [110] but may vary from 0.69 when Y = creatinine clearance [109] to 1.02 when Y = total blood volume [110]. The a in Equation (456) is a species-independent allometric constant. For cardiac output (CO) $a = 0.1017$ L/min and $b = 0.9988$ when W is in kg [107]. For hepatic blood flow (Q_L) $a = 0.0554$ L/min and $b = 0.894$ when W is in kg. Table 11.1 gives values of cardiac output and liver blood flow obtained with these parameter values.

TABLE 11.1. Cardiac Output and Liver Blood Flow Estimated with Equation (456).

Species	W (kg)	Q_L (L/min)	CO (L/min)
Human	70	2.47	7.08
Dog	10	0.434	1.01
Dog	20	0.807	2.02
Rabbit	3	0.15	0.305
Rabbit	4	0.19	0.406

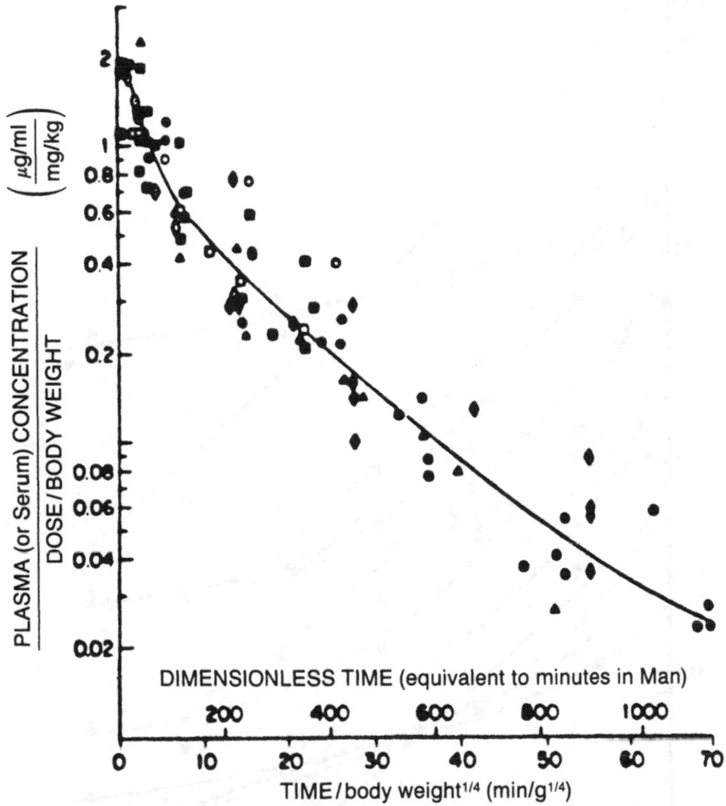

FIGURE 11.4. Correlation of methotrexate plasma concentrations using physiological time [102]. See Figure 11.5 for original data (with permission from the National Cancer Institute).

Another useful scaling concept is physiological time, t', defined by Equation (457) where t is chronological time and W is body weight.

$$t' = t/W^{0.25} \tag{457}$$

While chronological time is the same for all species, physiological time is different for each species. All species have approximately the same physiologic and metabolic rates when measured in the physiological time frame. Thus physiological time takes care of the fact that the mouse has a heart rate and a relative cardiac output that are about an order of magnitude higher than those of man.

An example of the use of physiological time is shown in Figure 11.4 [102] where methotrexate plasma concentrations per mg/kg in the mouse, rat, monkey, dog, and man are related via physiological time.

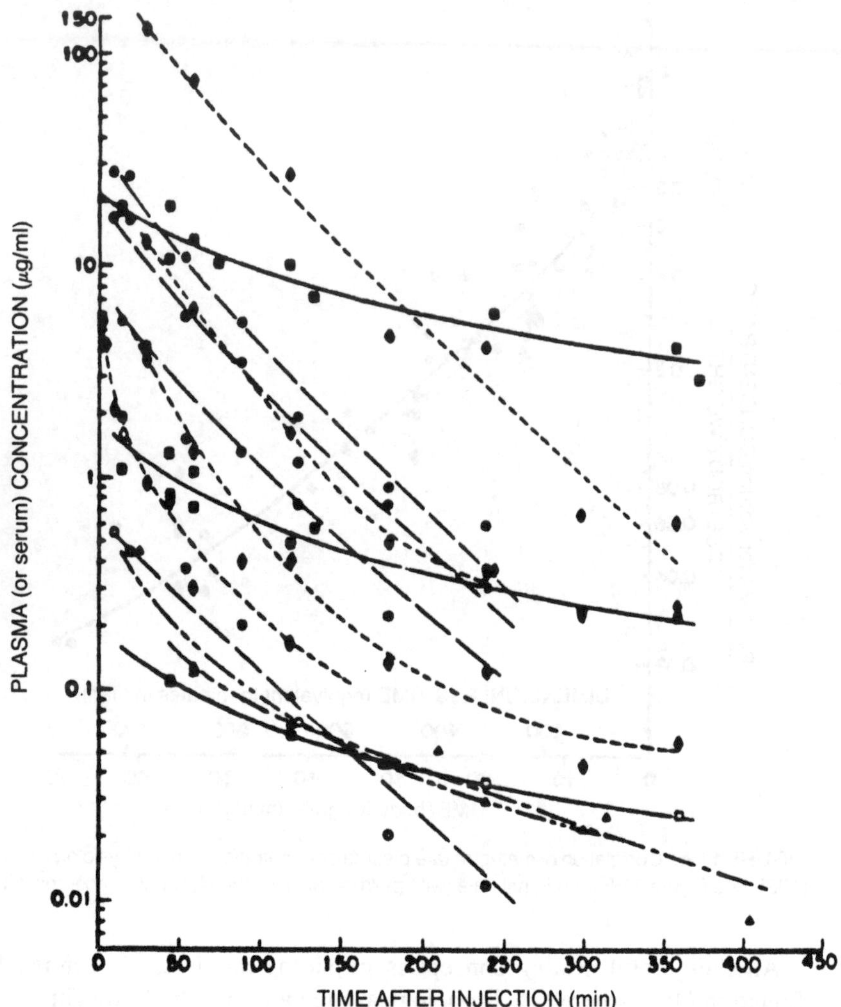

FIGURE 11.5. Plasma and serum concentrations of methotrexate after I.V. or I.P. injection of various doses in mice (diamonds), rats (circles), monkeys (closed triangles), dogs (open triangles), and human beings (squares). This is the original data from which Figure 11.4 was constructed (from Reference [102], with permission from the National Cancer Institute).

THE SIMPLEST PHYSIOLOGICALLY BASED MODELS

Introduction

So far in this chapter we have considered the type of physiologically based pharmacokinetics where real mean physiologically measured blood flows and pharmacokinetically measured R values are used in modelling, and one *simulates* blood and tissue concentration profiles and then shows plots of the simulated curves drawn through the observed blood and tissue concentrations. A quantitative measure of how good such correlations are, such as a coefficient of determination, has never been reported. Often "eye-ball" observations of such plots suggest that an r^2 value would be quite low.

There is another type of physiological pharamacokinetics where one can use simple models, with only two or three compartments (rather than the ten compartments of Figure 11.1), and pharmacokinetically estimate compartment model parameters that have physiological significance and meaning, such as liver blood flow and unbound tissue volumes.

Simple Physiologically Based Models That Involve Biexponential Disposition

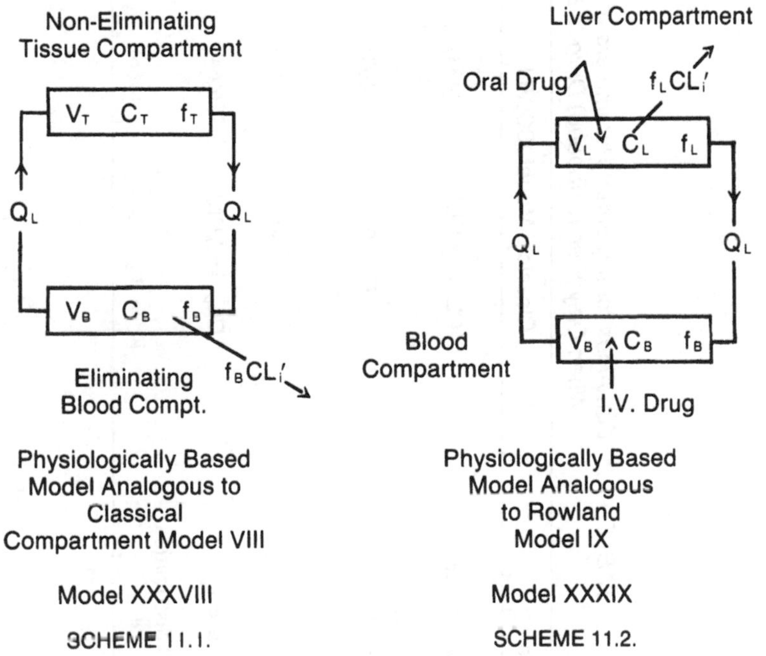

SCHEME 11.1.

SCHEME 11.2.

Equations for important parameters of Schemes 11.1 and 11.2 are shown in Table 11.2.

Derivations of the equations in Table 11.2 are given in this chapter, Chapter 12, or the Appendix.

For Model XXXVIII V_T, C_T, and f_T are the volume, concentration, and

TABLE 11.2. Equations for Important Parameters of Models XXXVIII and XXXIX When Only Free (Unbound) Drug Is Eliminated.

Parameter	Classical Model XXXVIII		Rowland Model XXXIX	
Systemic clearance	$CL_s = f_B CL_i'$	(458)	$CL_s = \dfrac{Q_L f_B CL_i'}{Q_L + f_B CL_i'}$	(459)
Extraction ratio	$E = 0$	(460)	$E = \dfrac{f_B CL_i'}{Q_L + f_B CL_i'}$	(461)
R_T or R_L	$R_T = C_{Tss}/C_{Bss}$	(462)	$R_L = \dfrac{C_{Lss}}{C_{Bss}} \dfrac{1}{F}$	(463)
Oral clearance	$CL_{po} = f_B CL_i'$	(464)	$CL_{po} = f_B CL_i'$	(465)
Volume of distribution steady state	$V_{ss} = V_B + R_L V_L$	(466)	$V_{ss} = V_B + V_L R_L F$	(467)

free fraction of the drug in the noneliminating peripheral compartment; and for Model XXXIX V_L, C_L, and f_L are the volume, concentration, and free fraction of the drug in the eliminating liver compartment. For both models, V_B, C_B, and f_B are the volume, concentration, and free fraction of drug in the central compartment. CL_i' is the intrinsic clearance of free (unbound) drug, F is the bioavailability, CL_m is the metabolic clearance, CL_s is the systemic clearance or $CL_{I.V.}$ = dose/(AUC)$_{I.V.}$, and CL_r = renal clearance. For Model XXXIX, E is the extraction ratio and Q_L is liver blood flow rate. For Model XXXVIII, Q_T is the pooled tissue blood flow rate. CL_{po} is the oral clearance = dose/(AUC)$_{po}$.

Unfortunately some of the equations in the literature do not agree with the correct equations in Table 11.2. Gibaldi and Koup [112] have $CL_B = Q_B f_B CL_i'/(Q_B + f_B CL_i')$ as their Equation (23) rather than the correct Equation (459) in Table 11.2 with Q_L rather than Q_B. Chen and Gross [113] have

$$R_L = \left[1 + \frac{CL_r}{Q_L}\right]\frac{C_{Lss}}{C_{Bss}}$$

for I.V. infusion at a constant rate to steady state and for an eliminating organ such as the liver, but the correct equation is Equation (463) in Table 11.2. This is shown later.

DERIVATIONS FOR MODEL XXXVIII WHERE ONLY FREE (UNBOUND) DRUG IS ELIMINATED

The differential equations when an I.V. infusion at the rate R_o is given to steady state are:

$$V_B\left[\frac{dC_{Bss}}{dt}\right] = R_o - Q_T C_{Bss} + \left[\frac{Q_T}{R_T}\right]C_{Tss} - f_B CL_i' C_{Bss} = 0 \quad (468)$$

$$V_T\left[\frac{dC_{Tss}}{dt}\right] = Q_T C_{Bss} - \left[\frac{Q_T}{R_T}\right]C_{Tss} = 0 \quad (469)$$

From Equation (469) one obtains:

$$R_T = \frac{C_{Tss}}{C_{Bss}} = \frac{f_B}{f_T} \quad (470)$$

for the noneliminating tissue at steady state. Thus the simplest of all physiological models provides the definition for R_T. From Equation (470) one obtains Equation (471).

$$f_T C_{Tss} = f_B C_{Bss} \tag{471}$$

Hence at steady state the free drug concentrations are equal in the tissue and central compartments. From Equation (470) one obtains Equation (472).

$$\frac{C_{Tss}}{R_T} = C_{Bss} \tag{472}$$

$$R_o = f_B CL_i' C_{Bss} \tag{473}$$

Now,

$$CL_s = \frac{R_o}{C_{Bss}} = f_B CL_i' = CL_m \tag{474}$$

If f_B is measured, then the intrinsic metabolic clearance of free drug may be obtained with Equation (475).

$$CL_i' = f_B CL_i' / f_B \tag{475}$$

If the plasma concentration, C_p, is measured rather than the whole blood concentration, C_B, then the blood/plasma ratio, C_B/C_p, should be measured and two concentrations related.

If a drug obeys this model then from Equation (474) one can see that a plot of $CL_s = D/(AUC)_{I.V.}$ *versus* free fraction, f_B, will give a straight line forced through the origin with slope equal to CL_i', providing there is no intrasubject variation in CL_i' or only one animal is studied. If there is intrasubject variation in CL_i' when a group of animals is studied, then one obtains an "average" value of CL_i' as the slope. Yacobi and Levy [115] collected data on warfarin in the rat; a plot as above with $r = 0.95$ was obtained. For the Rowland Model XXXIX a plot of metabolic clearance, $f_B CL_i'$, *versus* f_B should give a similar plot. But since warfarin is a low clearance drug with F approaching unity, CL_s is only slightly smaller than $f_B CL_i'$, and the true slope, CL_i'/F, is nearly the same as CL_i' if one plotted CL_s *versus* f_B. As seen later, warfarin obeys the Rowland Model XXXIX quite well. Hence, in the case of warfarin in the rat the data of Yacobi and Levy [115] most probably gave a slope equal to CL_i'/F not CL_i'.

The elimination half-life, $t_{1/2}$, is given exactly by Equation (476) for Model XXXVIII when only free drug is eliminated.

$$t_{1/2} = \frac{0.693V_\beta}{f_B \mathrm{CL}_i'} \tag{476}$$

Since $V_\beta = \mathrm{CL}_s/\lambda_1 = V_B k_{10}^*/\lambda_1$ and $k_{10}^* = \lambda_1\lambda_2/k_{21}^*$ for this model [see Equation (28) under Model VIII, Chapter 2] then $V_\beta = V_B\lambda_2/k_{21}^*$ and the predicted $t_{1/2}$ is also given by Equation (477). Here the asterisks on the $k_{i,j}$s refer to Model VIII to distinguish them from Model IX where there are no asterisks.

$$t_{1/2} = \frac{0.693V_B\lambda_2}{k_{21}^* f_B \mathrm{CL}_i'} \tag{477}$$

DERIVATIONS FOR MODEL XXXIX WHERE ONLY FREE (UNBOUND) DRUG IS ELIMINATED

Rowland et al. [114] were the first to publish on Model XXXIX but the treatment given in this book is more extensive.

When an intravenous infusion is given at the rate, R_o, to steady state the differential equations are as follows.

$$V_B\left[\frac{dC_{B\mathrm{ss}}}{dt}\right] = R_o - Q_L C_{B\mathrm{ss}} + \left[\frac{Q_L}{R_L}\right]C_{L\mathrm{ss}} = 0 \tag{478}$$

$$V_L\left[\frac{dC_{L\mathrm{ss}}}{dt}\right] = Q_L C_{B\mathrm{ss}} - \left[\frac{Q_L}{R_L}\right]C_{L\mathrm{ss}} - \mathrm{CL}_i' f_L C_{L\mathrm{ss}} = 0 \tag{479}$$

From Equation (479) we get:

$$R_o = Q_L\left(C_{B\mathrm{ss}} - \frac{C_{L\mathrm{ss}}}{R_L}\right) \tag{480}$$

and

$$Q_L\left(C_{B\mathrm{ss}} - \frac{C_{L\mathrm{ss}}}{R_L}\right) = \mathrm{CL}_i' f_L C_{L\mathrm{ss}} \tag{481}$$

Also,

$$f_L \mathrm{CL}_i' = \frac{f_B \mathrm{CL}_i'}{R_L} \tag{482}$$

Substituting from Equations (481) and (482) into Equation (480) gives:

$$R_o = \left[\frac{f_B \mathrm{CL}_i'}{R_L}\right] \tag{483}$$

Hence, the input rate, R_o, is equal to the rate of metabolism at steady state. From Equation (480) we get:

$$R_o = Q_L C_{Bss} - \left[\frac{Q_L}{R_L}\right] C_{Lss} \tag{484}$$

Now,

$$\mathrm{CL}_s = \frac{R_o}{C_{Bss}} = Q_L - \frac{Q_L C_{Lss}}{R_L C_{Bss}} \tag{485}$$

Rearranging Equation (485) gives:

$$\frac{Q_L C_{Lss}}{R_L C_{Bss}} = Q_L - \mathrm{CL}_s \tag{486}$$

Substituting for CL_s in Equation (486) then solving for R_L gives:

$$R_L = \left[\frac{C_{Lss}}{C_{Bss}}\right]\left[\frac{Q_L}{Q_L - \mathrm{CL}_s}\right] = \left[\frac{C_{Lss}}{C_{Bss}}\right]\left[\frac{Q_L + f_B \mathrm{CL}_i'}{Q_L}\right] = \left[\frac{C_{Lss}}{C_{Bss}}\right]\frac{1}{F} \tag{487}$$

ORAL ADMINISTRATION IN MODEL XXXIX

The differential equations for oral administration at a constant rate, R_o, to steady state are:

$$V_B\left[\frac{dC_{Bss}}{dt}\right] = -Q_L C_{Bss} + \left[\frac{Q_L}{R_L}\right] C_{Lss} = 0 \tag{488}$$

$$V_L\left[\frac{dC_{Lss}}{dt}\right] = Q_L C_{Bss} - \left[\frac{Q_L}{R_L}\right] C_{Lss} - f_L \mathrm{CL}_i' C_{Lss} + R_o = 0 \tag{489}$$

From Equations (470) and (489) one obtains Equations (490) and (491).

$$R_L = \frac{C_{Lss}}{C_{Bss}} = \frac{f_L}{f_B} \tag{490}$$

$$f_B C_{Bss} = f_L C_{Lss} \tag{491}$$

Hence at steady state during oral administration at a constant rate, the free drug concentration in the liver compartment is equal to the free drug concentration in the noneliminating central compartment.

From Equations (483), (490), and (491) one obtains Equation (492).

$$R_o = f_L \mathrm{CL}_i' C_{Lss} \left[\frac{f_B \mathrm{CL}_i'}{R_L} \right] C_{Lss} \tag{492}$$

Hence the rate of input is equal to the rate of metabolism.

If we let $R_{\mathrm{po}} = [C_{Lss}/C_{Bss}]$ oral and $R_{\mathrm{I.V.}} = [C_{Lss}/C_{Bss}]$ then $R_{\mathrm{po}}/R_{\mathrm{I.V.}} = F$, which supports Equation (487).

Gibaldi et al. [116] published Equation (493) where they had β in place of λ_1; they also indicated that the denominator was equal to the steady-state volume of distribution, V_{ss}.

$$\lambda_1 = \frac{f_B \mathrm{CL}_i'}{V_p + (f_p/f_T)V_T} = \frac{f_p \mathrm{CL}_i'}{V_{ss}} \tag{493}$$

Equation (493) really applies to Model XXXVI since T refers to the peripheral noneliminating compartment.

Thus one can predict the elimination half-life, $t_{1/2}$, with Equation (494) for Model XXXVIII.

$$t_{1/2} = \frac{0.693\, V_{ss}}{f_B \mathrm{CL}_i'} = \frac{0.693\, [V_p + R_T V_T]}{f_p \mathrm{CL}_i'} \tag{494}$$

For the Rowland Model XXXIX Equation (494) is written as Equation (495) since the peripheral compartment is the liver and the eliminating compartment with volume $= V_L$. The successful use of Equation (495) with warfarin in the rat is covered below.

$$t_{1/2} = \frac{0.693\, [V_p + R_L V_L]}{f_p \mathrm{CL}_i'} \tag{495}$$

EXAMPLE OF APPLICATION OF EQUATIONS FOR MODEL XXXIX

Data on warfarin in the rat [115–117] may be explained by Model XXXIX using an average liver plasma flow of 4.7 ml/min and a liver volume of 19.5 ml for a 500 g (0.5 kg) rat, as published by Gerlowski and Jain [118].

Yacobi and Levy [115] measured an average $R_L = 2.5$ for warfarin in the rat and their plot of clearance *versus* f_p, shown by Levy [116], gives a value of $CL_i'/F = 1.025$ L/kg/hr. Since F is nearly equal to unity for warfarin because it is a low clearance drug, then $CL_i' \sim 1.03$ L/kg/hr. Also, from Figure 3 of Levy [116] the median $f_p = 0.01$. Hence, $f_p CL_i' = (0.01)(1.03) = 0.0103$ L/kg/hr. Gerlowski and Jain [118] give $Q_L = 4/7/0.5 = 9.4$ ml/kg/min $= 564$ ml/kg/hr $= 0.654$ L/kg/hr. Hence, $CL_s = Q_L f_p CL_i'/(Q_L + f_p CL_i') = (0.564)(0.0103)/(0.564 + 0.0103) = 0.0101$ L/kg/hr. This gives $F = CL_s/CL_{po} = 0.0101/0.0103 = 0.981$. Gerlowski and Jain [114] give $V_L = 19.55/0.5 = 39.1$ ml/kg $= 0.0391$ L/kg. Hence, $R_L V_L = (2.5)(0.0391) = 0.0977$ L/kg. Benya and Wagner [118] found $V_p = 0.110$ L/kg for warfarin in the rat. Substituting these values into Equation (495) gives:

$$t_{1/2} = \frac{0.693\,[0.110 + 0.0975]}{0.0103} = 14.0 \text{ hr} \tag{496}$$

Levy [116] reported an elimination rate constant of 0.0507 hr^{-1}, corresponding to a half-life of 13.7 hr. Hence the agreement between observed and predicted half-lives is good, which supports the application of Model XXXIX to warfarin in the rat. In Chapter 12 it is shown that most of the parameters of Model XXXIX may be estimated without measuring R_L.

Derivation for Model XXXX Where There Is Urinary Excretion from the Central Compartment as Well as Metabolism in the Peripheral Compartment

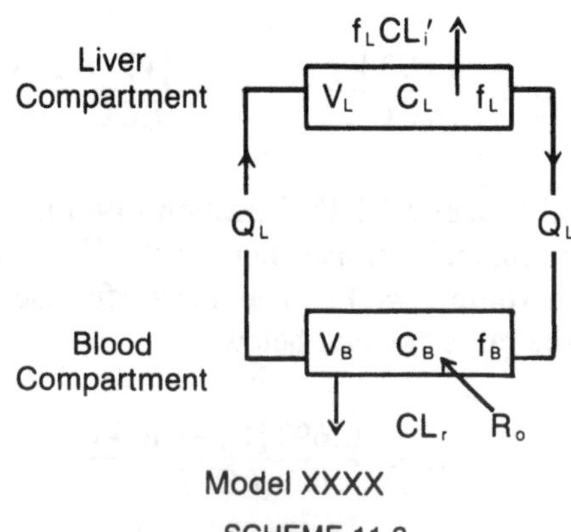

Model XXXX

SCHEME 11.3.

When an infusion at a constant rate, R_o, is given into the central compartment to steady state, the differential equations are as follows.

$$V_B\left[\frac{dC_{Bss}}{dt}\right] = R_o - Q_L C_{Bss} + \left[\frac{Q_L}{R_L}\right] C_{Lss} - \mathrm{CL}_r C_{Bss} = 0 \quad (497)$$

$$V_L\left[\frac{dC_{Lss}}{dt}\right] = Q_L C_{Bss} - \left[\frac{Q_L}{R_L}\right] C_{Lss} - f_L \mathrm{CL}_i' C_{Lss} = 0 \quad (498)$$

From Equation (497) one obtains:

$$R_o - \mathrm{CL}_r C_{Bss} = Q_L C_{Bss} - \left[\frac{Q_L}{R_L}\right] C_{Lss} \quad (499)$$

From Equation (498) one obtains:

$$f_L \mathrm{CL}_i' C_{Lss} = Q_L C_{Bss} - \left[\frac{Q_L}{R_L}\right] C_{Lss} \quad (500)$$

From Equations (499) and (500) one obtains:

$$R_o = \mathrm{CL}_r C_{Bss} + f_L \mathrm{CL}_i' C_{Lss} \quad (501)$$

Hence the rate of input, R_o, is equal to the sum of the rates of urinary excretion and the rate of metabolism as expected.

Dividing Equation (499) through by C_{Bss} gives:

$$\frac{R_o}{C_{Bss}} = \mathrm{CL}_r - \mathrm{CL}_s = Q_L - \frac{Q_L C_{Lss}}{R_L C_{Bss}} \quad (502)$$

But

$$\mathrm{CL}_s = \mathrm{CL}_{nr} + \mathrm{CL}_r \quad (503)$$

where CL_{nr} is the nonrenal clearance as a result of metabolism.

Also,

$$\mathrm{CL}_{nr} = \frac{Q_L f_B \mathrm{CL}_i'}{Q_L + f_B \mathrm{CL}_i'} \quad (504)$$

Substituting from Equations (503) and (504) into Equation (502) gives the previous Equation (487). Hence with or without urinary excretion the expression for R_L is the same. For Model XXXX the predicted elimination half-life is obtained with Equation (505).

$$t_{1/2} = \frac{0.693\,[V_B + R_L V_L]}{\mathrm{CL}_r + f_B \mathrm{CL}_i'} \tag{505}$$

In Chapter 12 it is shown that essentially all of the parameters of Models XXXIX and XXXX may be estimated without measuring R_L.

CHAPTER 12

Relationship between Physiologically Based Flow Models and Usual Compartment Models

INTRODUCTION

This chapter shows how one can relate the usual pharmacokinetic compartment models and physiologically based models. The method allows one to estimate most of the parameters of physiologically based models. The usual criticism of compartment modelling is that the estimated parameters are meaningless. However, with the new techniques described in this chapter the parameters now have physiological significance.

When writing differential equations for the usual compartment models, one uses amounts in each compartment and multiplies these amounts by first-order rate constants to obtain rates with dimensions of mass/time. When writing differential equations for physiologically based models, one multiplies concentrations by clearances to obtain rates with dimensions of mass/time. Hence, to compare the two kinds of models one must convert the clearances to first-order rate constants and the concentrations to amounts. Once this has been done, one equates the coefficients of the amounts and obtains expressions for the rate constants in terms of the physiologically based model parameters.

CLASSICAL TWO-COMPARTMENT OPEN MODEL

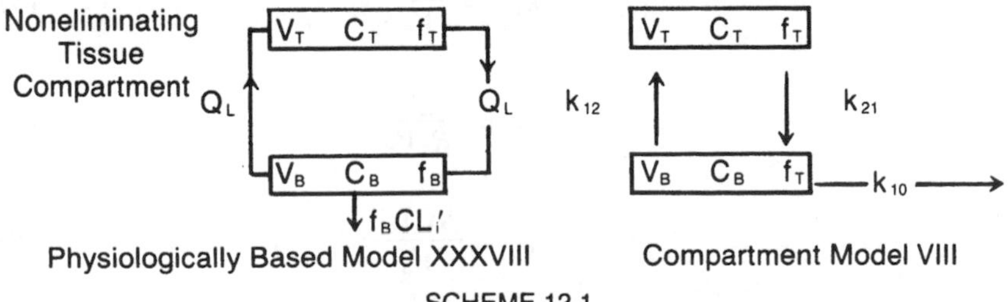

SCHEME 12.1.

MODEL XXXVIII WITH BOLUS I.V. ADMINISTRATION INTO THE CENTRAL COMPARTMENT AND ELIMINATION OF FREE DRUG ONLY

The differential equations are as follows.

$$V_B\left[\frac{dC_B}{dt}\right] = -Q_T C_B + \left[\frac{Q_T}{R_T}\right]C_T - f_B \mathrm{CL}_i' C_B \tag{506}$$

$$V_T\left[\frac{dC_T}{dt}\right] = Q_T C_B - \left[\frac{Q_T}{R_T}\right]C_T \tag{507}$$

Multiplying the numerators and denominators of the right-hand sides of Equations (506) and (507) by the appropriate volume and using $V_B C_B = A_B$ and $V_T C_T = A_T$ gives Equations (508) and (509).

$$\frac{dA_B}{dt} = -\left[\frac{Q_T}{V_B}\right]A_B + \left[\frac{Q_T}{R_T V_B}\right]A_T - \left[\frac{f_B \mathrm{CL}_i'}{V_B}\right]A_B \tag{508}$$

$$\frac{dA_T}{dt} = \left[\frac{Q_T}{V_B}\right]A_B - \left[\frac{Q_T}{R_T V_T}\right]A_T \tag{509}$$

For Model VIII the differential equations are:

$$\frac{dA_B}{dt} = -k_{12}A_B + k_{21}A_T - k_{10}A_B \tag{510}$$

$$\frac{dA_T}{dt} = k_{12}A_B - k_{21}A_T \tag{511}$$

Matching coefficients in Equations (508) and (509) with those in Equations (510) and (511) gives:

$$k_{12} = \frac{Q_T}{V_B} \tag{512}$$

$$k_{21} = \frac{Q_T}{R_T V_T} \tag{513}$$

$$k_{10} = \frac{f_B \mathrm{CL}_i'}{V_B} \tag{514}$$

To apply the above equations the procedure is to measure f_B (or f_p and C_B/C_p) and C_B (or C_p) as a function of time, fit a biexponential equation to the time course data, then estimate k_{12}, k_{21}, and k_{10} using Equations (26) through (29).

Note that the symbol V_B or V_p replaces Vc in those equations. With this model it is necessary to measure $R_T = C_{T\text{ss}}/C_{B\text{ss}}$ at steady state to obtain all the model parameters as follows.

$$Q_L = V_B k_{12} \tag{515}$$

$$V_T R_T = Q_L/k_{21} \tag{516}$$

$$f_B \text{CL}_i' = V_B k_{10} = \text{CL}_m = \text{CL}_s = D/(\text{AUC})_{\text{I.V.}} \tag{517}$$

Hence with this model the metabolic clearance, CL_m, is the same as the systemic clearance, CL_s.

$$V_T = V_T R_T / R_T \tag{518}$$

$$f_T = f_B R_T \tag{519}$$

ROWLAND MODELS IX AND XXXIX WHEN ONLY FREE DRUG IS ELIMINATED

Rowland et al. [114] were the first to show the relationship between the two-compartment open model with peripheral elimination (Model XIX) and a simple physiologically based flow model (Model XXXIX). This has been repeated below since the treatment is different in parts and has been extended.

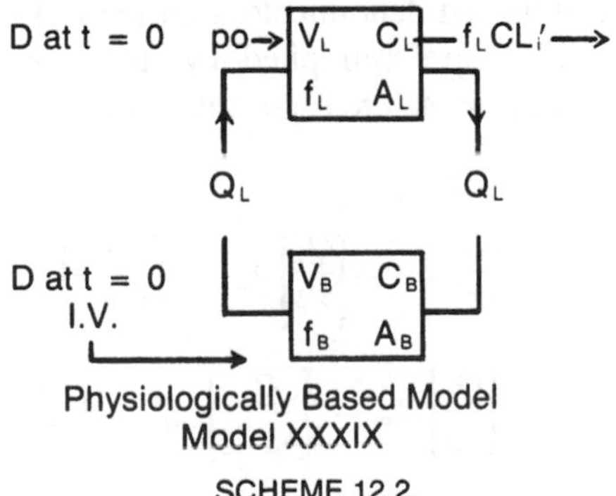

SCHEME 12.2.

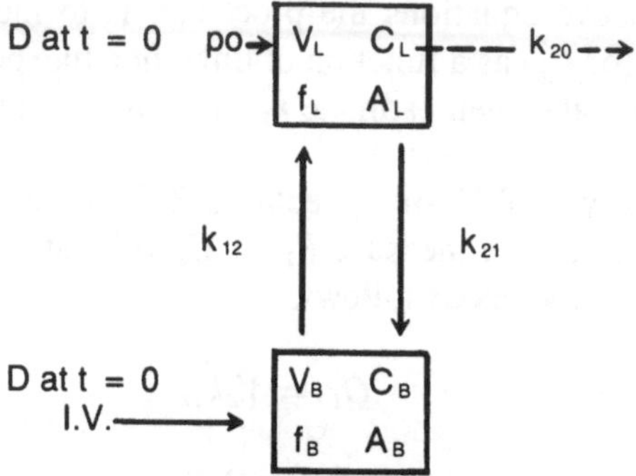

Usual Compartment Model Modified
for Free Fractions
Model IX

SCHEME 12.3.

The symbols are shown in Schemes 12.2 and 12.3. The *V*s are volumes, the *C*s are concentrations, the *A*s are amounts, the *f*s are free fractions. *L* refers to the peripheral liver compartment and *B* refers to the blood and rest of the body compartment. Q_L is the liver blood flow rate, $R_L = f_B/f_L = C_{Lss}/C_{Bss}$, and CL_i' is the intrinsic clearance of free (unbound) drug and the rate of metabolism $= f_L CL_i' C_L = f_B CL_i'/R_L$.

$$V_B\left[\frac{C_B}{dt}\right] = -Q_L C_B + \left[\frac{Q_L}{R_L}\right] C_L \tag{520}$$

$$V_L\left[\frac{dC_L}{dt}\right] = Q_L C_B - \left[\frac{Q_L}{R_L}\right] C_L - \left[\frac{f_B CL_i'}{R_L}\right] C_L \tag{521}$$

When the numerators and denominators on the right-hand sides of Equations (520) and (521) are multiplied by the appropriate volumes and $V_B C_B = A_B$ and $V_L C_L = A_L$ are used, then Equations (522) and (523) are obtained.

$$\frac{dA_B}{dt} = -\left[\frac{Q_L}{V_B}\right] A_B + \left[\frac{Q_L}{V_L R_L}\right] A_L \tag{522}$$

$$\frac{dA_L}{dt} = \left[\frac{Q_L}{V_B}\right] A_B - \left[\frac{Q_L}{V_L R_L}\right] A_L - \left[\frac{f_B CL_i'}{V_L R_L}\right] A_L \tag{523}$$

COMPARTMENT MODEL VIII: SINGLE-DOSE BOLUS I.V.

The differential equations are as follows.

$$\frac{dA_B}{dt} = -k_{12}A_B + k_{21}A_L \tag{524}$$

$$\frac{dA_L}{dt} = k_{12}A_B - k_{21}A_L - k_{20}A_L \tag{525}$$

Matching the bracketed coefficients of the A_B and A_L terms in Equation (522) with the corresponding coefficients in Equation (524) and doing similarly with Equations (523) and (525) gives:

$$k_{12} = Q_L/V_B \tag{526}$$

$$k_{21} = Q_L/V_LR_L \tag{527}$$

$$k_{20} = \frac{f_B\text{CL}_i'}{V_LR_L} = \frac{f_L\text{CL}_i'}{V_L} \tag{528}$$

It is possible to estimate parameters from I.V. data only since $\text{CL}_{po} = \text{CL}_s/F$, but it is preferred to obtain CL_s as $D/(\text{AUC})_{\text{I.V.}}$ from I.V. data and CL_{po} as $D/(\text{AUC})_{po}$ from po data. It is also important to estimate Q_L from CL_s and CL_{po} since the extraction ratio $E = f_B\text{CL}_i'/(Q_L + f_B\text{CL}_i')$. One must measure f_B (or f_p and C_B/C_p) and C_B or C_p as a function of time. The time course data are fitted with a biexponential Equation (21) then the microconstants estimated with Equations (36) to (39).

ESTIMATION OF MODEL XXXIX PARAMETERS IF ONE ASSUMES THAT ONLY FREE (UNBOUND) DRUG IS ELIMINATED

$$F = \frac{(\text{AUC 0–Inf})_{po}}{(\text{AUC 0–Inf})_{\text{I.V.}}} = \frac{Q_L}{Q_L + f_B\text{CL}_i'} \tag{529}$$

$$E = 1 - F = \frac{f_B\text{CL}_i'}{Q_L + f_B\text{CL}_i'} \tag{530}$$

$$\mathrm{CL}_s = \frac{D}{(\mathrm{AUC}\ 0\text{–Inf})_{\mathrm{I.V.}}} = \frac{Q_L f_B \mathrm{CL}_i'}{Q_L + f_B \mathrm{CL}_i'} \tag{531}$$

$$\mathrm{CL}_{\mathrm{po}} = \frac{D}{(\mathrm{AUC}\ 0\text{–Inf})_{\mathrm{po}}} = f_B \mathrm{CL}_i' \tag{532}$$

$$Q_L = \frac{\mathrm{CL}_s \mathrm{CL}_{\mathrm{po}}}{\mathrm{CL}_{\mathrm{po}} - \mathrm{CL}_s} = V_B k_{12} \tag{533}$$

$$V_L R_L = Q_L / k_{21} \tag{534}$$

$$V_B = Q_L / k_{12} \tag{535}$$

$$\frac{V_L}{f_L} = \frac{V_L R_L}{f_B} \tag{536}$$

$$\mathrm{CL}_i' = \frac{f_B \mathrm{CL}_i'}{f_B} = \frac{\mathrm{CL}_{\mathrm{po}}}{f_B} \tag{537}$$

$$\text{Observed } t_{1/2} = \frac{0.693}{\lambda_1} \tag{538}$$

$$\text{Predicted } t_{1/2} = \frac{0.693\,[V_B + R_L V_L]}{f_B \mathrm{CL}_i'} \tag{539}$$

The unbound volume, V_L/f_L, reflects the degree of protein binding in the liver compartment but if you measure R_L then V_L and f_L may each be estimated. There is more than one way to estimate most of the parameters and the methods outlined above are those preferred by the author based mainly on application of the equations to real data. Estimating $\mathrm{CL}_{\mathrm{po}}$ from oral area data assumes that absorption is complete and that F is due only to the first-pass effect. This is usually better than estimating $\mathrm{CL}_{\mathrm{po}}$ from I.V. data as CL_s/F.

Example #1

The diphenhydramine human data of Albert et al. [119] was used. Plasma concentrations of diphenhydramine were measured after administration of 50 mg of drug as an aqueous infusion over one hour and 50 mg orally in the form of an aqueous solution. Oral capsule data were also available but were not used in this analysis.

Values of free fraction in plasma, f_p, of 0.0014 and 0.00175 were measured for subjects #1 and #2, respectively. Average values of blood/plasma concentration ratio, C_B/C_p, of 0.825 and 0.82 were measured for subjects #1 and #2, respectively. These provide blood free fractions, f_B, of 0.00172 and 0.00213 for subjects #1 and #2, respectively. The plasma concentrations were multiplied by the appropriate C_B/C_p ratio to convert to whole blood concentrations, C_B. The two sets of post-infusion C_B, t data were fitted by nonlinear least squares regression with the program MINSQ [2] to Equation (540), then Equations (541) to (544) were applied (see Appendix for derivations).

Table 11.1 lists the input data and the results of the analysis of the diphenhydramine whole blood concentrations. Note that the bioavailabilities, systemic clearances, oral clearances, intrinsic clearances, and liver blood flow rates are each very comparable for the two subjects. Subject #2–with the longer elimination half-life of 13.1 hours compared with 4.08 hours for subject #1–has the higher values of V_L/R_L, as expected [116].

$$C_B = Y_1 e^{-\lambda_1 (t-T)} + Y_2 e^{-\lambda_2 (t-T)} \tag{540}$$

where t is time from the start of the infusion and T is the infusion time.

$$k_{20} + k_{21} = \frac{\lambda_1\lambda_2[Y_1[(1 - e^{-\lambda_2 T}) + Y_2(1 - e^{-\lambda_1 T})]}{Y_2\lambda_2(1 - e^{-\lambda_1 T}) + Y_2\lambda_1(1 - e^{-\lambda_2 T})} \tag{541}$$

$$k_{12} = \lambda_1 + \lambda_2 - (k_{20} + k_{21}) \tag{542}$$

$$k_{20} = \lambda_1\lambda_2/k_{12} \tag{543}$$

$$k_{21} = (k_{20} + k_{21}) - k_{20} \tag{544}$$

The liver blood flow rates of 2.14 and 1.95 L/min found by this new type of analysis in these two subjects are higher than the accepted average liver blood flow rate of 1.50 L/min in man; the subjects were graduate students and may very well have had higher than normal flow rates. The flow rates may also be high because of pooling of other tissues with liver in the analysis. It should be noted that the AUCs reported in the original article differ from those reported in Table 12.1 as a result of the C_B/C_p ratios and the dimensions of the areas. The fitted post-infusion data are shown in Figure 12.1 and the oral data in Figure 12.2. The assumption that only free drug was metabolized is supported by the next section.

TABLE 12.1. Diphenhydramine Parameters in Two Subjects.

Parameter	Equation	Subject #1	Subject #2
Y_1 (mg/L)		0.0186	0.02266
λ_1 (min^{-1})		0.002834	0.0008791
Y_2 (mg/L)		0.09303	0.1362
λ_2 (min^{-1})		0.05652	0.006521
f_B		0.00172	0.00213
k_{12} (min^{-1})	(542)	0.04142	0.005825
k_{21} (min^{-1})	(544)	0.01406	0.0005905
k_{20} (min^{-1})	(543)	0.003867	0.0009842
$(AUC\ 0\text{–}60\ min)_{I.V.}$ [mg × min/L]		8.8005	4.22775
$(AUC\ 60\text{–}Inf)_{I.V.}$	$Y_1/\lambda_1 + Y_2/\lambda_2$	43.5109	46.6627
$(AUC\ 0\text{–}Inf)_{I.V.}$ [mg × min/L]		52.3114	50.8905
$(AUC\ 0\text{–}Inf)_{po}$		28.9594	25.2020
CL_s (L/min)	(531)	0.9550	0.9825
F	(529)	0.554	0.495
E	(530)	0.446	0.505
Q_L (L/min)	(533)	2.14	1.95
$CL_{po} = f_B CL_i'$	(532)	1.727	1.984
$V_L R_L$ (L)	(534)	152.2	2,117.
V_B (L)	(535)	51.7	334.
V_L/f_L (L)	(536)	88,490.	993,900.
CL_i' (L/min)	(537)	1,004.	931.5
Observed $t_{1/2}$ (hr)	(538)	4.08	13.1
Predicted $t_{1/2}$ (hr)	(539)	1.36	14.0

Example #2

The meptazinol data of Norbury et al. [120] was used. Doses of 25 mg were administered by bolus intravenous injection two hours after a light breakfast to five volunteers. Similarly, after a light breakfast, 200 mg doses as tablets were given orally. Blood was collected at 0.25, 0.5, 0.75, 1, 1.5, 2, 3, 4, and 6 hours after I.V. administration and at 0.25, 0.5, 0.75, 1, 2, 3, 4, and 6 hours after oral administration. Meptazinol was measured in plasma by HPLC with fluorescence detection. Plasma protein binding was measured with radioactive drug; the binding was 27.1% hence $f_p = 0.729$. Results given in Table 12.2 are those of this author and are not the values given in the original article. The intravenous data were fitted with biexponential equations by nonlinear least squares regression using the program RSTRIP [2] and a Zenith 386 microcomputer. $(AUC)_{po}$ was mea-

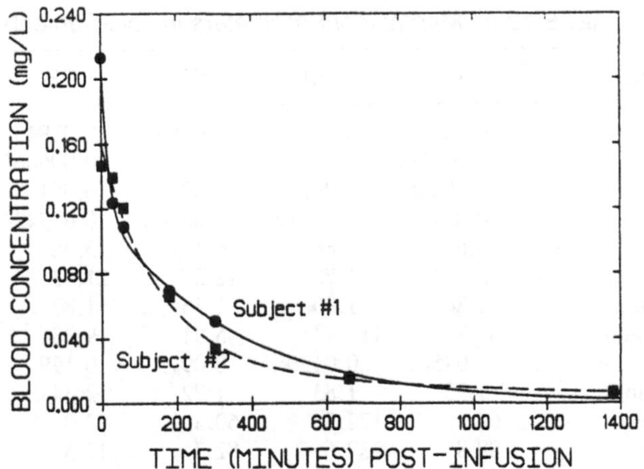

FIGURE 12.1. Post-infusion diphenhydramine blood concentration profiles of subjects #1 and #2 [119] fitted with Equation (540). The parameters are listed in Table 12.1.

sured by trapezoidal rule with extrapolation to infinity in the usual manner. Parameters of Model XXXIX were estimated with Equations (529) to (538) and are shown in Table 12.2.

The Q_L values of meptazinol in Table 12.2 are two to three times liver plasma flow of 0.8 L/min.

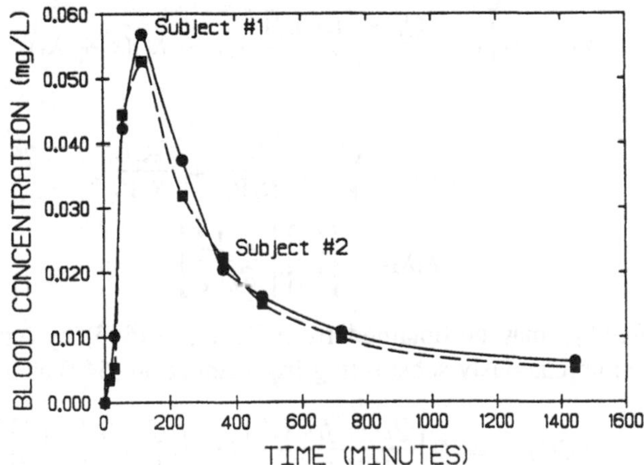

FIGURE 12.2. Diphenhydramine blood concentrations of subjects #1 and #2 following oral administration.

TABLE 12.2. Meptazinol Parameters in Five Subjects.

Parameter	DB	SM	DP	GR	GW
C_1 (mg/L)	0.06742	0.05454	0.07957	0.03943	0.07774
λ_1 (min^{-1})	0.005367	0.004573	0.005447	0.006066	0.006347
C_2 (mg/L)	0.08428	0.06697	0.09487	0.05103	0.1753
λ_2 (min^{-1})	0.08101	0.03472	0.06306	0.007423	0.1935
$(AUC)_{I.V.}$ (mg min/L)	13.60	13.86	16.11	13.38	13.15
$(AUC)_{po}$ (mg min/L)	5.291	1.722	12.57	21.33	10.24
CL_s (L/min)	1.84	1.80	1.55	1.87	1.90
CL_{po} (L/min)	37.8	116.1	5.91	9.38	19.52
F	0.0456	0.0150	0.0975	0.199	0.0973
Q_L (L/min)	1.93	1.83	1.72	2.34	2.10
V_LR_L (L)	64.7	172.5	60.3	82.0	38.3
V_L/f_L (L)	88.8	237.6	82.7	112.5	52.5
V_p (L)	40.7	86.4	137.6	342.5	15.4
CL_i' (L/min)	51.9	159.3	21.8	12.9	26.8
$t_{1/2}$ (hr)	2.15	2.53	2.12	1.90	1.82

EXPRESSIONS FOR CLEARANCES FOR MODELS IX AND XXXIX

Conversion of the differential Equations (524) and (525) to their Laplace transforms and application of matrix algebra [17d] provides the Laplace transform, a_B, of the amount of drug in the central compartment in the central (blood) compartment of Model XXXIX, shown as Equation (545) below (see Appendix for details).

$$a_B = D\left[s + \frac{Q_L + f_B CL_i'}{R_L V_L}\right] \Big/ [(s + \lambda_1)(s + \lambda_2)] \tag{545}$$

where

$$\lambda_1 + \lambda_2 = \frac{Q_L}{V_B} + \frac{Q_L}{R_L V_L} + \frac{f_B CL_i'}{R_L V_L} \tag{546}$$

$$\lambda_1 \lambda_2 = \left[\frac{Q_L}{V_B}\right]\left[\frac{f_B CL_i'}{R_L V_L}\right] \tag{547}$$

The $(AUC)_{I.V.}$ may be obtained from Equation (545) by letting $s = 0$ [121] and simultaneously substituting from Equation (547) as follows:

$$\begin{aligned}(AUC)_{I.V.} &= D\left[\frac{Q_L + f_B CL_i'}{R_L V_L}\right] \Big/ V_B\left[\left(\frac{Q_L}{V_B}\right)\left(\frac{f_B CL_i'}{R_L V_L}\right)\right] \\ &= D\left[\frac{Q_L + f_B CL_i'}{Q_L f_B CL_i'}\right]\end{aligned} \tag{548}$$

Hence,

$$\mathrm{CL}_s = \frac{D}{(\mathrm{AUC})_{\mathrm{I.V.}}} = \frac{Q_L f_B \mathrm{CL}_i'}{Q_L + f_B \mathrm{CL}_i'} \tag{549}$$

Equation (549) is the same as Equation (9.26) of Gibaldi and Perrier [24] and the same as Equation (504) in Chapter 11 if CL_S replaces CL_{nr}.

In a smiliar manner one can convert Equations (524) and (525) of Model IX to their Laplace transforms and apply matrix algebra; this leads to Equation (550).

$$(\mathrm{AUC})_{\mathrm{I.V.}} = \frac{D[k_{20} + k_{21}]}{V_B k_{12} k_{20}} \tag{550}$$

By using Equations (526) and (527) one can see that Equation (548) is identical to Equation (550).

The Rowland Model with first-order absorption into the peripheral compartment corresponding to oral administration is covered in the Appendix and leads to Equation (551).

$$(\mathrm{AUC})_{\mathrm{po}} = \frac{D}{f_B \mathrm{CL}_i'} \tag{551}$$

and

$$\mathrm{CL}_{\mathrm{po}} = \frac{D}{(\mathrm{AUC})_{\mathrm{po}}} = f_B \mathrm{CL}_i' \tag{552}$$

TWO-COMPARTMENT ROWLAND MODEL: SINGLE BOLUS INTRAVENOUS DOSE WHEN TOTAL DRUG IS METABOLIZED

The model is the same as Model XXXIX except CL_m replaces $f_L \mathrm{CL}_i'$ where CL_m is the metabolic clearance with reference to total drug in the liver compartment. The differential equations are as follows.

$$V_B \left[\frac{dC_B}{dt}\right] = -Q_L C_B + \left[\frac{Q_L}{R_L}\right] C_L \tag{553}$$

$$V_L \left[\frac{dC_L}{dt}\right] = Q_L C_B - \left[\frac{Q_L}{R_L}\right] \mathrm{CL} - \mathrm{CL}_m C_L \tag{554}$$

Converting to amounts gives:

$$\frac{dA_B}{dt} = -\left[\frac{Q_L}{V_B}\right]A_B + \left[\frac{Q_L}{R_L V_L}\right]C_L \tag{555}$$

$$\frac{dA_L}{dt} = \left[\frac{Q_L}{V_B}\right]A_B - \left[\frac{Q_L}{R_L V_L}\right]A_L - \left[\frac{\mathrm{CL}_m}{V_B}\right]A_L \tag{556}$$

Matching coefficients in Equation (553) with those of Equation (524) and those of Equation (554) with those of Equation (525) gives:

$$k_{12} = Q_L/V_B \tag{557}$$

$$k_{21} = Q_L/V_L R_L \tag{558}$$

$$k_{20} = \mathrm{CL}_m/V_L \tag{559}$$

Converting Equations (555) and (556) to their Laplace transforms and applying matrix algebra gives Equation (560).

$$(\mathrm{AUC})_{\mathrm{I.V.}} = D\left[\frac{Q_L + R_L\mathrm{CL}_m}{Q_L R_L \mathrm{CL}_m}\right] \tag{560}$$

Hence,

$$\mathrm{CL}_s = \frac{D}{(\mathrm{AUC})_{\mathrm{I.V.}}} = \frac{Q_L R_L \mathrm{CL}_m}{Q_L + R_L \mathrm{CL}_m} \tag{561}$$

Equation (561) is the same as Equation (9.28) of Gibaldi and Perrier [24] since their Cl_L is the same as CL_m in this book. Solving for R_L in Equation (561) gives Equation (562).

$$R_L = \frac{\mathrm{CL}_s Q_L}{\mathrm{CL}_m (Q_L - \mathrm{CL}_s)} \tag{562}$$

Example

Substituting values for diphenhydramine listed in Table 12.1 into Equation (562) gives $R_L = 1.000$ for subject #1 and $R_L = 0.998 \sim 1.000$ for subject #2. Hence Equation (561) reduces to Equation (581) which is correct for only free drug being metabolized. Thus this analysis suggests strongly that only free (unbound) diphenhydramine is metabolized.

TRIEXPONENTIAL FIRST-PASS MODEL WITH BOLUS INTRAVENOUS ADMINISTRATION

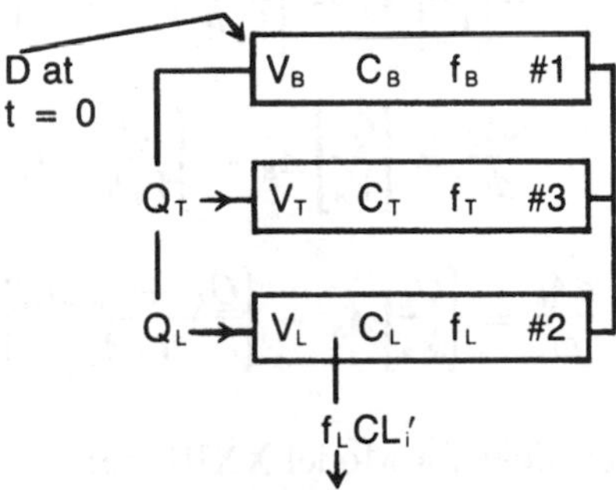

Physiologically Based Model XXXX

SCHEME 12.4.

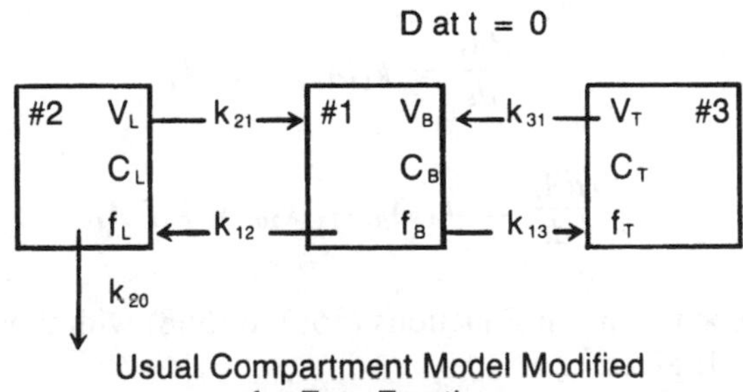

Usual Compartment Model Modified for Free Fractions Model XXIII

SCHEME 12.5.

Model XXXX was treated by Gibaldi and Koup [112] but the equation for systemic clearance was in error and consequently some of the examples were in error, hence the model and its corresponding compartment model, Model XXIII, are treated in detail here.

The differential equations for Model XXXX are as follows.

$$V_B\left[\frac{dC_B}{dt}\right] = -Q_BC_B + \left[\frac{Q_T}{R_T}\right]C_T + \left[\frac{Q_L}{R_L}\right]C_L \tag{563}$$

$$V_T\left[\frac{dC_T}{dt}\right] = Q_TC_B - \left[\frac{Q_T}{R_T}\right]C_T \tag{564}$$

$$V_L\left[\frac{dC_L}{dt}\right] = Q_LC_B - \left[\frac{Q_L}{R_L}\right]C_L - \left[\frac{f_B\mathrm{CL}_i'}{R_L}\right]C_L \tag{565}$$

Converting to amounts gives:

$$\frac{dA_B}{dt} = -\left[\frac{Q_B}{V_B}\right]A_B + \left[\frac{Q_T}{V_T R_T}\right]A_T + \left[\frac{Q_L}{V_L R_L}\right]A_L \tag{566}$$

$$\frac{dA_T}{dt} = \left[\frac{Q_T}{V_B}\right]A_B - \left[\frac{Q_T}{V_T R_T}\right]A_T \tag{567}$$

$$\frac{dA_L}{dt} = \left[\frac{Q_L}{V_B}\right]A_B - \left[\frac{Q_L + f_B \mathrm{CL}_i'}{V_L R_L}\right]A_L \tag{568}$$

The differential equations for Model XXIII are:

$$\frac{dA_B}{dt} = -(k_{12} + k_{13})A_B + k_{31}A_T + k_{21}A_L \tag{569}$$

$$\frac{dA_T}{dt} = k_{13}A_B - k_{31}A_T \tag{570}$$

$$\frac{dA_L}{dt} = k_{12}A_B - (k_{20} + k_{21})A_L \tag{571}$$

Matching coefficients in Equations (566) to (568) with those in Equations (569) to (571) gives:

$$k_{12} = Q_L/V_B \tag{572}$$

$$k_{13} = Q_T/V_B \tag{573}$$

$$Q_B = Q_L + Q_T \tag{574}$$

$$k_{21} = Q_L/R_L V_L \tag{575}$$

$$k_{31} = Q_T/R_T V_T \tag{576}$$

$$k_{20} = \frac{f_B \mathrm{CL}_i'}{R_L V_L} \tag{577}$$

$$k_{12} + k_{13} = Q_B/V_B \tag{578}$$

$$\mathrm{k}_{20} + k_{21} = \frac{Q_L + f_B \mathrm{CL}_i'}{R_L V_L} \tag{579}$$

By converting Equations (566) to (568) to their Laplace transforms followed by use of matrix algebra, one finds for the blood compartment #1:

$$(\mathrm{AUC})_{\mathrm{I.V.}} = D\left[\frac{Q_L + f_B\mathrm{CL}_i'}{Q_L f_B \mathrm{CL}_i'}\right] \tag{580}$$

and,

$$\mathrm{CL}_s = \frac{D}{(\mathrm{AUC})_{\mathrm{I.V.}}} = \frac{Q_L f_B \mathrm{CL}_i'}{Q_L + f_B\mathrm{CL}_i'} \tag{581}$$

Similarly, for oral administration:

$$(\mathrm{AUC})_{\mathrm{po}} = \frac{D}{f_B\mathrm{CL}_i'} \tag{582}$$

$$\mathrm{CL}_{\mathrm{po}} = f_B\mathrm{CL}_i' \tag{583}$$

Thus the clearance expressions for the three-compartment Model XXXX are exactly the same as for the Rowland Model XXXIX.

Similar Laplace transform treatment for Equations (569) to (571) gives:

$$(\mathrm{AUC})_{\mathrm{I.V.}} = D\left[\frac{k_{20} + k_{21}}{V_B k_{12} k_{20}}\right] \tag{584}$$

and

$$\mathrm{CL}_s = \frac{D}{(\mathrm{AUC})_{\mathrm{I.V.}}} = \frac{V_B k_{12} k_{20}}{k_{20} + k_{21}} = Q_L E \tag{585}$$

Similarly, for oral administration:

$$(\mathrm{AUC})_{\mathrm{po}} = \frac{D}{V_B}\left[\frac{k_{21}}{k_{20} + k_{21}}\right] \tag{586}$$

and,

$$\mathrm{CL}_{\mathrm{po}} = \frac{D}{(\mathrm{AUC})_{\mathrm{po}}} = \frac{V_B k_{12} k_{20}}{k_{21}} \tag{587}$$

Hence, from Equation (585):

$$Q_L = V_B k_{12} \tag{588}$$

and,

$$E = \frac{k_{20}}{k_{20} + k_{21}} \tag{589}$$

Thus,

$$F = 1 - E = \frac{k_{21}}{k_{20} + k_{21}} \tag{590}$$

ESTIMATION OF MODEL XXXX PARAMETERS WHEN IT IS ASSUMED THAT ONLY FREE (UNBOUND) DRUG IS ELIMINATED BY LIVER

It is assumed that you administer the drug intravenously and measure C_B as a function of a time, then fit a triexponential equation to the C_B, t data [Equation (140) under Model XXIII], then derive values of k_{12}, k_{21}, k_{13}, k_{31}, and k_{20}, using Equations (141) to (146) under Model XXIII. Also, one obtains $(\text{AUC})_{\text{I.V.}}$ by integrating the triexponential equation. It is also assumed that you administer the drug orally (preferably in solution) and measure $(\text{AUC})_{\text{po}}$ and f_B. Then, sequentially, one uses the following equations:

$$\text{CL}_s = \frac{D}{(\text{AUC})_{\text{I.V.}}} \tag{591}$$

$$\text{CL}_{\text{po}} = \frac{D}{(\text{AUC})_{\text{po}}} \tag{592}$$

$$\text{CL}_i' = \frac{\text{CL}_{\text{po}}}{f_B} = \frac{f_B \text{CL}_i'}{f_B} \tag{593}$$

$$Q_L = \frac{\text{CL}_s \text{CL}_{\text{po}}}{\text{CL}_{\text{po}} - \text{CL}_s} \tag{594}$$

$$V_B = Q_L/k_{12} \tag{595}$$

$$Q_T = V_B/k_{21} \tag{596}$$

$$V_L R_L = Q_L/k_{21} \tag{597}$$

$$V_T R_T = Q_T / k_{31} \tag{598}$$

$$\frac{V_L}{f_L} = V_L R_L / f_B \tag{599}$$

$$\frac{V_T}{f_T} = V_T R_T / f_B \tag{600}$$

If you measure R_T and R_L, then V_L, V_T, f_L, and f_T may be estimated individually. However, if these are not measured then the unbound volumes V_L/f_L and V_T/f_T are useful to indicate relative free fractions and hence binding since the smaller f_L and f_T, the larger the unbound volumes, V_L/f_L and V_T/f_T and the greater the binding.

In Chapter 13 simulations are done with Model XXXX to illustrate the effect of drug binding on pharmocokinetics.

The predicted elimination half-life may be obtained with Equation (601).

$$t_{1/2} = \frac{0.693\,[V_B + R_L V_L + R_T V_T]}{f_B \mathrm{CL}_i'} \tag{601}$$

In applying Model XXXIX to real data where there is biexponential disposition, the input variables are two coefficients and two exponents of the bioexponential disposition equation, the dose D, f_p or f_B, the C_B/C_p ratio, and $(\mathrm{AUC})_{\mathrm{po}}$ – a total of eight variables. The output variables are V_B, Q_L, CL_s, $\mathrm{CL}_{\mathrm{po}}$, CL_i', and V_L/f_L – a total of six variables.

In applying Model XXXX to real data where there is triexponential disposition the input variables are three coefficients and three exponents of the triexponential disposition equation, the dose D, f_p or f_B, the C_B/C_p ratio, and $(\mathrm{AUC})_{\mathrm{po}}$ – a total of ten variables. The output variables are V_B, Q_L, Q_T, CL_s, $\mathrm{CL}_{\mathrm{po}}$, CL_i', V_L/f_L, and V_T/f_T – a total of eight variables.

CHAPTER 13

Protein Binding

INTRODUCTION

Drugs bind to tissue proteins and to various proteins – such as albumin, α_1-acid glycoprotein, and transcortin – in plasma. The binding of a drug to a specific protein is a reversible, equilibrium process.

$$[D] + [P] \underset{k_2}{\overset{k_1}{\rightleftharpoons}} [DP] \tag{602}$$

where $[D]$ is the free (unbound) concentration, $[P]$ is the concentration of binding protein, and $[DP]$ is the concentration of drug-protein complex. The forward and reverse reactions are extremely rapid. Now,

$$K_a = \frac{k_1}{k_2} = \frac{[DP]}{[D][P]} \tag{603}$$

where K_a is the association constant, and

$$f_u = \frac{[D]}{[D] + [P]} \tag{604}$$

where f_u is the free fraction.

Albumin is the most important binding protein in plasma and interstitial fluid and it binds mostly acidic drugs. Its normal concentration is 39–55 mg/ml or 3.9–5.5 g/100 ml. Many basic drugs (e.g., quinidine, propranolol, lidocaine, and disopyramide) are bound to α_1-acid glycoprotein which has a normal concentration of 70–110 mg/100 ml or 0.07–1.1 mg/ml.

THE EFFECT OF PROTEIN BINDING ON PHARMACOKINETICS

The effect of protein binding on pharmacokinetics was simulated using Models XXIII and XXXX of Chapter 12. Similar to Gibaldi and Koup [112], common parameters used in the simulations were: $Q_B = 5.5$ L/min, $Q_T = 4$ L/min, $Q_L = 1.5$ L/min, $V_B = 5.4$ L, $V_L = 4$ L, $V_T = 40$ L, $D = 200$ mg, and $k_a = 0.04$ min^{-1}. Table 13.1 lists values of the parameters f_B, f_T, f_L, R_T, R_L, CL_i', and τ used in the simulations. Values of the constants k_{12}, k_{13}, k_{21}, k_{31}, and k_{20} of Model XXIII were obtained from these parameter values using Equations (572) to (577) of Chapter 12. The λ_1, λ_2, and λ_3 values were obtained as the cube roots of the Laplace transform Equation (605).

$$s^3 + a_2 s^2 + a_1 s + a_o = 0 \tag{605}$$

where

$$a_2 = k_{12} + k_{13} + k_{21} + k_{31} + k_{20} \tag{606}$$

$$a_1 = k_{12}(k_{20} + k_{31}) + (k_{13} + k_{31})(k_{20} + k_{21}) \tag{607}$$

$$a_o = k_{12} k_{20} k_{31} \tag{608}$$

Bolus intravenous C_B, t data were generated with Equation (609), which is the same as Equation (135) of Chapter 3.

TABLE 13.1. Parameter Values Used in Simulations.

Set No.	f_B	f_T	f_L	R_T	R_L	CL_i' (L/min)	τ (min)
1	0.25	0.8	0.25	0.3125	1.0	0.286	360
2	0.8	0.8	0.2	1.0	4.0	0.286	360
3	0.2	0.2	0.2	1.0	1.0	0.286	360
4	0.2	0.8	0.8	0.25	0.25	0.286	360
5	0.2	0.8	0.2	0.25	1.0	5.71	360
6	0.8	0.8	0.2	1.0	4.0	5.71	360
7	0.3	0.4	0.6	0.75	0.5	0.286	360
8	0.3	0.8	0.6	0.375	0.5	0.286	360
9	0.2	0.6	0.2	0.333	1.0	5.71	360

$$C_B = \left[\frac{D}{V_B}\right]\left[\frac{(k_{20} + k_{21} - \lambda_1)(k_{31} - \lambda_1)}{(\lambda_2 - \lambda_1)(\lambda_3 - \lambda_1)} e^{-\lambda_1 t}\right.$$

$$+ \frac{(k_{20} + k_{21} - \lambda_2)(k_{31} - \lambda_2)}{(\lambda_1 - \lambda_2)(\lambda_3 - \lambda_2)} e^{-\lambda_3 t}$$

$$\left.+ \frac{(k_{20} + k_{21} - \lambda_3)(k_{31} - \lambda_3)}{(\lambda_2 - \lambda_3)(\lambda_1 - \lambda_3)} e^{-\lambda_3 t}\right] \tag{609}$$

The model for oral administration with first-order absorption is Model XXXI shown in Scheme 13.1.

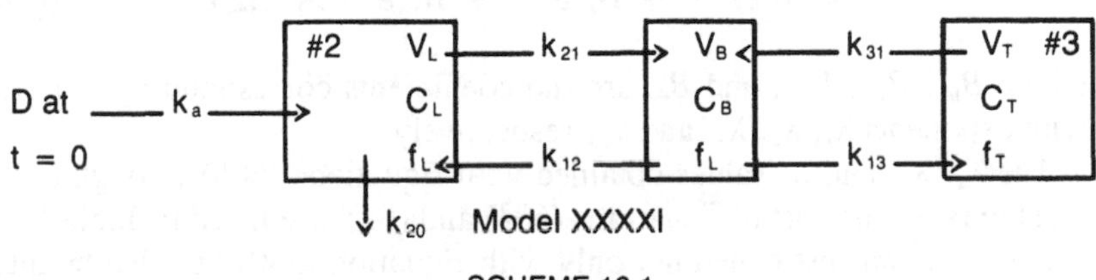

SCHEME 13.1.

Oral single-dose C_B, t data were generated with Equation (610).

$$C_B = \frac{D}{V_B}\left[\frac{k_a k_{21}(k_{31} - \lambda_1)e^{-\lambda_1 t}}{(k_a - \lambda_1)(\lambda_2 - \lambda_1)(\lambda_3 - \lambda_1)}\right.$$

$$+ \frac{k_a k_{21}(k_{31} - \lambda_2)e^{-\lambda_2 t}}{(k_a - \lambda_2)(\lambda_1 - \lambda_2)(\lambda_3 - \lambda_2)}$$

$$+ \frac{k_a k_{21}(k_{31} - \lambda_3)e^{-\lambda_3 t}}{(k_a - \lambda_3)(\lambda_1 - \lambda_3)(\lambda_2 - \lambda_3)}$$

$$\left.+ \frac{k_a k_{21}(k_{31} - k_a)e^{-k_a t}}{(\lambda_1 - k_a)(\lambda_2 - k_a)(\lambda_3 - k_a)}\right] \tag{610}$$

Equation (609) may be written as Equation (611).

$$C_B = C_1 e^{-\lambda_1 t} + C_2 e^{-\lambda_2 t} + C_3 e^{-\lambda_3 t} \tag{611}$$

where C_1, C_2, and C_3 are the coefficients in Equation (606) which have exponents λ_1, λ_2, and λ_3, respectively.

Equation (610) may be written as Equation (612).

$$C_B = B_1 e^{-\lambda_1 t} + B_2 e^{-\lambda_2 t} + B_3 e^{-\lambda_3 t} + B_4 e^{-k_a t} \tag{612}$$

where B_1, B_2, B_3, and B_4 are the coefficients of Equation (610), which have exponents λ_1, λ_2, λ_3, and k_a, respectively.

Oral steady-state data were generated with Equation (613).

$$C_{B\text{ss}} = \frac{B_1 e^{-\lambda_1 t}}{(1 - e^{-\lambda_1 \tau})} + \frac{B_2 e^{-\lambda_2 t}}{(1 - e^{-\lambda_2 \tau})} + \frac{B_3 e^{-\lambda_3 t}}{(1 - e^{-\lambda_3 \tau})} + \frac{B_4 e^{-k_a t}}{(1 - e^{-k_a \tau})} \tag{613}$$

Equation (613) may be written as Equation (614).

$$C_{B\text{ss}} = B_{1\text{ss}} e^{-\lambda_1 t} + B_{2\text{ss}} e^{-\lambda_2 t} + B_{3\text{ss}} e^{-\lambda_3 t} + B_{4\text{ss}} e^{-k_a t} \tag{614}$$

where $B_{1\text{ss}}$, $B_{2\text{ss}}$, $B_{3\text{ss}}$, and $B_{4\text{ss}}$ are the coefficients corresponding to terms with exponents λ_1, λ_2, λ_3, and k_a, respectively.

The λ_1, λ_2, and λ_3 values obtained with Equations (605) through (608) and the rate constants of Equations (604) and (610) are listed in Table 13.2.

$(\text{AUC})_{\text{I.V.}}$ was estimated not only with Equation (580) but also by integrating Equation (608). $(\text{AUC})_{\text{po}}$ was estimated not only with Equation (582) but also by integrating Equation (610). The area in a dosage interval at steady state after oral administration was shown to be the same as the single-dose area zero to infinity. All parameters were back-calculated from the time course data to support Equations (580) through (601). Results are shown in Table 13.3. In all cases the original parameter values were back-calculated accurately from time course data.

Emphasis graphically is on the oral time course data. Figure 13.1 shows

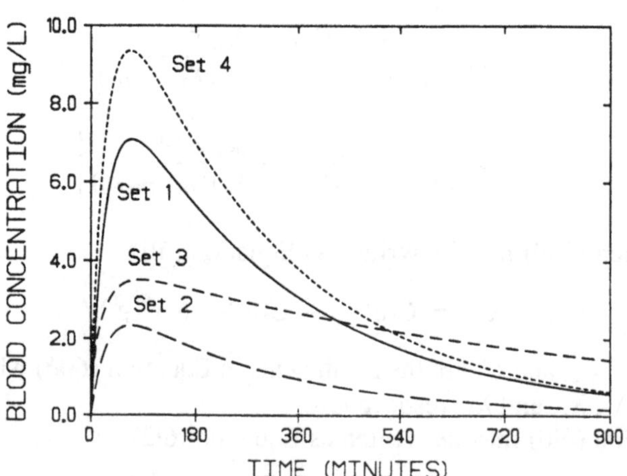

FIGURE 13.1. Single-dose C_B, t data for sets 1 to 4.

TABLE 13.2. Parameter Values Used in Simulations.

Set No.	k_{20} (min^{-1})	k_{21} (min^{-1})	k_{31} (min^{-1})	k_a (min^{-1})	$\lambda_1 \times 10^3$ (min^{-1})	λ_2 (min^{-1})	λ_3 (min^{-1})
1	0.017875	0.375	0.32	0.04	3.145	0.3726	1.3556
2	0.0143	0.09375	0.10	0.04	3.352	0.1061	1.1172
3	0.0143	0.375	0.10	0.04	1.112	0.2949	1.2118
4	0.0572	1.5	0.40	0.04	3.357	0.9246	2.0477
5	0.2855	0.375	0.40	0.04	36.49	0.6046	1.4380
6	0.2855	0.09375	0.10	0.04	20.001	0.3521	1.1256
7	0.0429	0.75	0.1333	0.04	2.153	0.5179	1.4247
8	0.0429	0.75	0.2667	0.04	3.622	0.5196	1.4829
9	0.2855	0.375	0.30	0.04	30.06	0.5768	1.3722

TABLE 13.3. Coefficients of Equations (611) to (613) Used in Simulations.

Coefficient	Set 1	Set 2	Set 3	Set 4	Set 5	Set 6	Set 7	Set 8	Set 9
C_1 (mg/L)	9.1521	3.2759	3.9970	12.1198	10.5525	2.8984	5.2365	8.8411	8.5897
C_2 (mg/L)	0.1086	0.004292	2.5294	11.8791	0.8952	0.9859	8.3754	4.6229	1.9732
C_3 (mg/L)	27.2762	33.7568	30.5106	13.0380	25.5893	32.1527	23.4251	23.5730	26.4740
B_1 (mg/L)	9.5578	3.2017	3.9916	12.7716	72.2704	1.5181	5.2493	9.2377	20.5622
B_2 (mg/L)	−0.2421	−0.1228	−1.5769	−1.2737	−0.4252	−0.4368	−1.9118	−1.24895	−0.6588
B_3 (mg/L)	0.3290	0.1165	0.4749	0.7043	0.3532	0.1534	0.8032	0.7103	0.4189
B_4 (mg/L)	−9.6447	−3.1954	−2.8695	−12.2922	−72.1984	−1.2347	−4.14075	−8.6991	−20.3223
B_{1ss} (mg/L)	14.1029	4.5682	12.0433	18.2105	72.2705	1.5192	9.7315	12.6791	21.1356
B_{2ss} (mg/L)	−0.2421	−0.1228	−1.5769	−1.2737	−0.4252	−0.4368	−1.9118	−1.24895	−0.6588
B_{3ss} (mg/L)	0.3290	0.1165	0.4749	0.7943	0.3532	0.1534	0.8093	0.7103	0.4189
B_{4ss} (mg/L)	−9.6447	−3.1954	−2.8695	−12.2922	−72.1984	−1.2347	−4.14075	−8.6991	−20.4910

single-dose C_B, t data for sets 1 to 4. The extreme results are given by sets 2 and 4. For these two sets the intrinsic metabolic clearance, CL_i', is the same, namely 0.286. But the f_B ratio for the two sets is $0.8/0.2 = 4$ and this caused the fourfold change in $CL_{po} = f_B CL_i' = 0.229/0.0572 = 4$ and a fourfold change in $(AUC)_{po}$ of $3497/873 = 4$. The fourfold change in free fraction in the liver compartment, f_L, from 0.2 to 0.8 had no effect on these oral $(AUC)_{po}$ results. Set 3 has the longest elimination half-life of 9.98 hours compared with 3.54, 3.10, and 3.31 hours for sets 1, 2, and 4, respectively, because tissue binding was greatest for set 3; since $f_T = 0.2$ for set 3 and 0.8 for sets 1, 2, and 4 then the fraction bound, namely $1 - f_T$, is 0.8 for set 3 and 0.2 for sets 1, 2 and 4.

Figure 13.2 shows the oral single-dose time courses of C_B *versus* t for data sets 5, 6, and 9. The intrinsic metabolic clearance, CL_i', is 5.71 L/min for each of these three sets. But $f_B = 0.8$ for set 6, hence $CL_{po} = f_B CL_i' = (0.8)(5.71) = 4.568$ L/min and the $(AUC)_{po}$ is only 43.78 mg × min/L, while for sets 5 and 9, $f_B = 0.2$ hence $CL_{po} = (0.2)(5.71) = 1.42$ and $(AUC)_{po} = 175.1$ mg × min/L. The $t_{1/2}$ for these three sets is short, namely 0.196, 0.155, and 0.230 hours for sets 5, 6, and 9, respectively; this is the result of no change in $f_L = 0.2$ for all three sets and $f_T = 0.8$, 0.8, and 0.6 for sets 5, 6, and 9, respectively.

Figure 13.3 shows the oral steady-state time courses of C_B *versus* t for data sets 3 and 4. The areas in a dosage interval at steady state are equal to the single-dose areas zero to infinity listed in Table 13.4 and both are equal to 3497 mg × min/L since the oral clearances are 0.0572 L/min for both sets.

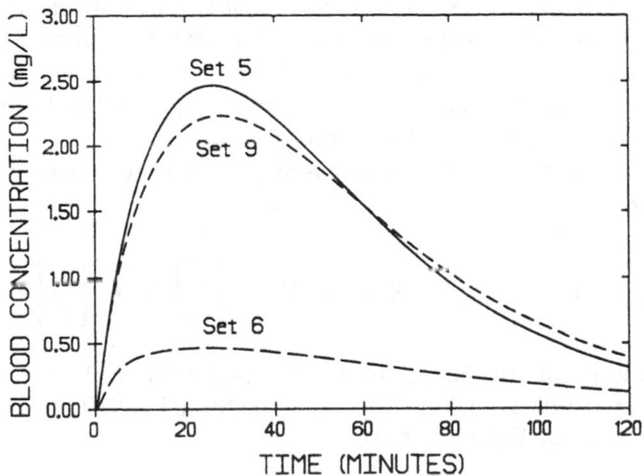

FIGURE 13.2. Single-dose C_B, t data for sets 5, 6, and 9.

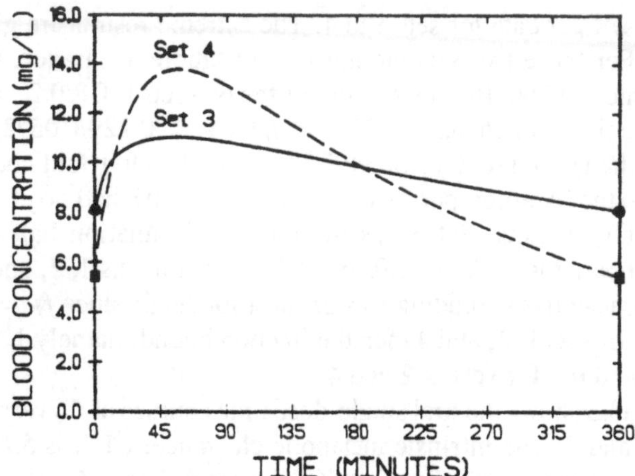

FIGURE 13.3. Oral steady-state C_B, t data for sets 3 and 4.

However, the elimination half-lives are 9.98 and 3.31 hours for sets 3 and 4, respectively, due to the higher binding of $1 - f_L = 1 - 0.2 = 0.8$ in the liver and $1 - f_T = 1 - 0.2 = 0.8$ in other tissues for set 3 compared with $1 - f_L = 1 - 0.8 = 0.2$ in liver and $1 - f_T = 1 - 0.8 = 0.2$ in other tissues for set 4.

Figure 13.4 shows the oral steady-state time courses for data sets 7 and 8. The areas in a dosage interval at steady state are equal to the single-dose areas zero to infinity listed in Table 13.4 and both are equal to 2331 mg × min/L since the oral clearances are the same, namely 0.0812 L/min. This gives $(C_{ss})_{av} = 2331/360 = 6.475$ mg/L for both sets as shown in Figure 13.4. However, the elimination half-life of set 7, namely 5.03 hour, is greater than $t_{1/2} = 3.02$ hour for set 8 since binding in the tissue compartment (fraction bound) $= 1 - f_T = 1 - 0.4 = 0.6$ for set 7 and $1 - f_T = 1 - 0.8 = -.2$ for set 8.

Gibaldi and Koup [112] gave Equation (615) for the apparent volume of distribution.

$$V_{app} = V_B + R_L V_L + R_T V_T = V_B + \left[\frac{f_b}{f_L}\right] V_L + \left[\frac{f_B}{f_T}\right] V_T \quad (615)$$

The volume of distribution steady state for the classical central compartment elimination model corresponding to Model XXIII, namely Model XIX, is given by Equation (616).

$$V_{ss} = V_B\left[1 + \frac{k_{12}}{k_{21}} + \frac{k_{13}}{k_{31}}\right] \quad (616)$$

TABLE 13.4. *Parameters Back-Calculated from Time Course Data.*

Parameter	Ecuation	Set 1	Set 2	Set 3	Set 4	Set 5	Set 6	Set 7	Set 8	Set 9
$(AUC)_{I.V.}$	(584)	2931	1007	3630	3630	308.5	177.1	2964	2464	308.5
$(AUC)_{po}$	(586)	2797	873	3497	3497	175.1	43.78	2331	2331	175.1
CL_s (L/min)	(585)	0.06824	0.1986	0.0551	0.0551	0.6483	1.129	0.0812	0.0812	0.6483
CL_{po} (L/min)	(587)	0.0715	0.229	0.0572	0.0572	1.142	4.568	0.0858	0.0858	1.142
CL_i' (L/min)	(593)	0.286	0.286	0.286	0.286	5.71	5.71	0.286	0.286	5.71
Q_L (L/min)	(594)	1.5	1.5	1.5	1.5	1.5	1.5	1.5	1.5	1.5
V_B (L)	(595)	5.4	5.4	5.4	5.4	5.4	5.4	5.4	5.4	5.4
Q_T (L/min)	(596)	4	4	4	4	4	4	4	4	4
$V_L R_L$ (L)	(597)	4	16	4	1	4	16	2	2	4
$V_T R_T$ (L)	(598)	12.5	40	40	10	10	40	30	15	13.33
V_L/f_L (L)	(599)	16	20	20	5	20	20	6.67	6.67	20
V_T/f_T (L)	(500)	50	50	200	50	50	50	100	50	66.7
$t_{1/2}$ (hr)										
$0.693/\lambda_1$		3.67	3.44	10.4	3.44	0.317	0.155	5.03	3.02	0.230
Est. $t_{1/2}$ (hr)	(601)	3.54	3.10	9.98	3.11	0.196	0.247	0.946	0.946	0.568
F	(590)	0.9545	0.868	0.963	0.963	0.568	0.247	0.946	0.946	0.568
V_{ss} (L)		306	268	864	287	17.0	13.4	436	261	19.9

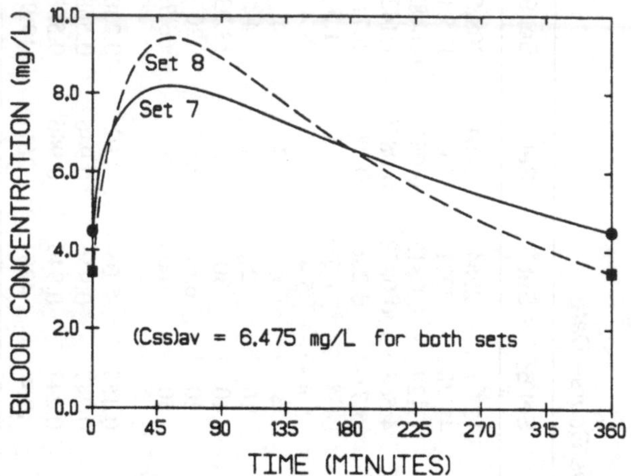

FIGURE 13.4. Oral steady-state C_B,t data for sets 7 and 8.

It is readily shown below that $V_{app} = V_{ss}$. A plot showing equivalence for data sets 1 to 9 is shown as Figure 13.5.

Since for Model XXIII:

$$V_B k_{12} = V_L R_L k_{21} \tag{617}$$

$$V_B k_{13} = V_T R_T k_{31} \tag{618}$$

Then

$$V_L = \frac{V_B k_{12}}{R_L k_{21}} \tag{619}$$

$$V_T = \frac{V_B k_{13}}{R_T k_{31}} \tag{620}$$

Substituting from Equations (619) and (620) into Equation (615) and cancelling the R_Ls and R_Ts gives Equation (616).

Figure 13.6 is a plot of the predicted $t_{1/2}$ from Equation (615) *versus* the observed half-life $= 0.693/60\lambda_1$ with $r = 0.999$ ($p < 0.001$). The slope, 0.981, is only slightly less than unity and there is a small negative intercept of -0.18.

Figure 13.7 is an ln-ln plot of $(AUC)_{po}$ *versus* CL_{po} with slope $= -1.000$. This illustrates that $(AUC)_{po}$ is inversely proportional to CL_{po} or $f_B CL_i'$.

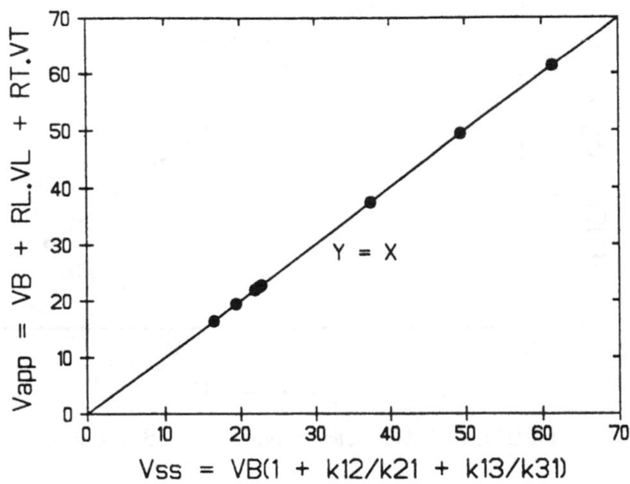

FIGURE 13.5. Showing the equivalence of V_{app} and V_{ss}.

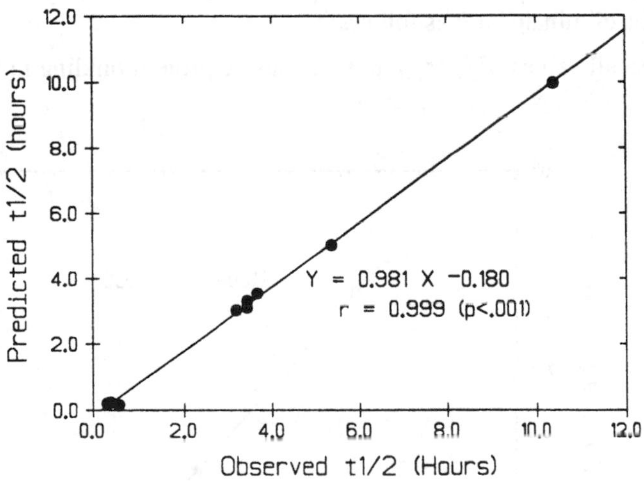

FIGURE 13.6. Plot of predicted *versus* observed half-life.

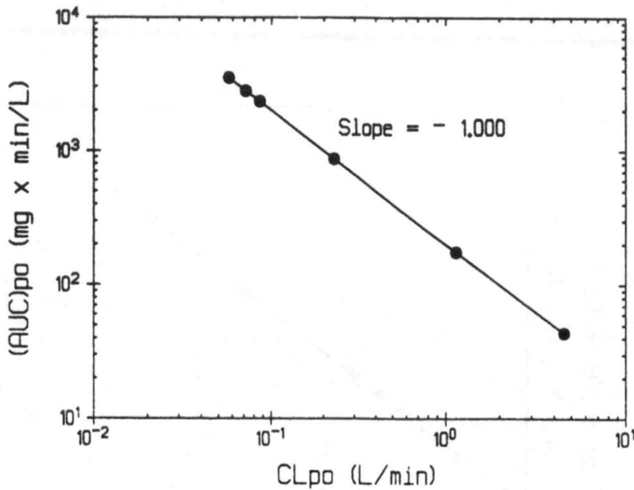

FIGURE 13.7. Correlation of oral AUCs and clearances.

Figure 13.8 is an ln-ln plot of $(AUC)_{I.V.}$ *versus* CL_s with slope = − 1.000. This illustrates that $(AUC)_{I.V.}$ is inversely proportional to CL_s.

The effects of protein binding on pharmacokinetics if only free (unbound) drug is eliminated are as follows:

(1) Small values of f_p or f_B mean extensive protein binding in blood.

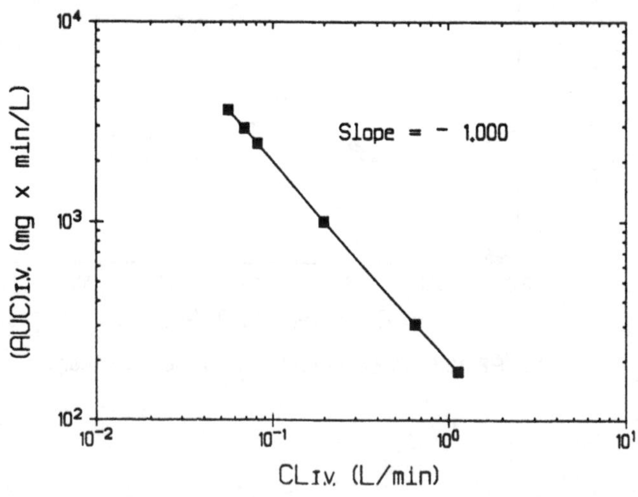

FIGURE 13.8. Correlation of I.V. AUCs and clearances.

(2) Oral clearance, CL_{po}, is directly proportional to f_B since $CL_{po} = f_B CL_i'$. $(AUC)_{po}$ is inversely proportional to CL_{po} and f_B since $(AUC)_{po} = D/f_B CL_i'$. Hence $(AUC)_{po}$ is dependent only on D, f_B, and CL_i' and is independent of liver and other tissue binding if such binding is linear.

(3) Systemic clearance $CL_s = Q_L f_B CL_i'/(Q_L + f_B CL_i')$, and hence is dependent upon liver blood flow, Q_L, and f_B and CL_i'. Thus, $(AUC)_{I.V.}$ is dependent upon D, Q_L, f_B, and CL_i' and is independent of liver and other tissue binding if such binding is linear.

(4) Small values of f_B, f_L, f_T, and CL_i' all favor longer elimination half-lives.

(5) Large values of V_B, V_L, and V_T all favor longer elimination half-lives.

(6) Oral and systemic clearances are related by $CL_{po} = CL_s/F$ where F is the bioavailability.

(7) Systemic clearance, $CL_s = Q_L E$, where E is the extraction ratio and $E = 1 - F = f_B CL_i'/(Q_L + f_B CL_i')$.

(8) $R_T < 1$ means $f_T > f_B$ and $R_L < 1$ means $f_L > f_B$. $R_T > 1$ means $f_T < f_B$ and $R_L > 1$ means $f_L < f_B$.

(9) For the nine sets of simulated data, systemic clearance is only weakly correlated with V_{ss} ($r = 0.700$, $p = 0.05$); the equation of the line is: $CL_s = 0.6075 - 0.00101\ V_{ss}$. Oral clearance was not significantly correlated with V_{ss} ($r = 0.548$, $p > 0.10$).

MEASUREMENT OF PLASMA PROTEIN BINDING

Ultrafiltration

Judd and Pesce wrote an excellent article [122] on ultrafiltration. Assumptions that should be validated for each drug and apparatus are: (1) the drug does not bind to the membrane; (2) there is no significant leakage of protein through the membrane; (3) the equilibrium constant does not change as the protein is concentrated; and (4) the membrane is equally permeable to the drug and water.

The pressure needed for ultrafiltration can be produced in most laboratory centrifuges and 15 μL of filtrate can be produced from a 1 ml sample in fifteen minutes. The temperature during ultrafiltration needs to be maintained at a constant.

Judd and Pesce [121] found that the free drug concentration remains remarkably constant in serial aliquots of ultrafiltrate generated from a single sample. This is based on the following:

$$\text{Free drug} + \text{Free protein} \underset{K}{\rightleftharpoons} \text{Protein-bound drug} \tag{621}$$

$$K = \frac{[\text{Protein-bound drug}]}{[\text{Free drug}][\text{Free protein}]} \tag{622}$$

$$[\text{Free drug}] = \frac{[\text{Protein-bound drug}]}{[\text{Free protein}]} \times \frac{1}{K} \tag{623}$$

Since protein-bound drug and free protein are both concentrated at the same rate, this ratio remains constant if K is a constant. Hence, the concentration of free drug should not change during ultrafiltration.

Equilibrium Dialysis

During equilibrium dialysis an osmotic water shift occurs such that the protein in the plasma compartment becomes diluted and the volume on the buffer side of the semipermeable membrane decreases. If a correction is not made for this volume shift (averaging 15–20%) then the concentrations of free and bound drug and the free fraction in *nonlinear* binding situations refer to lower protein and drug concentrations in the original plasma or serum. Correct equations were reported by Tozer et al. [123]. In cases of *linear* binding there is no problem and Equation (624) applies.

$$f_u = \frac{1}{(C_p/C_f) - R} \tag{624}$$

where f_u is the free fraction, C_p is the drug concentration in plasma *before* dialysis, C_f is the free drug concentration in the buffer compartment at dialysis equilibrium, and R = volume of buffer/volume of plasma, both *before* dialysis.

FITTING OF DATA TO BINDING EQUATIONS

Bound-free concentration data should be fitted by least squares regression on a computer. Linear transforms should not be used since they yield inaccurate parameter estimates.

For linear binding, Equation (625) may be used.

$$C_b = nK_aPC_f \tag{625}$$

In Equation (625), C_b is the bound concentration, C_f is the free drug concentration, n is the number of binding sites in a class, K_a is the association constant, and P is the binding protein concentration. If n, K_a, and P are all constants then they may be combined to give one constant relating bound and free drug concentrations.

If binding is nonlinear then the free fraction is not constant but varies with the drug concentration. Types of nonlinear binding equations with examples are shown below.

$$C_b = \frac{nK_aPC_f}{1 + K_aC_f} = \frac{nPC_f}{1/K_a + C_f} = \frac{nPC_f}{K_d + C_f} \tag{626}$$

In Equation (626), K_d is the dissociation constant, equal to $1/K_a$, and the other symbols have been defined above. Ibuprofen plasma protein binding obeys Equation (626).

$$C_b = \frac{n_TK_TP_TC_f}{1 + K_TC_f} + n_AK_AP_AC_f \tag{627}$$

Prednisolone and cortisol binding obeys Equation (627) where T refers to the specific binding protein transcortin and A refers to albumin.

$$C_b = \frac{n_1K_{a1}PC_f}{1 + K_{a1}C_f} + \frac{n_2K_{a2}PC_f}{1 + K_{a2}C_f} \tag{628}$$

Tolmetin binding obeys Equation (628) and the "1" and "2" refer to two different classes of binding sites.

Behm and Wagner [124] indicate how the nonlinear binding Equations (625) to (627) may be rearranged to allow estimation of the free drug concentration corresponding to the measured total drug concentration in the original plasma.

EFFECT OF DISEASE STATES AND OTHER FACTORS

Wilkinson's review [125] with ninety-two references is excellent. Some disease state effects are as follows. The plasma concentration of α_1-acid glycoprotein is elevated four- to fivefold in a variety of inflammatory and other disease states, but reduced in the neonate, by pregnancy, and following administration of oral contraceptives. Renal failure leads to the accumulation of unidentified endogenous substances that act as competitive inhibitors of binding. Administration of heparin causes release of both lipoprotein lipase from the capillary endothelium and hepatic lipase into the blood. This leads to an increase in the circulating concentration of nonesterified fatty acids by hydrolysis of triglycerides. Several studies have demonstrated a significant relationship between these heparin-induced fatty acid changes and decreases in *in vitro* binding of several drugs.

Many studies have documented that the stoppers of certain commercial blood collection tubes (Vacutainer in particular) contain binding inhibitors that may leak out in contact with blood.

CHAPTER 14

Pharmacokinetic-Pharmacodynamic Modelling

INTRODUCTION

Pharmacodynamics is the experimental science of the actions and effects of drugs. *Pharmacokinetics* is the study of the time courses of drug and metabolite amounts and concentrations in biological fluids, tissues, and excreta and also of pharmacologic effects and responses and the construction of models to interpret such data. Models are really not the schematic diagrams we draw, but rather the mathematical equations or set of equations.

PHARMACODYNAMIC MODELS

The linear model is

$$E = E_o + SC \tag{629}$$

where E is the effect at concentration, C, E_o is the baseline effect, and S is the slope of the straight line when E is plotted *versus* C.

The log-linear model is

$$E = S \log C + I \tag{630}$$

where E is the effect at log C concentration, S is the slope, and I is the intercept of the straight line when E is plotted *versus* log C.

The E_{max} model is

$$E = \frac{E_{max} C}{EC_{50} + C} \tag{631}$$

This is a nonlinear function with E_{max} being equal to E when the concentration, C, approaches infinity, and EC_{50} is the concentration when $E = E_{max}/2$.

The sigmoid E_{max} model is:

$$E = \frac{E_{max} C^s}{EC_{50}^s + C^s} \tag{632}$$

This is a nonlinear function with E_{max} being equal to E when the concentration, C^s, approaches infinity, and EC_{50}^s is the concentration when $E = E_{max}/2$. This function has a sigmoid shape and s is a power of the concentration.

In pharmacokinetic-pharmacodynamic modelling there have been two basic approaches. The first type is illustrated in Figures 14.1 and 14.2 where the effect is assumed to be related to the amount or concentration of the drug in one of the compartments of a compartment model. In this particular case the effect, namely scores on arithmetic tests, was assumed to be related to the concentration of the drug in the peripheral compartment of the classical two-compartment open model when LSD was given by bolus intravenous injection to five subjects [126]. Figure 14.2 shows the mean plasma concentrations of LSD (triangles), the mean performance scores (open circles), and the estimated concentration of LSD in the peripheral compartment (closed circles). Inset is a plot of performance score *versus* "tissue" concentration of LSD. To produce the plots, the plasma concentration-time data were first computer-fitted then the inset figure was produced. There is hysteresis in that the points have a time dependency about the line drawn in

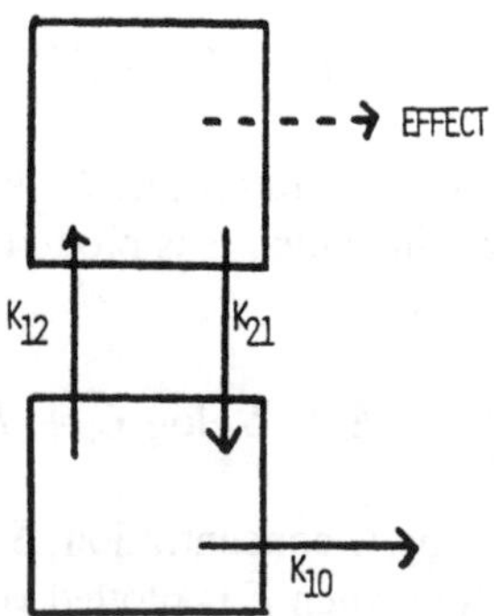

FIGURE 14.1. LSD example of Wagner et al. [126] where the pharmacodynamic effect is related to the amount or concentration of drug in the peripheral compartment of the classical two-compartment open model.

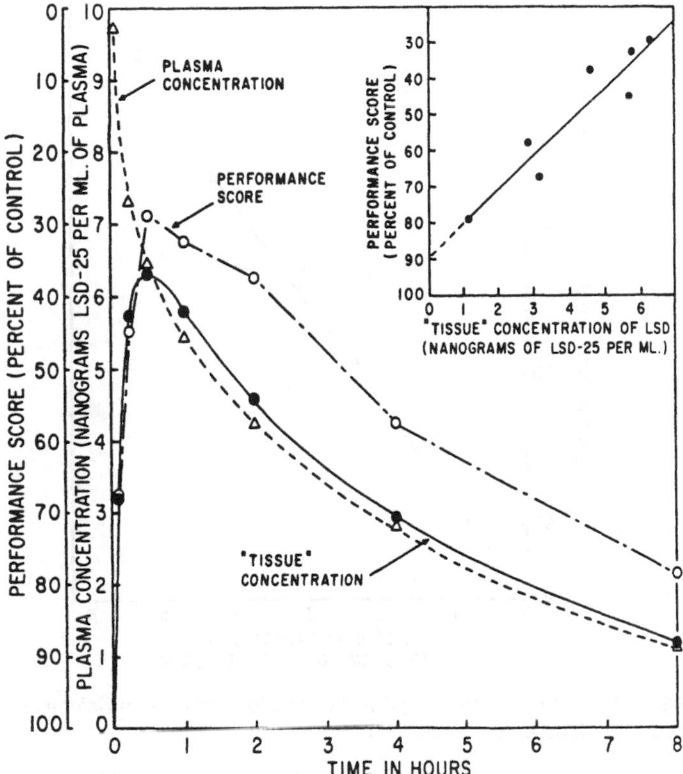

FIGURE 14.2. LSD example of Wagner et al. [126]; mean plasma concentrations of LSD (triangles), mean performance scores (open circles). Inset is a plot of performance score *versus* "tissue" concentration of LSD (with permission from Mosby-Year Book, Inc.).

the inset figure. Metzler [127] simultaneously fitted the plasma concentrations, performance scores, and time data on the computer and produced Figure 14.3, where the numbers refer to the time sequence of the points. The hysteresis has disappeared. The sigmoid E_{max} model was first published by Hill [128] to explain the oxygen dissociation of hemoglobin. In 1968 the author published a paper [129] which suggested that the equation be used as a pharmacodynamic model and illustrated its use.

The second type of modelling was introduced by Sheiner [130] and is illustrated in Figure 14.4. This model has an effect compartment connected to a pharmacokinetic model. Drug is assumed to enter and leave the effect compartment according to first-order kinetics – in this case with rate constants k_{1e} and k_{eo}. The sigmoid E_{max} pharmacodynamic model was used to relate the intensity of effect, E, to the amount, A_e, or concentration of drug

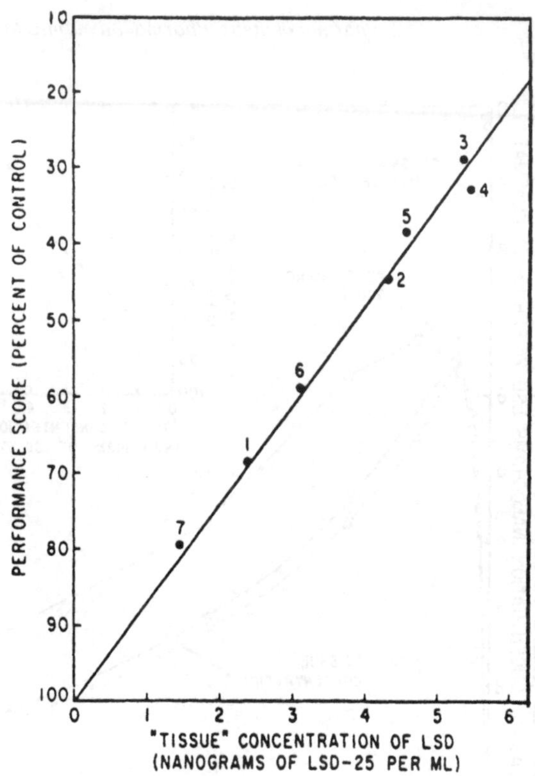

FIGURE 14.3. LSD plot of Metzler [127]. See text for details (from Reference [126], with permission from Mosby-Year Book, Inc.).

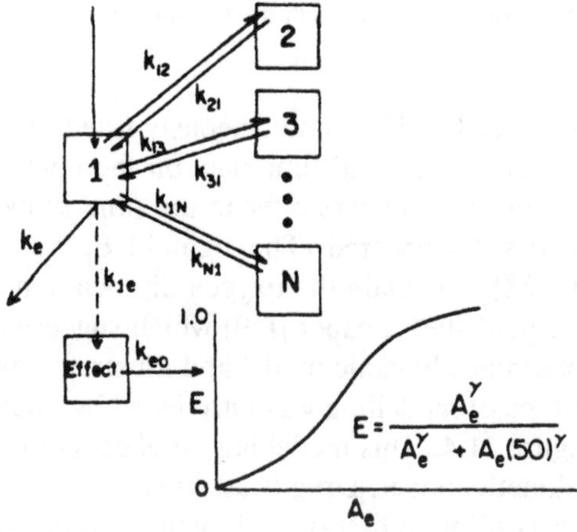

FIGURE 14.4. Effect model of Sheiner et al. [130]. See text for details (with permission from Mosby-Year Book, Inc.).

in the effect compartment; here γ replaces the s power in Equation (632). Sheiner proposed the model for *d*-tubocurarine in man and it has been subsequently used for several drugs and pharmacologic effects.

Figure 14.5 shows results of the *d*-tubocurarine study. The effect measurement was the force of thumb adduction measured with a force displacement transducer. In Figure 14.5 this effect is plotted *versus* time as the open circles, while the plasma concentration, expressed as a fraction of the peak concentration, is plotted *versus* time as the solid circles. The solid line represents the best fit to the model.

Figure 14.6 shows another application of the effect compartment where the pharmacokinetic model is the simple one-compartment open model with first-order elimination and first-order input (shown inset in the figure). The drug in this case was ergotamine tartrate and the effect endpoint was the systolic blood pressure gradient between toe and arm. The drug was administered by intramuscular injection. The dotted line gives the time course of plasma concentrations of ergotamine. The solid circles give the effect and the solid line is the model-predicted effect.

Multiple receptors may also be accommodated by combining sigmoid

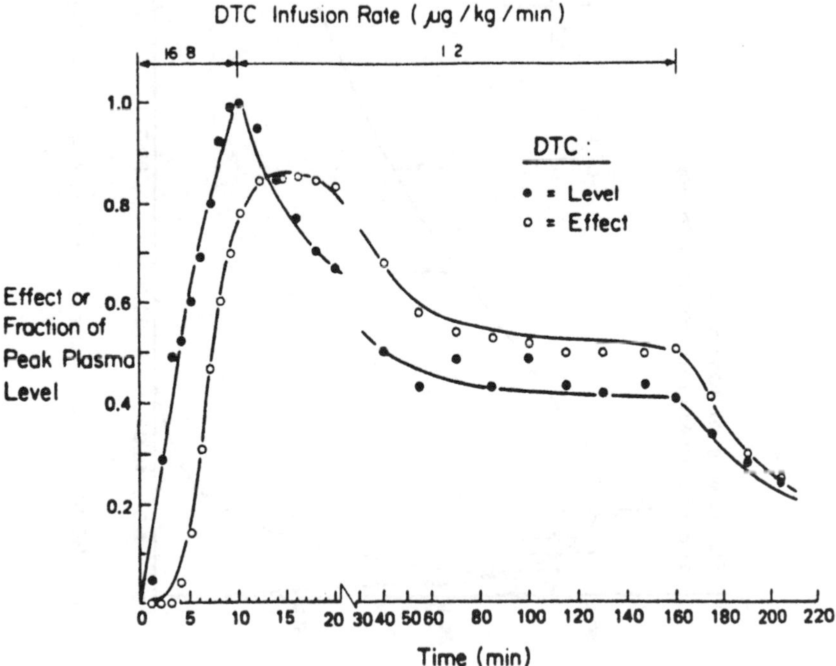

FIGURE 14.5. *d*-Tubocurarine results of Sheiner [130]. See text for details (with permission from Mosby-Year Book, Inc.).

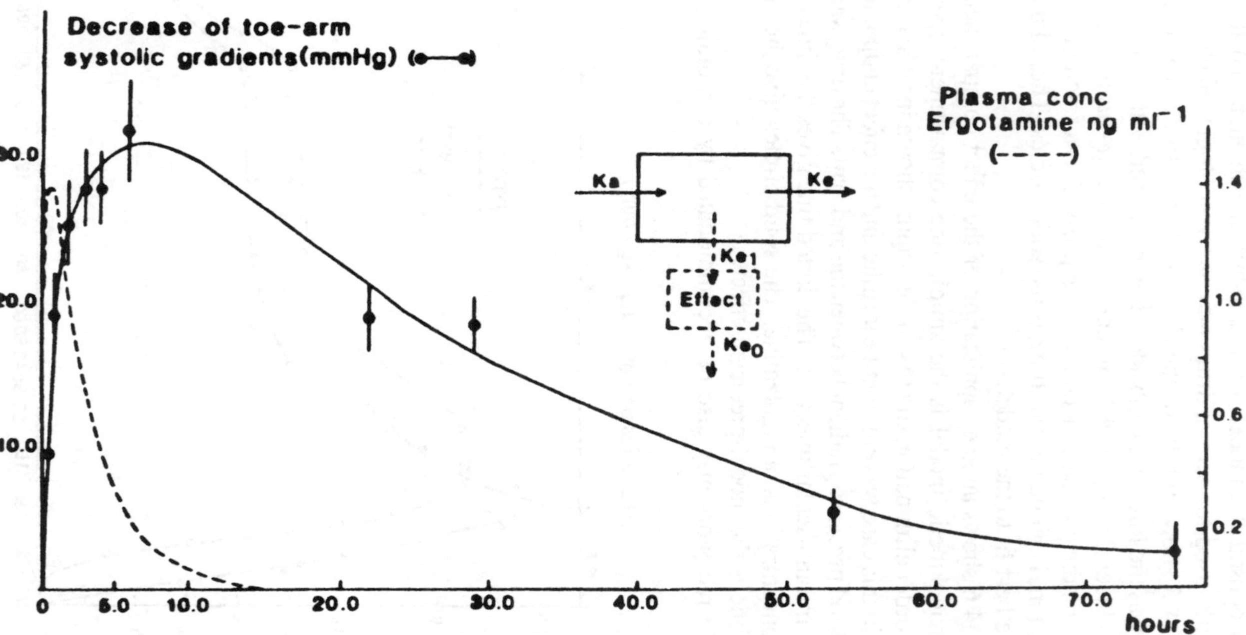

FIGURE 14.6. Ergotamine tartrate results of Paalzow et al. [131]. See text for details (with permission from Plenum Press).

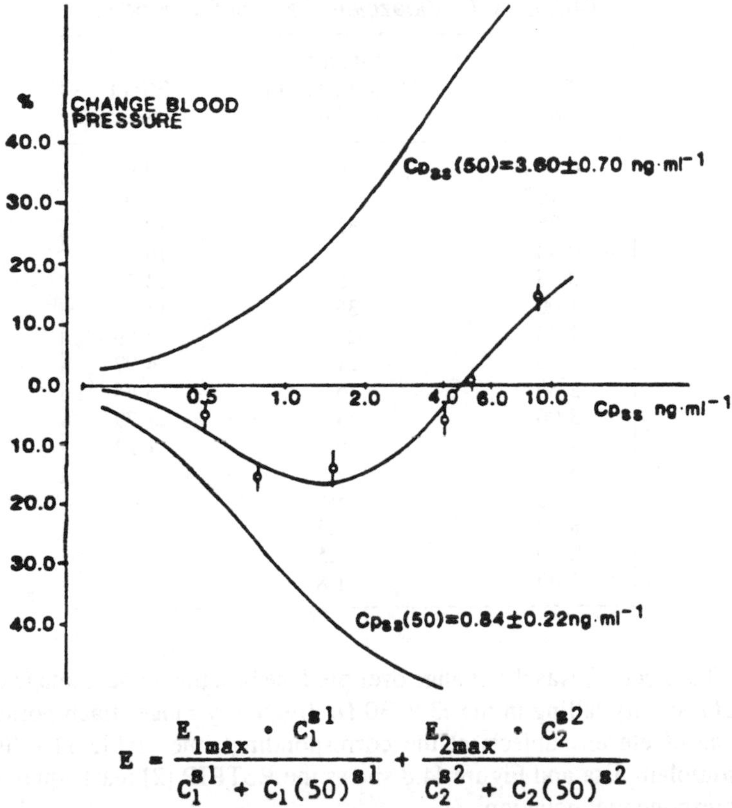

$$E = \frac{E_{1max} \cdot C_1^{s1}}{C_1^{s1} + C_1(50)^{s1}} + \frac{E_{2max} \cdot C_2^{s2}}{C_2^{s2} + C_2(50)^{s2}}$$

FIGURE 14.7. Multiple receptors accommodated by combining sigmoid E_{max} equations and applied to clonidine by Paalzow et al. [131] (with permission from Plenum Press).

E_{max} equations. This was applied to clonidine—which at low concentrations activates α_2-receptors—which leads to an inhibition of the neuronal activity of noradrenergic neurons. Higher concentrations activate α_1-receptors leading to an effect similar to increased noradrenergic activity. Paalzow et al. [131] show in Figure 14.7 the effects of clonidine on blood pressure at different steady-state concentrations in the normotensive rat. The observed effect was modelled using the equation shown, where there is a sum of two sigmoid E_{max} equations. The solid lines are the model-predicted blood pressures.

PROCEDURE FOR FITTING THE E_{max} MODEL

The data of Greenblatt et al. [132] in their Figures 2 and 4 were used. These data were plasma concentrations of midazolam after an infused intravenous dose of the drug given over a one-minute period.

TABLE 14.1. Midazolam Data for E_{max} Model.

Time (hr)	Plasma Concentration (ng/ml)	Effect (%)
0.0833	150.	19.9
0.167	100.	20.0
0.25	85.	19.3
0.50	60.	16.1
0.75	45.	13.2
1.00	35.	11.4
1.50	25.	7.59
2.00	21.	4.97
2.50	15.	3.44
3.00	14.	1.79
4.00	9.	1.24
5.00	7.	–
6.00	5.	–
8.00	3.5	
10.00	2.5	
12.00	1.8	

The effect, *E*, was the change over predose baseline in percentage of total EEG activity falling in the 13 to 30 Hz frequency range. Each point is the mean of eleven subjects at the corresponding time. Table 14.1 lists the midazolam data and Figure 14.8 shows the RSTRIP [2] least squares fit to a biexponential equation.

The plasma concentrations were then fitted again using the program MINSQ [2] to the biexponential Equation (75) under Model XII (Chapter 2) using $\lambda_1 = 0.308$ hr^{-1} and $\lambda_2 = 3.05$ hr^{-1} estimated in the original fit (Figure 14.8), the input zero-order rate, $k_o = (0.1 \times 10^3)/0.01667$ μg/hr (since the dose was 0.1 mg/kg given over one minute or 0.01667 hr) and the infusion time of 0.01667 hr as constants and estimating Vc and k_{21}. The fit was essentially identical to that shown in Figure 14.8 and the estimated parameters were $Vc = 0.58883$ L/kg and $k_{21} = 0.881$ hr^{-1}.

The effect-time data (Table 14.1) were then fitted to the appropriate effect compartment model using the program MINSQ [2] and the computer program shown below:

```
A: =KEO*KO*VC
B: =(K21-L2)*(1-EXP((-L2)*TI))*EXP((-L2)*T)/L2*
(L1-L2)*(KEO-L2))
C: =(K21-KEO)*(1-EXP((-L1)*TI))*EXP((-L1)*T)/(L1*
(L2-L1)*(KEO-L1))
D: =(K21-KEO)*(1-EXP((-KEO)*TI))*EXP((-KEO)*T)/
```

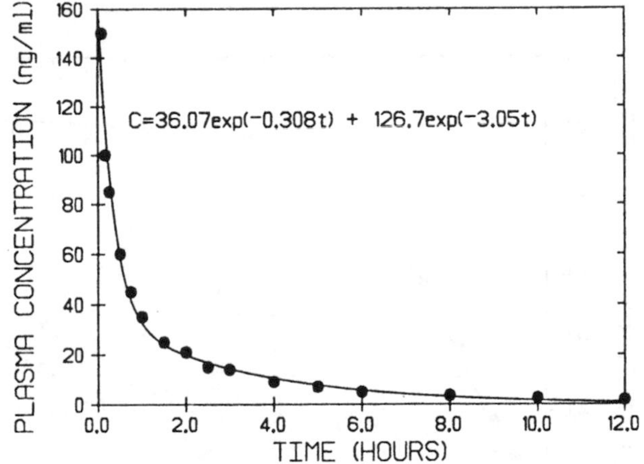

FIGURE 14.8. Midazolam plasma concentrations shown in Table 14.1 fitted to a biexponential equation.

```
(KEO*(L2-KEO)*(L1-KEO)
CE: =A*(B+C+D)
E: =EMAX*CE^G/(CE^G+C50^G)
```

During this fitting, KO (k_o), VC, K21, L1 (λ_1), L2 (λ_2), and TI (infusion time) were used as constants. Estimated parameters were KEO = 16,651 hr^{-1}, G (s) = 2.414, C_{50} = 31.1 ng/ml, and E_{max} = 20.5. Here KEO (k_{eo}) is the elimination rate constant from effect compartment (see Figure 14.4). This fit is shown in Figure 14.9. Since KEO was very large the plasma com-

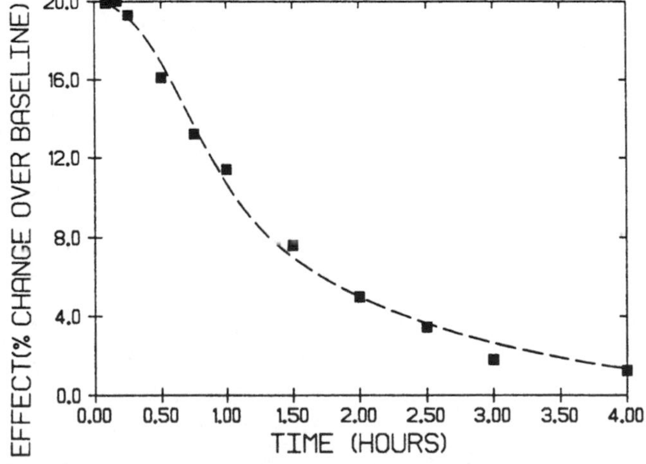

FIGURE 14.9. Midazolam effect-time data in Table 14.1 fitted to the appropriate effect compartment model using the computer program shown in text.

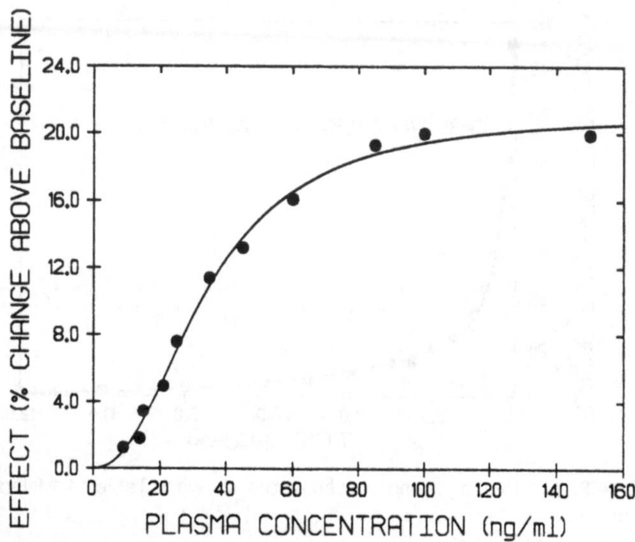

FIGURE 14.10. Midazolam effect-plasma concentration data in Table 14.1 fitted with the sigmoid E_{max} model. See text for parameters.

partment may be considered to be the effect compartment and the fit described below was done.

The effect-plasma concentration data were fitted to Equation (632) using MINSQ [2]. The estimated parameters were $E_{max} = 21.2$, $C_{50} = 34.0$ ng/ml, and $s = 2.24$. This fit is shown in Figure 14.10.

CHAPTER 15

Metabolic Kinetics

METABOLIC PATHWAYS

Metabolic pathways are those termed Phase I reactions which include oxidation, reduction, and hydrolysis, and Phase II reactions which are conjugation reactions such as sulfo-conjugation, glucuronidation, glutathione and amino acid conjugation, methylation, acetylation, glucosidation and hydration.

Table 15.1, taken from Rowland and Tozer [133], lists some therapeutically important metabolites.

METABOLITE KINETICS

Steady State in a Simple Linear System

$$C_{Dss} \text{ —— } CL_D \longrightarrow C_{Mss} \text{ —— } CL_M \longrightarrow \text{Eliminated Metabolite}$$

Model XXXXII

SCHEME 15.1.

In Scheme 15.1 $C_{D\text{ss}}$ and $C_{M\text{ss}}$ are the steady-state concentrations of drug and metabolite, respectively, and CL_D and CL_M are the clearances of drug and metabolite, respectively. If the drug is infused at a constant rate, R_o, until steady-state concentrations of both drug and metabolite are obtained, then:

$$R_o = \text{CL}_D C_{D\text{ss}} \tag{633}$$

$$\frac{dA_{M\text{ss}}}{dt} = f_m \text{CL}_D C_{D\text{ss}} - \text{CL}_M C_{M\text{ss}} = 0 \tag{634}$$

TABLE 15.1. Representative Therapeutically Important Metabolites.

Compound Administered	Metabolite	Compound Administered	Metabolite
Acetylsalicylic acid	Salicylic acid	Lidocaine	Desethyllidocaine
Meperidine	Normeperidine	Amitriptyline	Nortriptyline
Phenacetin	Acetaminophen	Verapamil	Norverapamil
Carbamazepine	Carbamazepine 10,11-epoxide	Phenylbutazone	Oxyphenbutazone
Prednisone	Prednisolone	Chlordiazepoxide	Desmethylchlordiazepoxide
Primidone	Phenobarbital	Codeine	Morphine
Procainamide	*N*-Acetyl procainamide	Diazepam	Desmethyldiazepam
Propranolol	4-Hydroxypropranolol	Glutethamide	4-Hydroxyglutethamide
Sulindac	Sulindac sulfide	Imipramine	Desipramine

In Equations (633) and (634), $A_{M_{ss}}$ is the amount of metabolite in the "body" at steady state, R_o is the infusion rate (mass/time), and f_m is the fraction of the drug that is converted to the metabolite.

Rearrangement of Equation (634) gives Equation (635).

$$C_{M_{ss}} = f_m \left[\frac{CL_D}{CL_M}\right] C_{D_{ss}} \tag{635}$$

Equation (635) indicates that a plot of $C_{M_{ss}}$ *versus* $C_{D_{ss}}$ should yield a straight line with slope equal to $f_m[CL_D/CL_M]$. Figure 15.1 at the bottom is an illustration of such a plot for subject #2, taken from Wagner et al. [134], when loading doses were given at zero time followed by 1, 2, or 3 mg/hr of adinazolam mesylate from one to twelve hours; both the drug and the metabolite, *N*-desmethyladinazolam, were measured in plasma. Figure 15.1 at the top shows a plot of R_o *versus* $C_{D_{ss}}$ with slope equal to CL_D. From Equations (633) and (635) one obtains:

$$\frac{\text{Slope of } R_o \textit{ versus } C_{D_{ss}} \text{ plot}}{\text{Slope of } C_{M_{ss}} \textit{ versus } C_{D_{ss}} \text{ plot}} = \frac{CL_M}{f_m} \tag{636}$$

Steady State in a Nonlinear System

Here the drug is assumed to be metabolized in the liver according to Michaelis Menten kinetics.

$$R_o = \frac{V_m C_{D_{ss}}}{K_m + C_{D_{ss}}} \tag{637}$$

$$CL_D = \frac{V_m}{K_m + C_{D_{ss}}} = \frac{V_m - R_o}{K_m} \tag{638}$$

$$\frac{dA_{M_{ss}}}{dt} = \frac{f_m V_m C_{D_{ss}}}{K_m + C_{D_{ss}}} - CL_M C_{M_{ss}} = 0 \tag{639}$$

$$C_{M_{ss}} = \frac{(f_m V_m / CL_M) C_{D_{ss}}}{K_m + C_{D_{ss}}} \tag{640}$$

From Equations (637) and (640) one obtains Equation (641).

$$\frac{\text{Asymptote of } R_o \textit{ versus } C_{D_{ss}} \text{ plot}}{\text{Asymptote of } C_{M_{ss}} \textit{ versus } C_{D_{ss}} \text{ plot}} = \frac{V_m}{\{f_m V_m / CL_M\}} \tag{641}$$

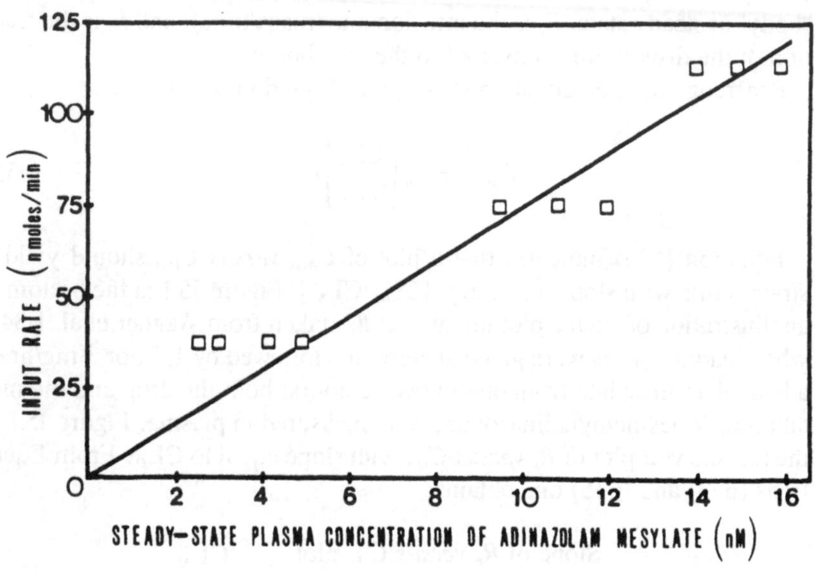

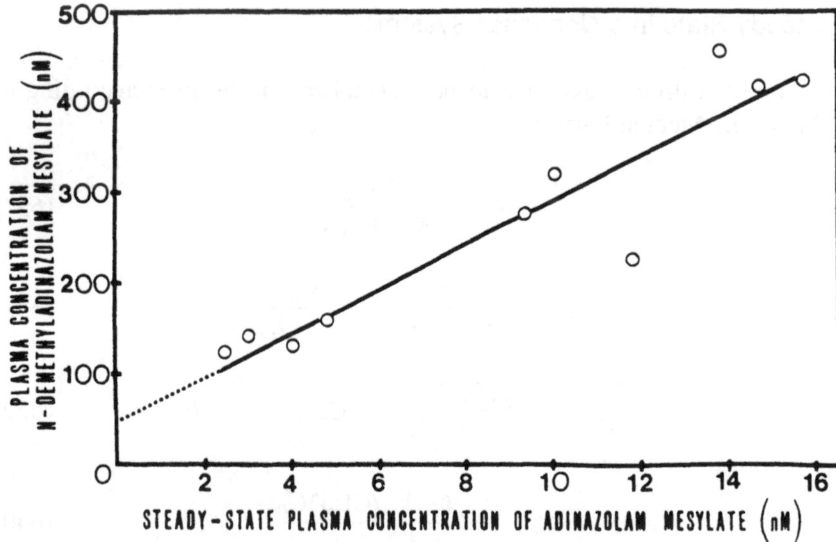

FIGURE 15.1. Application of Equations (633) and (635) based on a human study with adinazolam mesylate [133] (with permission from John Wiley & Sons, Ltd.).

Figure 15.2 at the bottom is an example of a plot of $C_{M\text{ss}}$ *versus* $C_{D\text{ss}}$; this is for subject #5 given adinazolam mesylate as above; the plot is based on Equation (640). The equation of the line is: $C_{M\text{ss}} = 5929\ C_{D\text{ss}}/(980 + C_{D\text{ss}})$. Figure 15.2 at the top is a plot of R_o *versus* $C_{D\text{ss}}$ based on Equation (637). The equation of the line is: $R_o = 316\ C_{D\text{ss}}/(2724 + C_{D\text{ss}})$. Hence, from Equation (641) we have:

$$\frac{V_m}{\{f_m V_m/\mathrm{CL}_M\}} = \frac{316}{5929} = 0.0533$$

It is interesting that in the adinazolam mesylate human study, four subjects exhibited linear kinetics and four subjects exhibited nonlinear kinetics.

Single Doses of Drug

When single doses of drug are administered and kinetics are linear, then:

$$\frac{dA_M}{dt} = f_m \mathrm{CL}_D C_D - \mathrm{CL}_M C_M \tag{642}$$

where C_D and C_M are concentrations of drug and metabolite at time t, respectively.

Integration of Equation (642), recognizing that $A_M = 0$ at $t = 0$ and at $t =$ infinity gives:

$$\frac{(\mathrm{AUC})_M}{(\mathrm{AUC})_D} = f_m \left[\frac{\mathrm{CL}_D}{\mathrm{CL}_M}\right] = \frac{\mathrm{CL}_D}{\{\mathrm{CL}_M/f_m\}} \tag{643}$$

where

$$\mathrm{CL}_D = \frac{D_D}{(\mathrm{AUC})_D} \tag{644}$$

and D_D is the dose of the drug and $(\mathrm{AUC})_D$ and $(\mathrm{AUC})_M$ are the areas under the drug and metabolite concentration-time curves, respectively. Substituting from Equation (644) into Equation (643), cancelling the $(\mathrm{AUC})_D$s and rearranging gives Equation (645).

$$\frac{\mathrm{CL}_M}{f_m} = \frac{D_D}{(\mathrm{AUC})_M} \tag{645}$$

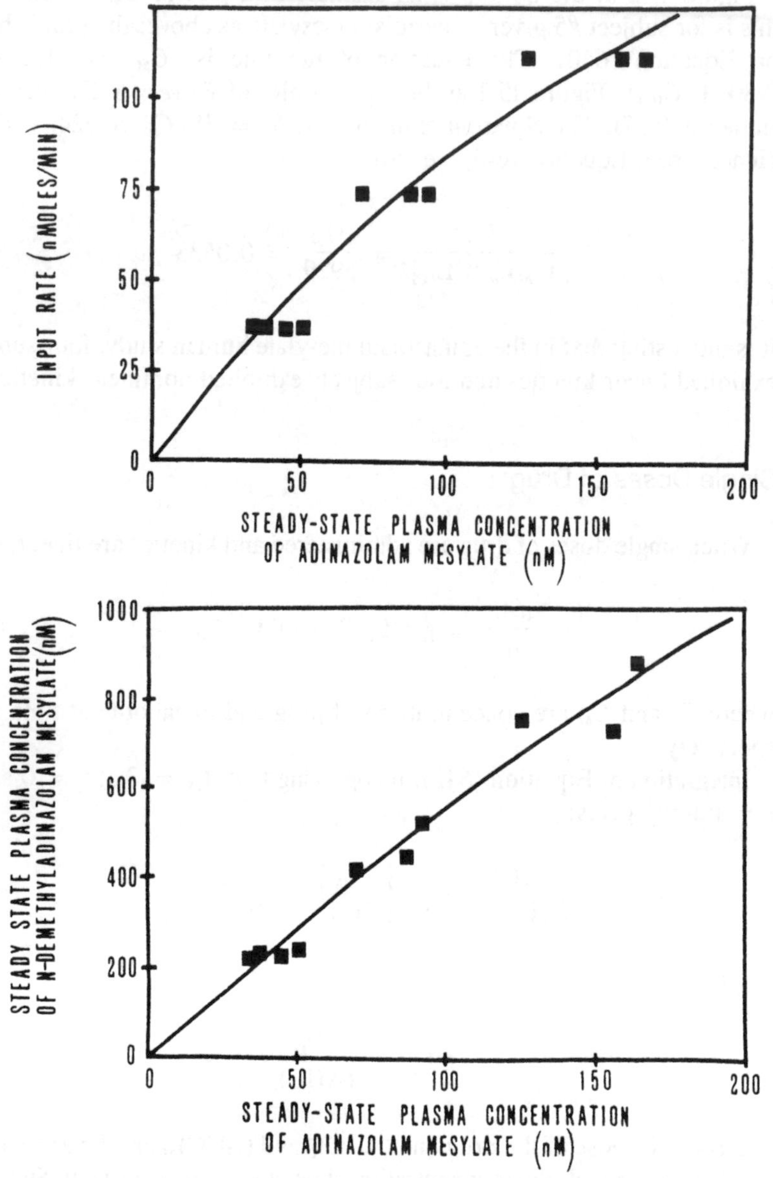

FIGURE 15.2. Application of Equations (637) and (640) based on a human study with adinazolam mesylate [133] (with permission from John Wiley & Sons, Ltd.).

Corresponding to the clearance Equation (642) we can also write Equation (646).

$$\frac{dA_M}{dt} = k_f A_D - k_{me} A_M \tag{646}$$

where k_f is the formation rate constant of the metabolite and k_{me} is the elimination rate constant of the metabolite.

For the drug we may write:

$$\frac{dA_D}{dt} = -KA_D \tag{647}$$

If formation of the metabolite is the only way that the drug is eliminated then $k_f = K$. If the metabolite is formed and unchanged drug is eliminated in the urine with rate constant, k_d, then:

$$K = k_f + k_d \tag{648}$$

If $K < k_{me}$ then the semilogarithmic fall-off curve of the metabolite declines in parallel with the drug, as illustrated in Figure 15.3. But if $k_{me} < K$, then the semilogarithmic fall-off curve of the metabolite has a smaller slope than that of the drug, as illustrated in Figure 15.4. Thus, the rate-limiting

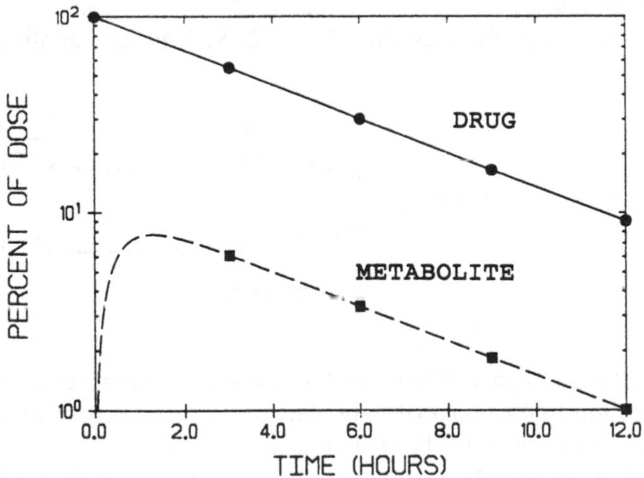

FIGURE 15.3. Parallel decline of drug and metabolite concentrations when $K < k_{me}$. See Equations (646) and (648) in text.

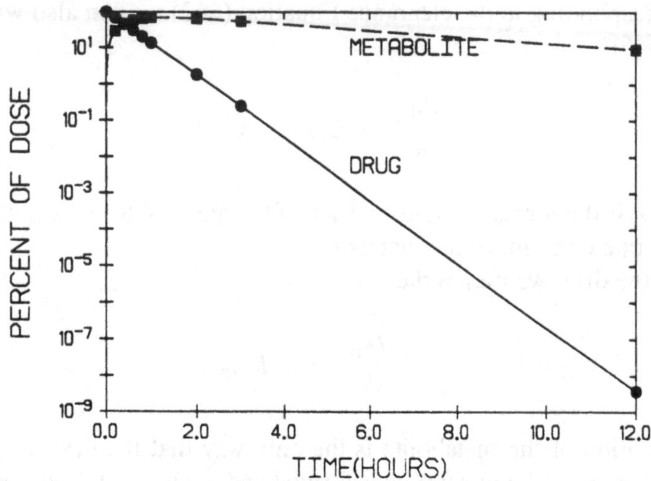

FIGURE 15.4. Metabolite and drug decline at different rates when $k_{me} < K$.

step (smaller first-order rate constant) determines the relative shapes of the drug and metabolite concentration-time curves. Also, a metabolite accumulates in the body more than the drug only when its elimination half-life is longer than that of the drug.

Table 15.2, taken from Rowland and Tozer [133], lists representative drugs undergoing extensive first-pass hepatic metabolism and forming active metabolites.

In cases such as the drugs in Table 15.2, Scheme 15.2 applies.

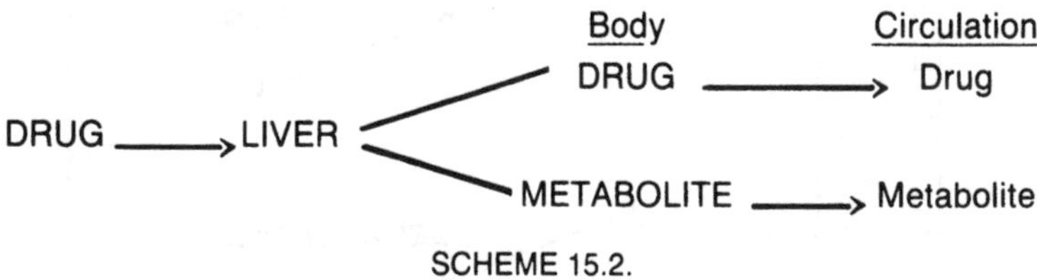

SCHEME 15.2.

Here the major part of the dose of drug is converted to metabolite(s) in the liver and this (these) enters the circulation. A minor fraction of the dose enters the circulation as unchanged drug. Hence, usually, the time of the peak metabolite concentration is about the same as the time of the peak drug concentration.

TABLE 15.2. Representative Drugs Undergoing Extensive First-Pass Hepatic Elimination and Forming Active Metabolites.

Drug	Active Metabolite*	Drug	Active Metabolite*
Alprenolol	4-Hydroxyalprenolol	Lorcainide	Norlorcainide
Amitriptyline	Nortiptyline	Meperidine	Normeperidine
Codeine	Morphine	Metoprolol	α-Hydroxymetoprolol
Dextropropoxyphene	Norpropoxyphene	Naloxone	6-β-Hydroxynaloxone
Encainide	*O*-Demethylencainide	Propranolol	4-Hydroxypropranolol
Imipramine	Desipramine	Quinidine	3-(*S*)-Hydroxyquinidine
Isosorbide dinitrate	Isosorbide 5-mononitrate	Verapamil	Norverapamil

*Some drugs have more than one active metabolite.

VOLUMES OF DISTRIBUTION

The volume of distribution of many basic drugs is frequently larger than 100 L, whereas their acidic metabolites have volumes of distribution of only about 10 to 20 L. The reasons are: (1) the drugs are highly tissue-bound while the metabolites are not; (2) the metabolites are more polar than the drug and hence get excreted in the urine more readily; and (3) the metabolites are highly albumin-bound, which restricts their distribution.

The volume of distribution of a metabolite may be directly estimated by administering the metabolite intravenously and indirectly estimated by dividing the metabolite clearance by the metabolite elimination rate constant [Equation (649)].

$$V_{dm} = \frac{CL_M}{k_{me}} \tag{649}$$

STEREOSELECTIVE DRUG METABOLISM

Only 41 out of 266 racemic drugs are marketed as drug preparations containing only a single isomer [135]. If only quantitative differences in therapeutic activity exist in the two isomers and the less active isomer is not predominantly responsible for side effects, the therapeutic benefit gained by using only the more active enantiomer is only marginal and does not justify the substantial increase in cost involved in manufacturing only the active isomer. However, if stereoselectivity in therapeutic activity is pronounced and adverse drug reactions are caused mainly by the less active isomer, then an isomeric drug preparation should be used therapeutically [135].

1-Verapamil is eight to ten times more potent than *d*-verapamil as a negative dromotropic agent. Systemic and oral clearances of 1-verapamil are substantially higher than that of *d*-verapamil. The bioavailability of *d*-verapamil (ca. 50%) is 2.5 times greater than that of 1-verapamil (ca. 20%). The concentration ratio of *d*-verapamil/1-verapamil in plasma is ca. 2 following I.V. administration and ca. 5 following oral administration [135].

CATENARY CHAINS

Frequently the drug and metabolite(s) form a catenary chain such as illustrated in Scheme 15.3, Model XXXXIII. Here D_g represents drug in the gut, k_a is the absorption rate constant, D_b represents drug in the blood, k_f is the metabolite formation rate constant, k_d is the rate constant for urinary excretion of unchanged drug, M_b is metabolite in blood, k_{me} is the urinary excretion rate constant of the metabolite, M_u is the metabolite in urine, and D_u is the unchanged drug in the urine. The values $k_a = 0.5\ \text{hr}^{-1}$, $k_f = 0.15\ \text{hr}^{-1}$, $k_d = 0.10\ \text{hr}^{-1}$, $K = k_f + k_d = 0.25\ \text{hr}^{-1}$, $k_{me} = 0.75\ \text{hr}^{-1}$, and $D = 100$ were assigned to the parameters of Scheme 15.3.

$$D_g \xrightarrow{k_a} D_b \xrightarrow{k_f} M_b \xrightarrow{k_{me}} M_u \qquad D_b \xrightarrow{k_d} D_u$$

Model XXXXIII

SCHEME 15.3.

The equations for the variables are as follows:

$$D_g = De^{-k_a t} \tag{650}$$

$$D_b = D\left\{\frac{k_a}{k_a - K}\right\}[e^{-Kt} - e^{-k_a t}] \tag{651}$$

$$D_u = D\left\{\frac{k_d}{K}\right\}\left[1 - \left\{\frac{1}{k_a - K}\right\}\{k_a e^{-Kt} - Ke^{-k_a t}\}\right] \tag{652}$$

$$M_b = k_a k_f D\left[\frac{e^{-Kt}}{(k_a - K)(k_{me} - K)} + \frac{e^{-k_a t}}{(K - k_a)(k_{me} - k_a)} + \frac{e^{-k_{me} t}}{(K - k_{me})(k_a - k_{me})}\right] \tag{653}$$

$$M_u = \left\{\frac{k_f}{K}\right\} D\left[1 - \left\{\frac{k_a k_{me} e^{-Kt}}{(k_a - K)(k_{me} - K)} + \frac{k_{me} K e^{-k_a t}}{(K - k_a)(k_{me} - k_a)} + \frac{k_a K e^{-k_{me} t}}{(k_a - k_{me})(K - k_{me})}\right\}\right] \tag{654}$$

The curves for D_g, D_b, D_u, and M_b are shown in Figure 15.5. Note that D_g drops rapidly with increase in time so that after four hours only 13.5% of the dose remains in the gut. D_b reaches a peak at about three hours then falls off. M_b is a low flat curve with a peak at about four hours – delayed with respect to the drug D_b curve. D_u is s-shaped and asymptotes at 40% since the asymptote is $(k_d/K)D = (0.1/0.25)(100) = 40\%$. The M_u

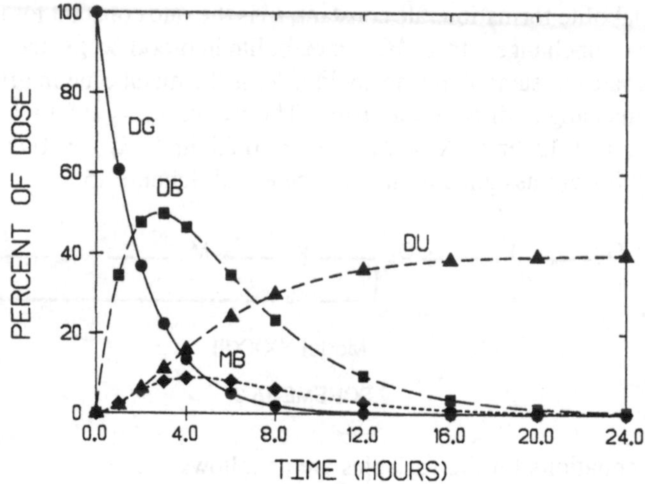

FIGURE 15.5. Catenary chain example based on Model XXXXIII with parameter values listed in the text.

curve is not shown but rises slowly and has an asymptote $= (k_f/K)D =$ $(0.15/0.25)(100) = 60\%$.

CHRONOPHARMACOKINETICS

Sometimes the clearance or protein binding has a diurnal cycle and hence the magnitude depends upon the time of the day. If this is so, then there may be an optimum time of the day to dose that particular drug.

Examples which have been reported are as follows [135]:

(1) Demethylation of indomethacin was reported to be greater at night than in the daytime.
(2) Lithium clearance in the rat was higher during the dark period than during the light period.
(3) The differences in the blood clearance of cyclosporin between night and day were 39% and 42% in two patients studied.
(4) Patients receiving cisplatin at 6 P.M. had a mean peak urinary concentration of free platinum (29.6 μg/ml) approximately 50% lower than that achieved (43.8 μg/ml) by the same patients after 6 A.M. administration. It was also shown that protein binding of cisplatin varies with the time of day.
(5) In freely fed rats there was a significant increase (57%) in V_{max} of UDP-glucuronyltransferase at 9 P.M. compared with that measured at 9 A.M.

CHAPTER 16

Laplace Derivation of Linear Pharmacokinetic Equations

INTRODUCTION

In my previous textbook [17d] a more extensive treatment of the use of Laplace transforms was published. In this chapter only the bare essentials are given to allow the reader to use Laplace transforms to obtain integrated linear pharmacokinetic equations. In the method of Laplace transforms the derivative in a linear differential equation is replaced by its Laplace transform and the resulting algebraic equation(s) is (are) solved by simple methods. The solution is obtained by taking the inverse Laplace transform or antitransform.

The Laplace transform, $L\{F(t)\}$, of a function, $F(t)$, is defined as:

$$L\{F(t)\} = \int_0^{\mathrm{Inf}} F(t)e^{-st}dt \qquad (655)$$

DETERMINANT

The definition of a determinant is quite complicated. But to use determinants to solve simple sets of linear equations and to obtain integrated equations for linear pharmacokinetic models, one has only to have the expansion of determinants of 2 × 2, 3 × 3, and 4 × 4 square matrices. These are given below. The symbol for a determinant is Δ.

The determinant of a 2 × 2 square matrix is:

$$\Delta = \begin{vmatrix} a_{11} & a_{12} \\ a_{21} & a_{22} \end{vmatrix} = a_{11}a_{22} - a_{12}a_{21} \qquad (656)$$

The determinant of a 3 × 3 square matrix is:

$$\Delta = \begin{vmatrix} a_{11} & a_{12} & a_{13} \\ a_{21} & a_{22} & a_{23} \\ a_{31} & a_{32} & a_{33} \end{vmatrix} = a_{11}a_{22}a_{33} - a_{11}a_{32}a_{23} - a_{12}a_{21}a_{33}$$

$$+ a_{12}a_{31}a_{23} + a_{13}a_{21}a_{32} - a_{13}a_{22}a_{31} \tag{657}$$

The determinant of a 4 × 4 square matrix is:

$$\Delta = \begin{vmatrix} a_{11} & a_{12} & a_{13} & a_{14} \\ a_{21} & a_{22} & a_{23} & a_{24} \\ a_{31} & a_{32} & a_{33} & a_{34} \\ a_{41} & a_{42} & a_{43} & a_{44} \end{vmatrix} = a_{11}a_{22}a_{33}a_{44} - a_{12}a_{21}a_{33}a_{44}$$

$$- a_{11}a_{22}a_{34}a_{43} + a_{12}a_{21}a_{34}a_{43} + a_{13}a_{21}a_{32}a_{44} - a_{11}a_{23}a_{32}a_{44}$$

$$- a_{13}a_{21}a_{34}a_{42} + a_{11}a_{23}a_{34}a_{42} + a_{11}a_{24}a_{32}a_{43} - a_{14}a_{21}a_{32}a_{43}$$

$$- a_{11}a_{24}a_{33}a_{42} + a_{14}a_{21}a_{33}a_{42} + a_{12}a_{23}a_{31}a_{44} - a_{13}a_{22}a_{31}a_{44}$$

$$- a_{12}a_{23}a_{34}a_{41} + a_{13}a_{22}a_{34}a_{41} + a_{14}a_{22}a_{31}a_{43} - a_{12}a_{24}a_{31}a_{43}$$

$$- a_{14}a_{22}a_{33}a_{41} + a_{12}a_{24}a_{33}a_{41} + a_{13}a_{24}a_{31}a_{42} - a_{14}a_{23}a_{31}a_{42}$$

$$- a_{13}a_{24}a_{32}a_{41} + a_{14}a_{23}a_{32}a_{41} \tag{658}$$

PROCEDURE FOR FINDING THE INTEGRATED EQUATION FOR THE AMOUNT IN ANY COMPARTMENT OF A LINEAR MODEL BY USE OF LAPLACE TRANSFORMS

1. Write the differential equations of the model.
2. Convert the differential equations to their Laplace transforms.
3. Rearrange the Laplace equations so that the Laplace variables are all on the left-hand sides and the constant terms are on the right-hand sides.
4. Write the coefficient matrix so that the coefficients for the central compartment are in column 1.
5. Write the determinant, Δ, of the coefficient matrix.

6. Expand Δ to obtain a polynomial in the Laplace variable s.

$$\Delta = s^n + a_{n-1}s^{n-1} + \ldots + a_o = 0 \qquad (659)$$

The exponents, $n, n - 1$, etc. are numbers such as $1,2,\ldots,n$. The coefficients, $a_{n-1},\ldots,a_o$, are functions of the rate constants of the model.

7. The exponents, $\lambda_1,\lambda_2,\ldots,\lambda_n$, are the absolute values of the negative roots of the n degree polynomial in s [Equation (659)] so that:

$$\Delta = (s + \lambda_1)(s + \lambda_2),\ldots,(s + \lambda_n) \qquad (660)$$

8. Obtain the Laplace transform of the amount in the central compartment by making a ratio of the matrix of coefficients with the coefficients of column 1 replaced by the column of constant terms to the determinant then obtain the solution.

9. For each compartment do as in step **8** but replace the appropriate column by the column of constant terms.

10. Apply the Heaviside Expansion Formula to obtain the antitransform. This formula is:

$$L^{-1}\{f(t)\} = L^{-1}\left\{\frac{P(s)}{Q(s)}\right\} = \sum_{n=1}^{m} \frac{P(\lambda_i)}{Q'(\lambda_i)} e^{-\lambda_i t} \qquad (661)$$

The theorem may be stated:

> If $f(t)$ is the quotient $P(s)/Q(s)$ of two polynomials such that $Q(s)$ has the higher degree and contains the factor $s + \lambda_i$, which is not repeated, then the term in $F(t)$ corresponding to that factor can be written in the form
>
> $$\frac{P(\lambda_i)}{Q'(\lambda_i)} e^{-\lambda_i t}$$
>
> Here $Q'(s)$ is the derivative of $Q(s)$ or $Q'(\lambda_i)$ is the derivative of $Q(\lambda_i)$.

You do not have to multiply out factors such as $(s + \lambda_1)(s + \lambda_2)$ to obtain the derivative. Instead you use the product rule:

$$\frac{duv}{dx} = u\frac{dv}{dx} + v\frac{du}{dv} \qquad (662)$$

Applying Equation (662) to Equation (663) gives:

$$Q(s) = (s + \lambda_1)(s + \lambda_2) \tag{663}$$

$$Q'(s) = (s + \lambda_2)\frac{d(s + \lambda_1)}{ds} + (s + \lambda_1)\frac{d(s + \lambda_2)}{ds}$$

$$= (s + \lambda_1) + (s + \lambda_2) \tag{664}$$

RELATIONSHIPS BETWEEN THE λ_i AND RATE CONSTANTS

In item **6** above, the $a_{n-1}, \ldots, a_o$ are functions of the rate constants. For the second degree polynomial in s, we have:

$$\Delta = s^2 + a_1 s + a_o = 0 \tag{665}$$

$$\lambda_1 + \lambda_2 = a_1 \tag{666}$$

$$\lambda_1 \lambda_2 = a_o \tag{667}$$

$$\lambda_1 = \frac{1}{2}\left[a_1 - \sqrt{a_1^2 - 4a_o}\right] \tag{668}$$

$$\lambda_2 = \frac{1}{2}\left[a_1 + \sqrt{a_1^2 - 4a_o}\right] \tag{669}$$

For the third degree polynomial in s, we have:

$$\Delta = s^3 + a_2 s^2 + a_1 s + a_o = (s + \lambda_1)(s + \lambda_2)(s + \lambda_3) = 0 \tag{670}$$

$$\lambda_1 + \lambda_2 + \lambda_3 = a_2 \tag{671}$$

$$\lambda_1\lambda_2 + \lambda_1\lambda_3 + \lambda_2\lambda_3 = a_1 \tag{672}$$

$$\lambda_1\lambda_2\lambda_3 = a_o \tag{673}$$

You have to find the roots λ_1, λ_2, and λ_3 of the polynomial with a computer program.

SOME LAPLACE TRANSFORMS

$F(t)$	$L\{F(t)\} = f(t)$
1	$1/s$
k_o	k_o/s
t	$1/s^2$
$C_1e^{-\lambda_i t}$	$C_1/(s + \lambda_1)$
$\frac{dA_1}{dt}$	$sa_1 - A_1^o$

where $a_1 = L\{A_1\}$ and A_1^o is the initial condition at $t = 0$.

EXAMPLES

I. Model VIII

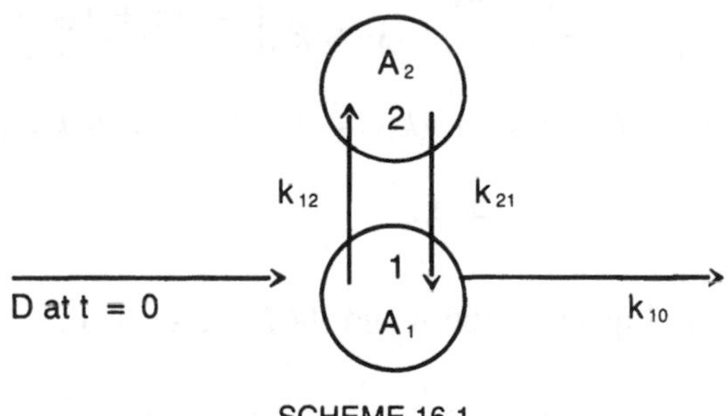

SCHEME 16.1.

The step numbers below are the same as in the Procedure list above.

1.

$$\frac{dA_1}{dt} = -(k_{12} + k_{10}) + k_{21}A_2 \tag{674}$$

$$\frac{dA_2}{dt} = k_{12}A_1 - k_{21}A_2 \tag{675}$$

2. Letting a_1 and a_2 be the Laplace transforms of the amounts A_1 and A_2, respectively, and converting Equations (674) and (675) to their Laplace transforms yields:

$$sa_1 - D = -(k_{12} + k_{10})a_1 + k_{21}a_2 \tag{676}$$

$$sa_2 - 0 = k_{12}a_1 - k_{21}a_2 \tag{677}$$

3. Rearranging Equations (676) and (677) gives:

$$(s + k_{12} + k_{10})a_1 - k_{21}a_2 = D \tag{678}$$

$$-k_{12}a_1 + (s + k_{21})a_2 = 0 \tag{679}$$

4. The coefficient matrix from Equations (678) and (679) is:

$$\begin{bmatrix} s + k_{12} + k_{10} & -k_{21} \\ -k_{12} & s + k_{21} \end{bmatrix} \begin{bmatrix} D \\ 0 \end{bmatrix} \tag{680}$$

5. to 7. The determinant is:

$$\Delta = \begin{vmatrix} s + k_{12} + k_{10} & -k_{21} \\ -k_{12} & s + k_{21} \end{vmatrix} = (s + k_{12} + k_{10})(s + k_{21})$$

$$- \quad k_{12}k_{21} = s^2 + (k_{12} + k_{21} + k_{10})s + k_{21}k_{10}$$

$$= (s + \lambda_1)(s + \lambda_2) \tag{681}$$

Applying Equations (666) and (667) we see that:

$$\lambda_1 + \lambda_2 = a_1 = k_{12} + k_{21} + k_{10} \tag{682}$$

$$\lambda_1\lambda_2 = a_o = k_{21}k_{10} \tag{683}$$

Applying Equations (668) and (669) we see that:

$$\lambda_1 = \frac{1}{2}\left[(k_{12} + k_{21} + k_{10}) - \sqrt{(k_{12} + k_{21} + k_{10})^2 - 4k_{21}k_{10}}\right] \tag{684}$$

$$\lambda_2 = \frac{1}{2}\left[(k_{12} + k_{21} + k_{10}) + \sqrt{(k_{12} + k_{21} + k_{10})^2 - 4k_{21}k_{10}}\right] \tag{685}$$

8.

$$a_1 = \frac{\begin{vmatrix} D & -k_{21} \\ 0 & s + k_{21} \end{vmatrix}}{\Delta} = \frac{D(s + k_{21})}{(s + \lambda_1)(s + \lambda_2)} \tag{686}$$

9.

$$a_2 = \frac{\begin{vmatrix} s + k_{21} + k_{10} & D \\ -k_{12} & 0 \end{vmatrix}}{\Delta} = \frac{k_{12}D}{(s + \lambda_1)(s + \lambda_2)} \tag{687}$$

10. From step **8** above using the Heaviside Expansion Formula we see that:

$$P(s) = D(s + k_{21}) \tag{688}$$

$$Q(s) = (s + \lambda_1)(s + \lambda_2) \tag{689}$$

From Equation (664) we have:

$$Q'(s) = (s + \lambda_1) + (s + \lambda_2) \tag{690}$$

$$L^{-1}\left[\frac{D(s + k_{21})}{(s + \lambda_1)(s + \lambda_2)}\right] = \frac{D(k_{21} - \lambda_1)}{(\lambda_2 - \lambda_1)(-\lambda_1 + \lambda_1)} e^{-\lambda_1 t}$$

$$+ \frac{D(k_{21} - \lambda_2)}{(-\lambda_2 + \lambda_2)(\lambda_1 - \lambda_2)} e^{-\lambda_2 t}$$

$$= \frac{D}{\lambda_2 - \lambda_1}[(k_{21} - \lambda_1)e^{-\lambda_2 t}$$

$$- (k_{21} - \lambda_2)e^{-\lambda_2 t}] \tag{691}$$

In practice you don't write the $-\lambda_1 + \lambda_1$, etc. terms. You just first write $e^{-\lambda_1 t}$ and for this term $s = -\lambda_1$ hence the numerator becomes $D(k_{21} - \lambda_1)$ and the denominator becomes $\lambda_2 - \lambda_1$, but you omit the $s + \lambda_i$ terms. Since $C = A_1/Vc$ for the central compartment then:

$$C = \frac{D}{Vc}[(k_{21} - \lambda_1)e^{-\lambda_1 t} - (k_{21} - \lambda_2)e^{-\lambda_2 t}] \tag{692}$$

II. Model II

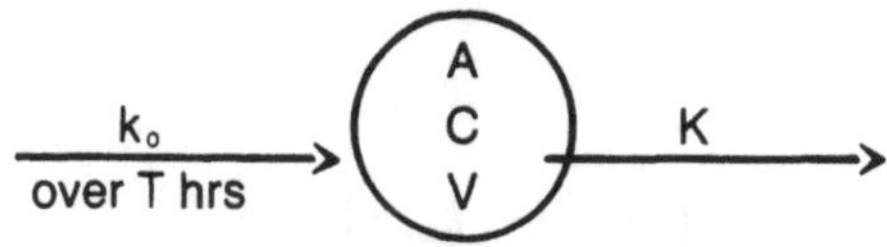

SCHEME 16.2.

1.

$$\frac{dA}{dt} = k_o - KA \tag{693}$$

2.

$$sa - 0 = \frac{k_o}{s} - KA \tag{694}$$

3.

$$(s + K)a = \frac{k_o}{s} \tag{695}$$

$$a = \frac{k_o}{s(s + K)} \tag{696}$$

There are no steps **4** to **8** in this case.

9. The roots are 0 and $-K$, hence:

$$A = L^{-1}\{a\} = \frac{k_o}{-K} e^{-Kt} + \frac{k_o}{K} e^{-0t} = \frac{k_o}{K}[1 - e^{-Kt}] \tag{697}$$

Since $C = A/V$ then

$$C = \frac{k_o}{VK}[1 - e^{-Kt}] \tag{698}$$

III.

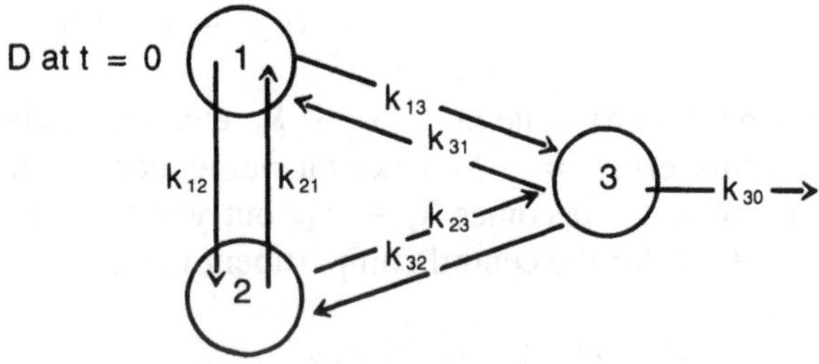

SCHEME 16.3.

Let $E_1 = k_{12} + k_{13}$, $E_2 = k_{21} + k_{23}$, $E_3 = k_{31} + k_{32} + k_{30}$, then:

1.

$$\frac{dA_1}{dt} = -E_1A_1 + k_{21}A_2 + k_{31}A_3 \tag{699}$$

$$\frac{dA_2}{dt} = k_{12}A_1 - E_2A_2 + k_{32}A_3 \tag{700}$$

$$\frac{dA_3}{dt} = k_{13}A_1 + k_{23}A_2 - E_3A_3 \tag{701}$$

2.

$$sa_1 - D = -E_1a_1 + k_{21}a_2 + k_{31}a_3 \tag{702}$$

$$sa_2 - 0 = k_{12}a_1 - E_2a_2 + k_{32}a_3 \tag{703}$$

$$sa_3 - 0 = k_{13}a_1 + k_{23}a_2 - E_3a_3 \tag{704}$$

3.

$$(s + E_1)a_1 - k_{21}a_2 - k_{31}a_3 = D \tag{705}$$

$$-k_{12}a_1 + (s + E_2)a_2 - k_{32}a_3 = 0 \tag{706}$$

$$-k_{13}a_1 = k_{23}a_2 + (s + E_3)a_3 = 0 \tag{707}$$

4. to **7.** Including applying Equations (657) and (670) to (673).

$$\Delta = \begin{vmatrix} s + E_1 & -k_{21} & -k_{31} \\ -k_{12} & s + E_2 & -k_{32} \\ -k_{13} & -k_{23} & s + E_3 \end{vmatrix} = (s + E_1)(s + E_2)(s + E_3)$$

$$- (s + E_1)(-k_{21})(-k_{32}) - (-k_{21})(-k_{12})(s + E_3)$$

$$+ (-k_{21})(-k_{13})(-k_{32}) + (-k_{31})(-k_{12})(-k_{23})$$

$$- (-k_{31})(s + E_3)(-k_{13}) = s^3 + (E_1 + E_2 + E_3)s^2$$

$$+ (E_1E_2 + E_1E_3 + E_2E_3 - k_{23}k_{32} - k_{12}k_{21} - k_{13}k_{31})s$$

$$+ E_1E_2E_3 - k_{23}k_{32}E_1 - k_{12}k_{21}E_3 - k_{13}k_{21}k_{32} - k_{12}k_{23}k_{31}$$

$$- k_{13}k_{31}E_2 = (s + \lambda_1)(s + \lambda_2)(s + \lambda_3) \tag{708}$$

where λ_1, λ_2, and λ_3 are the absolute values of the negative roots of the cubic Equation (709).

$$s^3 + a_2 s^2 + a_1 s + a_o = 0 \tag{709}$$

where

$$a_2 = E_1 + E_2 + E_3 \tag{710}$$

$$a_1 = E_1E_2 + E_1E_3 + E_2E_3 - k_{23}k_{32} - k_{12}k_{21} - k_{13}k_{31} \tag{711}$$

$$a_o = E_1E_2E_3 - k_{23}k_{32}E_1 - k_{12}k_{21}E_3 - k_{13}k_{21}k_{32} - k_{12}k_{23}k_{32} - k_{13}k_{31}E_2 \tag{712}$$

8. In this model, compartment #3 is the central compartment.

$$a_3 = \frac{\begin{vmatrix} s + E_1 & -k_{21} & D \\ -k_{12} & s + E_2 & 0 \\ -k_{13} & -k_{21} & 0 \end{vmatrix}}{\Delta}$$

$$= \frac{D(-k_{12})(-k_{23}) - D(s + E_2)(-k_{13})}{(s + \lambda_1)(s + \lambda_2)(s + \lambda_3)}$$

$$= \frac{D[k_{13}(s + E_2) + k_{12}k_{23}]}{(s + \lambda_1)(s + \lambda_2)(s + \lambda_3)} \tag{713}$$

9. Taking the antitransform of Equation (713), and, at the same time using $C_3 = A_3/V_3$, we get:

$$C_3 = \frac{D}{V_3}\left[\left\{\frac{k_{13}(E_2 - \lambda_1) + k_{12}k_{23}}{(\lambda_2 - \lambda_1)(\lambda_3 - \lambda_1)}\right\} e^{-\lambda_1 t}\right.$$

$$+ \left\{\frac{k_{13}(E_2 - \lambda_2) + k_{12}k_{23}}{(\lambda_1 - \lambda_2)(\lambda_3 - \lambda_2)}\right\} e^{-\lambda_2 t}$$

$$\left. + \left\{\frac{k_{13}(E_2 - \lambda_3) + k_{12}k_{23}}{(\lambda_1 - \lambda_3)(\lambda_2 - \lambda_3)}\right\} e^{-\lambda_3 t}\right] \tag{714}$$

Note in these integrated equations that the term with λ_1 as the exponential exponent has λ_1 at the *end* of the denominator terms; the term with λ_2 as the exponential exponent has λ_2 at the *end* of each denominator term. This is true in all such expressions. With practice you can immediately write the antitransform from the transform.

Also you can immediately write the expression for the AUC once the Laplace transform of the amount in a given compartment is known. This is done by letting $s = 0$ [121]. Hence for Equation (714) we have:

$$\mathrm{AUC} = \frac{a_3}{V_3}(s = 0) = \frac{D[k_{13}E_2 + k_{12}k_{23}]}{V_3\lambda_1\lambda_2\lambda_3} \qquad (715)$$

CHAPTER 17

Measures of Fit

INTRODUCTION

When data are fitted with a model using a program such as MINSQ [2] or NONLIN [68] measures of fit are needed to assess how good the fit is. These measures indicate the closeness of the observed data points to the model-predicted line and the confidence one should have in the parameter estimates. Some measures of fit are discussed below.

STANDARD DEVIATION (S.D.) OF ESTIMATED PARAMETER

$$\text{S.D.} = \sqrt{s^2 C_{ii}} \tag{716}$$

where $s^2 = \Sigma\, \text{dev}^2/(N - P)$, $\Sigma\, \text{dev}^2 = \Sigma\, (Y - \hat{Y})^2$, N = number of data points, P = number of parameters estimated, $N - P$ = number of degrees of freedom, and C_{ii} is ith diagonal element of the variance-covariance matrix of the elements.

For a reasonably good fit, the standard deviations should be about 10% of the estimated parameters.

CORRELATION (Corr)

The correlation coefficient between the model-predicted ($\hat{Y}$) and observed (Y) data points is given by Equation (717), where $\bar{Y}$ and $\bar{\hat{Y}}$ are average values.

$$\text{Corr} = \frac{\left[\sum_{i=1}^{n} w_i(Y_i - \overline{Y})(\hat{Y} - \overline{\hat{Y}})\right]}{\left[\sqrt{w_i \sum_{i=1}^{n} (Y - \overline{Y})^2}\right]\left[\sqrt{w_i \int_{i=1}^{n} (\hat{Y} - \overline{\hat{Y}})^2}\right]} \tag{717}$$

COEFFICIENT OF DETERMINATION (r^2)

$$r^2 = 1 - \left[\sum_{i=1}^{n} w_i(Y - \hat{Y})^2/s_y^2\right] \tag{718}$$

where

$$s_y^2 = \sum_{i=1}^{n} Y^2 - (\Sigma\, Y)^2/N$$

$$\sum_{i=1}^{n} w_i(Y - \hat{Y})^2 = \text{sum of weighted squared deviations}$$

$$w_i = \text{weights} = 1/Y_i \text{ for reciprocal weighting}$$

AKAIKE INFORMATION CRITERION (AIC)

$$\text{AIC} = N \ln \left[\sum_{i=1}^{n} w_i(Y_i - \hat{Y}_i)^2\right] + 2NP \tag{719}$$

The square bracketed part of Equation (719) is the sum of weighted squared deviations. The more appropriate model has the smallest AIC. The AIC may be used to decide whether data are best described by a biexponential equation or a triexponential equation. An example is shown below.

F-TEST FOR BEST FITTING MODEL

$$F = \left[\frac{S_m - S_{m+1}}{S_{m+1}}\right] \times \left[\frac{N - P_{m+1}}{(N - P_m) - (N - P_{m+1})}\right] \tag{720}$$

with degrees of freedom in the numerator of $(N - P_m) - (N - P_{m+1})$ and degrees of freedom in the denominator of $N - P_{m+1}$. Here s_m is the sum of weighted squared deviations for the fit with "m" terms and s_{m+1} is the sum of weighted squared deviations of the fit with "$m + 1$" terms. One compares the experimental F value given by Equation (720) with the tabled F found in a statistics book. If the experimental F is greater than the tabled F, then the fit with "$m + 1$" terms is significantly better than the fit with "m" terms.

The AIC and F-test were applied to the intravenous diazepam data of Kaplan et al. [25] fitted by NONLIN [68] with reciprocal weighting to both the biexponential (m terms) and triexponential ($m + 1$ terms) equation. Results are given in Table 17.1. For each of the four subjects, both tests indicated that the triexponential equation was better than the biexponential equation, except for AIC with subject #4.

MODEL SELECTION CRITERION (MSC)

$$\text{MSC} = \ln [\{\Sigma\, w_i(Y_i - \bar{Y})^2\}/\{\Sigma\, w_i(Y_i - \hat{Y})^2\}] - 2\frac{P}{N} \qquad (721)$$

where the symbols have been defined above.

When there is equal weighting this is equivalent to:

$$\text{MSC} = \ln [s_y^2/\Sigma\ \text{dev}^2] - 2\frac{P}{N} \qquad (722)$$

where the symbols have been defined above.

SUM OF WEIGHTED OR UNWEIGHTED DEVIATIONS

$$\text{Sum} - \sum_{i=1}^{n} w_i(Y_i - \hat{Y})^2 \qquad (723)$$

For equal weighting $w_i = 1$.

STANDARD DEVIATION OF THE DATA (S.D.)

$$\text{S.D.} = [\Sigma\, w_i(Y - \hat{Y})^2/(N - P)]^{1/2} \qquad (724)$$

TABLE 17.1. Application of AIC and F-Test to I.V. Data of Kaplan et al. [25] Fitted with NONLIN with Reciprocal Weighting.

Subject	$s_m \times 10^4$	$s_{m+1} \times 10^4$	F	AIC Biexp.	AIC Triexp.
1	11.55	3.813	$\left[\frac{11.55 - 3.813}{3.813}\right]\left[\frac{16 - 6}{12 - 10}\right] = 10.15$ ($p < .005$)	−100	−114
2	4.399	0.4423	$\left[\frac{4.339 - 0.4423}{0.4423}\right]\left[\frac{16 - 6}{12 - 10}\right] = 44.1$ ($p < 0.001$)	−116	−148
3	18.76	7.663	$\left[\frac{18.76 - 7.663}{7.663}\right]\left[\frac{16 - 6}{12 - 10}\right] = 7.24$ ($p < 0.05$)	−92.5	−103
4	28.80	7.629	$\left[\frac{28.80 - 7.629}{7.629}\right]\left[\frac{16 - 6}{12 - 10}\right] = 13.88$ ($p < .005$)	−85.6	−66.0

TABLE 17.2. Concentration-Time Data to Illustrate Weighting.

Time (Hours)	Observed Concentrations	Model-Predicted Concentrations: Equal Weights	Model-Predicted Concentrations: Weighting $1/Y_i$
0.05	2.78	2.4450	2.3944
0.3	2.08	2.2034	2.1671
0.5	1.83	2.0273	2.0007
0.8	1.59	1.7893	1.7748
2.0	1.23	1.0857	1.0989
4.0	0.58	0.47220	0.49433
5.0	0.32	0.31140	0.33154
7.0	0.13	0.13543	0.14914
10.0	0.041	0.038844	0.044993

COMPARISON OF FITS WITH DIFFERENT WEIGHTING

Usually when data are fitted using different weighting schemes, the various measures of fit are compared directly. But using this method, weighting will always make the fit look better if $1/Y^n$ weights are used. The author thinks one should compare the measures of fit calculated from unweighted data even though weighting was used in fitting the data. A set of data to illustrate this is shown in Table 17.2.

The data in Table 17.2 above were fitted to the equation: $C = C_oe^{-Kt}$ with both equal weights and reciprocal weighting using the program MINSQ [2]. Estimates of C_o were 2.50 for equal weights and 2.44 for $1/Y_i$ weighting. Estimates of K were 0.416 for equal weights and 0.399 for $1/Y_i$ weighting.

If you compare the measures of fit in columns 2 and 3 of Table 17.3 you conclude that the $1/Y_i$ fit is considerably superior to the equal weight fit since the sum of squared deviations and S.D. of data are lower, and r^2, Corr, and MSC are higher for the $1/Y_i$ fit compared with the equal weight fit.

But if you compare the measures of fit in columns 3 and 4 of Table 17.3

TABLE 17.3. Measures of Fit of Data in Table 17.2.

Measure of Fit	$1/Y_i$ Fit: Using $w_i = 1/Y_i$	$1/Y_i$ Fit: Using $w_i = 1$	Equal Wt. Fit
Sum of squared dev.	0.12469	0.24436	0.23862
S.D. of data	0.13347	0.18684	0.18463
r^2	0.99117	0.96725	0.96802
Corr	0.99558	0.98370	0.98399
MSC	4.2848	2.9745	2.9983

the two fits are about the same but the equal weight fit is slightly superior to the $1/Y_i$ fit. The author thinks that comparison of the measures of fit in columns 3 and 4 is the correct way to decide which weighting scheme is best. That method quantitatively compares the degree of closeness of the data points to the fitted model-predicted line.

ESTIMATION OF MICROSCOPIC RATE CONSTANTS OF ROWLAND MODEL IX FROM POST-INFUSION CONCENTRATION-TIME DATA

Fit post-infusion C_B, t data to the equation:

$$C_B = Y_1 e^{-\lambda_1 (t-T)} + Y_2 e^{-\lambda_2 (t-T)} \tag{725}$$

where Y_1 and Y_2 arc the coefficients corresponding to time T, T is the infusion time and t is time measured from the start of the infusion. Since,

$$C_B = \frac{D}{TV_B}\left[\frac{(k_{20} + k_{21} - \lambda_1)(1 - e^{+\lambda_1 T})e^{-\lambda_1 t}}{-\lambda_1(\lambda_2 - \lambda_1)} + \frac{(k_{20} + k_{21} - \lambda_2)(1 - e^{+\lambda_2 T})e^{-\lambda_2 t}}{-\lambda_2(\lambda_1 - \lambda_2)}\right. \tag{726}$$

then

$$Y_1 = \frac{D}{TV_B}\left[\frac{(k_{20} + k_{21} - \lambda_1)(1 - e^{-\lambda_1 T})}{\lambda_1(\lambda_2 - \lambda_1)}\right] \tag{727}$$

and

$$Y_2 = \frac{D}{TV_B}\left[\frac{\lambda_2 - (k_{20} + k_{21})(1 - e^{-\lambda_2 T})}{\lambda_2(\lambda_2 - \lambda_1)}\right] \tag{728}$$

After simplification:

$$\frac{Y_2}{Y_1} = \frac{\lambda_1\lambda_2(1 - e^{-\lambda_2 T}) - (k_{20} + k_{21})\lambda_1(1 - e^{-\lambda_2 T})}{(k_{20} + k_{21})\lambda_2(1 - e^{-\lambda_1 T}) - \lambda_1\lambda_2(1 - e^{-\lambda_1 T})} \quad (729)$$

whence

$$k_{20} + k_{21} = \frac{\lambda_1\lambda_2(Y_1(1 - e^{-\lambda_2 T}) + Y_2(1 - e^{-\lambda_1 T}))}{Y_2\lambda_2(1 - e^{-\lambda_1 T}) + Y_1\lambda_1(1 - e^{-\lambda_2 T})} \quad (730)$$

Let

$$P = Y_1(1 - e^{-\lambda_2 T}) \quad (731)$$

and

$$R = Y_2(1 - e^{-\lambda_1 T}) \quad (732)$$

Then

$$k_{20} + k_{21} = \frac{\lambda_1\lambda_2(P + R)}{\lambda_2 R + \lambda_1 P} \quad (733)$$

$$k_{12} = \lambda_1 + \lambda_2 - (k_{20} + k_{21}) \quad (734)$$

$$k_{20} = \lambda_1\lambda_2/k_{12} \quad (735)$$

$$k_{21} = (k_{20} + k_{21}) - k_{20} \quad (736)$$

APPLICATION OF LAPLACE TRANSFORMS TO THE DIFFERENTIAL EQUATIONS (522) AND (523) OF MODEL XXXIX

Taking the Laplace transforms of Equations (522) and (523) in Chapter 12 gives:

$$sa_B - D = -\left[\frac{Q_L}{V_B}\right]a_B + \left[\frac{Q_L}{R_L V_L}\right]a_L \quad (737)$$

$$sa_L = \left[\frac{Q_L}{V_B}\right]a_B - \left[\frac{Q_L}{V_L R_L}\right]a_L - \left[\frac{f_B \mathrm{CL}_i'}{V_L R_L}\right]a_L \quad (738)$$

The determinant, Δ, is:

$$\Delta = \begin{vmatrix} s + \dfrac{Q_L}{V_B} & -\dfrac{Q_L}{R_L V_L} \\ -\dfrac{Q_L}{V_B} & s + \dfrac{Q_L + f_B \mathrm{CL}_i'}{R_L V_L} \end{vmatrix}$$

$$= \left[s + \frac{Q_L}{V_B}\right]\left[s + \frac{Q_L + f_B \mathrm{CL}_i'}{R_L V_L}\right] - \left[\frac{Q_L}{V_B}\right]\left[\frac{Q_L}{R_L V_L}\right]$$

$$= s^2 + \left[\frac{Q_L}{V_B} + \frac{Q_L}{R_L V_L} + \frac{f_B \mathrm{CL}_i'}{R_L V_L}\right] s + \left[\frac{Q_L}{V_B}\right]\left[\frac{f_B \mathrm{CL}_i'}{V_L R_L}\right]$$

$$= (s + \lambda_1)(s + \lambda_2) \tag{739}$$

where

$$\lambda_1 + \lambda_2 = \frac{Q_L}{V_B} + \frac{Q_L}{R_L V_L} + \frac{f_B \mathrm{CL}_i'}{R_L V_L} \tag{740}$$

$$\lambda_1 \lambda_2 = \left[\frac{Q_L}{V_B}\right]\left[\frac{f_B \mathrm{CL}_i'}{V_L R_L}\right] \tag{741}$$

The Laplace transform of the amount in the blood compartment, a_B, is:

$$a_B = \frac{\begin{vmatrix} D & -\dfrac{Q_L}{R_L V_L} \\ 0 & s + \dfrac{Q_L + f_B \mathrm{CL}_i'}{R_L V_L} \end{vmatrix}}{\Delta} = \frac{D\left[s + \dfrac{Q_L + f_B \mathrm{CL}_i'}{R_L V_L}\right]}{(s + \lambda_1)(s + \lambda_2)} \tag{742}$$

$$(\mathrm{AUC})_{\mathrm{I.V.}} = \underset{s \to 0}{a_B / V_B} = D\left[\frac{Q_L + f_B \mathrm{CL}_i'}{Q_L f_B \mathrm{CL}_i'}\right] \tag{743}$$

$$\mathrm{CL}_s = \frac{D}{(\mathrm{AUC})_{\mathrm{I.V}}} = \frac{Q_L f_B \mathrm{CL}_i'}{Q_L + f_B \mathrm{CL}_i'} \tag{744}$$

THE ROWLAND MODEL WITH FIRST-ORDER ABSORPTION

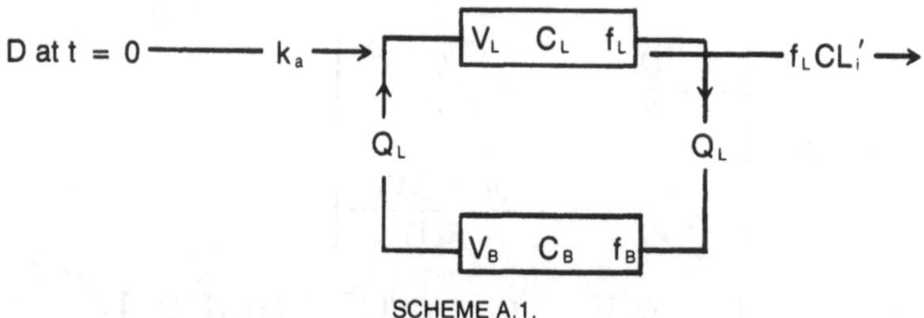

SCHEME A.1.

The differential equations are:

$$\frac{dA_a}{dt} = -k_a A_a \tag{745}$$

$$\frac{dA_B}{dt} = -\left[\frac{Q_L}{V_B}\right] A_B + \left[\frac{Q_L}{R_L V_L}\right] A_L \tag{746}$$

$$\frac{dA_L}{dt} = k_a A_a + \left[\frac{Q_L}{V_B}\right] A_B - \left[\frac{Q_L}{R_L V_L}\right] A_L - \left[\frac{f_B \text{CL}_i'}{R_L V_L}\right] A_L \tag{747}$$

where A_a, A_B, and A_L are the amounts of drug at the absorption site, in the central (blood) compartment and peripheral (liver) compartment at time t, respectively.

Converting to Laplace transforms gives:

$$sa_a - D = -k_a a_a \tag{748}$$

$$sa_B = k_a a_a + \left[\frac{Q_L}{V_B}\right] a_B + \left[\frac{Q_L}{R_L V_L}\right] a_L \tag{749}$$

$$sa_L = k_a a_a + \left[\frac{Q_L}{V_B}\right] a_B - \left[\frac{Q_L + f_B \text{CL}_i'}{R_L V_L}\right] a_L \tag{750}$$

Rearrangement of Equations (748) to (750) gives:

$$(s + k_a) a_a = D \tag{751}$$

$$\left(s + \frac{Q_L}{V_B}\right) a_B - \left(\frac{Q_L}{R_L V_L}\right) a_L = 0 \tag{752}$$

$$-k_a a_a - \left[\frac{Q_L}{V_B}\right] a_B + \left[\frac{s + Q_L + f_B \text{CL}_i'}{V_L R_L}\right] a_L = 0 \tag{753}$$

The determinant, Δ, of the matrix from Equations (751) to (753) is:

$$(s + k_a)\left[s + \frac{Q_L}{V_B}\right]\left[s + \frac{Q_L + f_B \mathrm{CL}_i'}{V_L R_L}\right] - \left[\frac{Q_L}{V_B}\right]\left[\frac{Q_L}{R_L V_L}\right]$$

$$= (s + k_a)(s + \lambda_1)(s + \lambda_2) \quad (754)$$

where $\lambda_1 + \lambda_2$ and $\lambda_1 \lambda_2$ are defined in Equations (740) and (741) above.

The Laplace transform of the amount in the central (blood) compartment, a_B, is:

$$a_B = \frac{\begin{vmatrix} s + k_a & D & 0 \\ 0 & 0 & -\dfrac{Q_L}{R_L V_L} \\ -ka & 0 & s + \dfrac{Q_L + f_B \mathrm{CL}_i'}{R_L V_L} \end{vmatrix}}{\Delta}$$

$$= \frac{Dk_a (Q_L / R_L V_L)}{(s + \lambda_1)(s + \lambda_2)(s + k_a)} \quad (755)$$

$$(\mathrm{AUC})_{\mathrm{po}} = \underset{s \to 0}{a_B / V_B} = \frac{D}{f_B \mathrm{CL}_i'} \quad (756)$$

ROWLAND MODEL WITH INTRAVENOUS INFUSION TO STEADY STATE AND VOLUME OF DISTRIBUTION STEADY STATE

The differential equations are:

$$V_B\left[\frac{dC_{B\mathrm{ss}}}{dt}\right] = -Q_L C_{B\mathrm{ss}}\left[\frac{Q_L}{R_L}\right] C_{L\mathrm{ss}} + R_o = 0 \quad (757)$$

$$V_L\left[\frac{dC_{L\mathrm{ss}}}{dt}\right] = Q_L C_{B\mathrm{ss}} - \left[\frac{Q_L}{R_L}\right] C_{L\mathrm{ss}} - \mathrm{CL}_i' f_L C_{L\mathrm{ss}} = 0 \quad (758)$$

Converting to amounts gives:

$$\left[\frac{dA_{B\mathrm{ss}}}{dt}\right] = -\left[\frac{Q_L}{V_B}\right] A_{B\mathrm{ss}} + \left[\frac{Q_L}{R_L V_L}\right] A_{L\mathrm{ss}} + R_o = 0 \quad (759)$$

$$\left[\frac{dA_{L\mathrm{ss}}}{dt}\right] = \left[\frac{Q_L}{V_B}\right] A_{B\mathrm{ss}} - \left[\frac{Q_L}{R_L V_L}\right] A_{L\mathrm{ss}} - \left[\frac{\mathrm{CL}_i'}{V_L}\right] f_L A_{L\mathrm{ss}} = 0 \quad (760)$$

Equation (759) gives:

$$R_o = \left[\frac{Q_L}{V_B}\right] A_{Bss} - \left[\frac{Q_L}{R_L V_L}\right] A_{Lss} \qquad (761)$$

Substituting from Equation (761) into Equation (759) and using $f_L \mathrm{CL}_i' = f_B \mathrm{CL}_i'/R_L$ gives:

$$R_o = \left[\frac{f_B \mathrm{CL}_i'}{V_L R_L}\right] A_{Lss} \qquad (762)$$

Solving for A_{Lss} in Equation (762) gives:

$$A_{Lss} = \frac{V_L R_L R_o}{f_B \mathrm{CL}_i'} \qquad (763)$$

Substituting from Equation (763) into Equation (759) gives:

$$R_o = \left[\frac{Q_L}{V_B}\right] A_{Bss} + \frac{Q_L R_o}{f_B \mathrm{CL}_i'} \qquad (764)$$

Solving for A_{Bss} in Equation (764) gives:

$$A_{Bss} = R_o V_B \left[\frac{1}{Q_L} + \frac{1}{f_B \mathrm{CL}_i'}\right] \qquad (765)$$

Now,

$$C_{Bss} = \frac{A_{Bss}}{V_B} = R_o \left[\frac{1}{Q_L} + \frac{1}{f_B \mathrm{CL}_i'}\right] \qquad (766)$$

and by substituting from Equations (763), (765), and (766) we get after simplification:

$$V_{ss} = \frac{A_{Bss} + A_{Lss}}{C_{Bss}} = V_B + V_L R_L \left[\frac{Q_L}{Q_L + f_B \mathrm{CL}_i'}\right] = V_B + V_L R_L F \qquad (767)$$

where F is the first-pass bioavailability.

REFERENCES

1. Smith, D. L., J. G. Wagner and G. C. Gerritsen. 1967. "Absorption, Metabolism, and Excretion of 5-Methylpyrazole-3-Carboxylic Acid in the Rat, Dog, and Human," *J. Pharm. Sci.*, 56:1150–1157.
2. MicroMath, P.O. Box 21550, Salt Lake City, UT 84121.
3. Fox, I., A. Dawson, P. Loynds, J. Eisner, K. Findlen, E. Levin, D. Hanson, T. Mant, J. Wagner and J. Maraganore. 1993. "Anticoagulant Activity of Hirulog a Direct Thrombin Inhibitor in Humans," *Thrombosis and Hemostasis*, 69:157–163.
4. Ayres, J. W., D. J. Weidler, J. McKichan and J. G. Wagner. 1977. "Pharmacokinetics of Tolmetin with and without Concomitant Administration of Antacid in Man," *Eur. J. Clin. Pharmacol.*, 12:421–428.
5. Wagner, J. G. and C. D. Alway. 1964. "Serum Levels of 'Lincocin' from Single Dose Serum Levels When 'Lincocin' (as the Hydrochloride) Was Administered by Constant Rate Intravenous Infusion," *Nature*, 201:1101–1103.
6. Wagner, J. G. and E. Nelson. 1963. "Per Cent Absorbed Time Plots Derived from Blood Level and/or Urinary Excretion Data," *J. Pharm. Sci.*, 52:610–611.
7. O'Reilly, R. A., P. G. Welling and J. G. Wagner. 1971. "Pharmacokinetics of Warfarin Following Intravenous Administration to Man," *Thromb. Diath. Haemorrh.*, 25:178–186.
8. Wagner, J. G., E. Novak, L. G. Leslie and C. M. Metzler. 1968. "Absorption, Distribution, and Elimination of Spectinomycin Dihydrochloride in Man," *Int. J. Clin. Pharmacol.*, 1:261–285.
9. Rowland, M. and S. Riegelman. 1968. "Pharmacokinetics of Acetylsalicylic Acid and Salicylic Acid after Intravenous Administration in Man," *J. Pharm. Sci.*, 57:1313–1319.
10. Wagner, J. G. 1983. "Pharmacokinetic Absorption Plots from Oral Data Alone or Oral/Intravenous Data and an Exact Loo-Riegelman Equation," *J. Pharm. Sci.*, 72:838–842.
11. Proost, J. H. 1985. "Wagner's Exact Loo-Riegelman Equation: The Need for a Criterion to Choose between the Linear and Logarithmic Trapezoidal Rule," *J. Pharm. Sci.*, 74:793–794.
12. Gonzalez, I. 1989. "Absorption of Flurbiprofen and Some Other Drugs through Oral Mucosa," Ph.D. dissertation, The University of Michigan.

13. Wagner, J. G. 1988. "Types of Mean Residence Times," *Biopharm. Drug Dispos.*, 9:41–57.

14. Wagner, J. G., D. A. Ganes, K. K. Midha, I. Gonzalez-Younes, J. C. Sackellares, L. D. Olsen, M. B. Affrime and J. E. Patrick. 1990. "Stepwise Determination of Multicompartment Disposition and Absorption Parameters from Extravascular Concentration-Time Data. Application to Mesoridazine, Flurbiprofen, Flunarizine, Labetalol and Diazepam," *J. Pharmacokin. Biopharm.*, 19:413–455.

15. Albert, K. S., M. R. Hallmark, E. Sakmar, D. J. Weidler and J. G. Wagner. 1975. "Pharmacokinetics of Diphenhydramine in Man," *J. Pharmacokin. Biopharm.*, 3:159–170.

16. Dressman, J. B., R. R. Berardi, G. H. Elta, T. M. Gray, P. A. Montgomery, H. S. Lau, K. L. Pelekoudas, G. J. Szpunar and J. G. Wagner. 1991. "Absorption of Flurbiprofen in the Fed and Fasted States," *Pharm. Res.*, 7:75–83.

17. Wagner, J. G. 1975. *Fundamentals of Clinical Pharmacokinetics, First Edition*. Hamilton, IL: Drug Intelligence Publications, Inc., pp. 107–108; (a) pp. 91–92; (b) pp. 93–101; (c) pp. 136–144; (d) pp. 231–246.

18. Wagner, J. G., P. G. Welling, S. B. Roth, E. Sakmar, K. P. Lee and J. E. Walker. 1972. "Plasma Concentrations of Propoxyphene in Man. I. Following Oral Administration of the Drug in Solution and Capsule Forms," *Int. J. Clin. Pharmacol.*, 5:371–380.

19. Wagner, J. G., P. G. Welling and A. J. Sedman. 1972. "Plasma Concentrations of Propoxyphene in Man. II. Pharmacokinetics," *Int. J. Clin. Pharmacol.*, 5:381–388.

20. Mitenko, P. A. and R. I. Ogilvie. 1972. "Rapidly Achieved Plasma Concentration Plateaus with Observations on Theophylline Kinetics," *Clin. Pharmacol. Ther.*, 13:329–335.

21. Boyes, R. N., D. B. Scott, R. J. Jebson, M. J. Goodman and D. G. Julian. 1971. "Pharmacokinetics of Lidocaine in Man," *Clin. Pharmacol. Ther.*, 12:105–115.

22. Wagner, J. G. 1974. "A Safe Method of Rapidly Achieving Plasma Concentration Plateaus," *Clin. Pharmacol. Ther.*, 16:691–700.

23. Wagner, J. G., A. R. DiSanto, W. R. Gillespie and K. S. Albert. 1981. "Reversible Metabolism and Pharmacokinetics: Application to Prednisone-Prednisolone," *Res. Comm. Chem. Path. Pharmacol.*, 32:387–405.

24. Gibaldi, M. and D. Perrier. 1082. *Pharmacokinetics, Second Edition*. New York, NY: Marcel Dekker, Inc., pp. 92–97.

25. Kaplan, S. A., M. L. Jack, K. Alexander and R. E. Weinfeld. 1973. "Pharmacokinetic Profile of Diazepam in Man Following Single Intravenous and Oral and Chronic Oral Administration," *J. Pharm. Sci.*, 62:1789–1796.

26. Nagashima, R. N., G. Levy and R. A. O'Reilly. 1968. "Comparative Pharmacokinetics of Coumarin Anticoagulants IV. Application of a Three-Compartment Model to the Analysis of Dose-Dependent Kinetics of Bishydroxycoumarin Elimination," *J. Pharm. Sci.*, 57:1888–1895.

27. Loo, J. and S. Riegelman. 1968. "New Method for Calculating the Intrinsic Absorption Rate of Drugs," *J. Pharm. Sci.*, 57:918–928.

28. Veng-Pedersen, P. 1980. "An Algorithm and Computer Program for Deconvolution in Linear Pharmacokinetics," *J. Pharmacokin. Biopharm.*, 8:463–481.

29. Lee, K. M. 1991. "The Role of Time-Dependent Gastrointestinal Parameters in the Oral Absorption of Drugs," dissertation, The University of Michigan.

30. Veng-Pedersen. P. 1988. "Linear and Nonlinear System Approaches in Pharmacokinetics. I. General Considerations," *J. Pharmacokin. Biopharm.*, 16:413–472.

31. DiStefano, J. J., III. 1982. "Noncompartmental vs. Compartmental Analysis: Some Bases for Choice," *Am. J. Physiol.*, 243:R1–R6.

32. Matis, J. H. 1987. "An Introduction to Stochastic Compartmental Models," *Pharmacokinetics. Mathematical and Statistical Approaches to Metabolism and Distribution of Chemicals and Drugs*, A. Pecile and A. Rescigno, eds., NATO ASI Series, New York, NY: Plenum Press, pp. 113–128.

33. Veng-Pedersen, P. 1989. "Mean Time Parameters in Pharmacokinetics. Definition, Computation and Clinical Implications (Part I)," *Clin. Pharmacokin.*, 17:345–366.

34. Perrier, D. and M. Mayersohn. 1982. "Noncompartmental Determination of the Steady-State Volume of Distribution for Any Mode of Administration," *J. Pharm. Sci.*, 713:72–373.

35. Veng-Pedersen, P. and W. Gillespie. 1985. "The Mean Residence Time of Drugs in the Systemic Circulation," *J. Pharm. Sci.*, 74:791–792.

36. Wagner, J. G. 1983. "Significance of Ratios of Different Volumes of Distribution in Pharmacokinetics," *Biopharm. Drug Disp.*, 4:263–270.

37. Veng-Pedersen, P. 1988. "System Approaches in Pharmacokinetics: I. Basic Concepts," *J. Clin. Pharmacol.*, 28:1–5.

38. Benet, L. Z. and C.-W. Chiang. 1972. "The Use and Application of Deconvolution Methods in Pharmacokinetics," abstracts of papers presented at *The 13th National Meeting of the American Pharmaceutical Association Academy of Pharmaceutical Sciences, Chicago, IL, November 5–9, 1972*, Vol. 2, No. 2.

39. Kiwada, H., K. Morita, M. Kayaski, S. Awazu and M. Hanano. 1977. "A Numerical Calculation Method for Deconvolution in Linear Compartment Analysis in Pharmacokinetics," *Chem. Pharm. Bull.*, 25:1312–1318.

40. Rescigno, A. and G. Segre. 1966. *Drug and Tracer Kinetics.* Waltham, MA: Blaisdell Publishing.

41. Cutler, D. J. 1978. "Numerical Deconvolution by Least Squares: Use of Prescribed Input Functions," *J. Pharmacokin. Biopharm.*, 6:227–263.

42. Veng-Pedersen, P. 1980. "Model-Independent Method of Analyzing Input in Linear Pharmacokinetic Systems Having Polyexponential Impulse Response. I. Theoretical Analysis," *J. Pharm. Sci.*, 69:298–305.

43. Veng-Pedersen, P. 1980. "Model-Independent Method of Analyzing Input in Linear Pharmacokinetic Systems Having Polyexponential Response. II. Numerical Evaluation," *J. Pharm. Sci.*, 69:305–312.

44. Veng-Pedersen, P. 1980. "Novel Deconvolution Method for Linear Pharmacokinetic Systems with Polyexponential Impulse Response," *J. Pharm. Sci.*, 69:312–318.

45. Veng-Pedersen, P. 1980. "Novel Approach to Bioavailability Testing: Statistical Method for Comparing Drug Input Calculated by a Least-Squares Deconvolution Technique," *J. Pharm. Sci.*, 69:318–324.

46. Gillespie, W. R. and P. Veng-Pedersen. 1985. "A Polyexponential Deconvolution Method. Evaluation of the 'Gastrointestinal Bioavailability' and Mean *in vivo* Dissolution Time of Some Ibuprofen Dosage Forms," *J. Pharmacokin. Biopharm.*, 13:289–309.

47. Veng-Pedersen. P. 1984. "Theorems and Implications of a Model-Independent Elimination/Distribution Function Decomposition of Linear and Some Nonlinear Drug Dispositions. I. Derivation and Theoretical Analyses," *J. Pharmacokin. Biopharm.*, 12:627–648.

48. Veng-Pedersen, P. and W. R. Gillespie. 1987. "Theorems and Implications of a Model

Independent Elimination/Distribution Function Decomposition of Linear and Some Nonlinear Drug Dispositions: IV. Exact Relationship between Terminal Log-Linear Slope Parameter Beta and Drug Clearance," *J. Pharmacokin. Biopharm.*, 15:305–325.

49. Wagner, J. G. 1976. "Rapid Method of Obtaining Area under Curve for Any Compartment of Any Linear Pharmacokinetic Model in Terms of Rate Constants," *J. Pharmacokin. Biopharm.*, 4:281–285.

50. Engberg-Pedersen, H., P. Morch and L. Tybring. 1964. "Kinetics of Serum Aminosalicylic Acid Levels," *Brit. J. Pharmacol.*, 23:1–13.

51. Niebergall, P. J., E. T. Sugita and R. L. Schnaare. 1974. "Calculation of Plasma Level *versus* Time Profiles for Variable Dosing Regimens," *J. Pharm. Sci.*, 63:100–105.

52. Howell, J. R. 1975. "Mathematical Formulation for Nonuniform Dosing," *J. Pharm. Sci.*, 64:464–466.

53. Ng, P. K. 1981. "Prediction of Multiple-Dose Blood Level Curves of Drugs Administered Four Times Daily at Non-Uniform Dosing Intervals," *Int. J. Bio-Medical Computing*, 12:217–226.

54. Michaelis, L. and M. L. Menten. 1913. "Die Kinetik der Invertinwirkung," *Biochem. Z.*, 49:333–369.

55. Henri, V. 1902. "Theorie Generale de l'Action de Quelques Diastases," *Compt. Rend. Hebd. Seanc. Acad. Sci. (Paris)*, 135:916–919.

56. Wagner, J. G. 1988. "Effect of First-Pass Michaelis-Menten Metabolism on Performance of Controlled Release Dosage Forms," *Oral Sustained Release Formulations. Design and Evaluation*, A. Yacobi and E. Halperin-Walega, eds., Maxwell House, Fairview Park, Elmsford, NY: Pergamon Books, Inc., pp. 95–124.

57. Wagner, J. G., G. J. Szpunar and J. J. Ferry. 1985. "Michaelis-Menten Elimination Kinetics: Areas under Curves, Steady-State Concentrations and Clearances for Compartment Models with Different Types of Input," *Biopharm. Drug Dispos.*, 6:177–200.

58. Wagner, J. G., P. K. Wilkinson, A. J. Sedman, D. R. Kay and D. J. Weidler. 1976. "Elimination of Alcohol from Human Blood," *J. Pharm. Sci.*, 65:152–154.

59. Wilkinson, P. K., A. J. Sedman, E. Sakmar, R. H. Earhart, D. J. Weidler and J. G. Wagner. 1976. "Blood Ethanol Concentrations during and Following Constant-Rate Intravenous Infusion of Ethanol," *Clin. Pharmacol. Ther.*, 19:213–223.

60. Wagner, J. G. 1985. "New and Simple Method to Predict Dosage of Drugs Obeying Simple Michaelis-Menten Elimination Kinetics and to Distinguish Such Kinetics from Simple First Order Kinetics," *Ther. Drug Monit.*, 7:377–386.

61. Wagner, J. G. 1986. "Lack of First-Pass Metabolism of Ethanol at Blood Concentrations in the Social Drinking Range," *Life Sci.*, 39:407–414.

62. Wilkinson, P. K., A. J. Sedman, E. Sakmar, D. R. Kay and J. G. Wagner. 1977. "Pharmacokinetics of Ethanol after Oral Administration in the Fasting State," *J. Pharmacokin. Biopharm.*, 5:207–224.

63. Wilkinson, G. N. 1961. "Statistical Estimation in Enzyme Kinetics," *Biochem. J.*, 80:324–332.

64. Eisenthal, R. and A. Cornish-Bowden. 1974. "The Direct Linear Plot. A New Graphical Procedure for Estimating Enzyme Kinetic Parameters," *Biochem. J.*, 139:715–720.

65. Wagner, J. G. 1973. "Properties of the Michaelis-Menten Equation and Its Integrated Form Which Are Useful in Pharmacokinetics," *J. Pharmacokin. Biopharm.*, 1:103–121; 1:337–338.

66. Thomas, G., J. C. Thalabard and C. A. A. Girre. 1987. "Program for the Integrated Michaelis-Menten Equation," *TIPS*, 8:292–294.

67. Wilkinson, P. K. 1975. "Effect of Food on Blood Levels of Ethanol," Ph.D. Dissertation, The University of Michigan, 167 pp.

68. Metzler, C. M. 1969. Biostatistical Technical Report 7292/69/7292/005, The Upjohn Company, Kalamazoo, MI, Nov. 25.

69. Wagner, J. G. 1985. "Propranolol: Pooled Michaelis-Menten Parameters and the Effect of Input Rate on Bioavailability," *Clin. Pharmacol. Ther.*, 37:481–487.

70. Sedman, A. J. and J. G. Wagner. 1974. "Quantitative Pooling of Michaelis-Menten Equations in Models with Parallel Metabolite Formation Paths," *J. Pharmacokin. Biopharm.*, 2:149–160.

71. Gerber, N. and J. G. Wagner. 1972. "Explanation of Dose-Dependent Decline of Diphenylhydantoin Plasma Levels by Fitting to the Integrated Form of the Michaelis-Menten Equation," *Res. Com. Chem. Pathol. Pharm.*, 3:455–466.

72. Wagner, J. G. 1984. "Commentary: Predictability of Verapamil Steady-State Plasma Levels from Single-Dose Data Explained," *Clin. Pharmacol. Ther.*, 36:1–4.

73. Wagner, J. G., J. C. Sackellares, P. D. Donofrio, S. Berent and E. Sakmar. 1984. "Nonlinear Pharmacokinetics of CI-912 in Adult Epileptic Patients," *Ther. Drug Monit.*, 6:277-283.

74. Ferry, J. J. and J. G. Wagner. 1985. "A Pharmacokinetic Model for Prednisone after Infusion to Steady-State in the Rabbit," *Biopharm. Drug Dispos.*, 6:335–339.

75. Wagner, J. G. 1985. "Theophylline: Pooled Michaelis-Menten Parameters (V_{max} and K_m) and Implications," *Clin. Pharmacokin.*, 10:432–442.

76. Wagner, J. G., J. W. Gyves, P. L. Stetson, S. C. Walker-Andrews, I. S. Wollner, M. K. Cochran and W. D. Ensminger. 1986. "Steady-State Nonlinear Pharmacokinetics of 5-Fluorouracil during Hepatic Arterial and Intravenous Infusions in Cancer Patients," *Cancer Res.*, 46:1499–1506.

77. Robinson, P. J., L. Bass, S. M. Pond, M. S. Roberts and J. G. Wagner. 1988. "Clinical Applicability of Current Pharmacokinetic Models: Splanchic Elimination of 5-Fluorouracil in Cancer Patients," *J. Pharmacokin. Biopharm.*, 16:229–249.

78. Wagner, J. G., P. L. Stetson, J. A. Knol, J. C. Andrews, S. Walker-Andrews, C. A. Knutsen, N. Johnson, D. Prieskorn, P. Terrio, Z. Yang, D. Ganes and W. D. Ensminger. 1989. "Steady-State Arterial and Hepatic Venous Plasma Concentrations of 5-Bromo-2′-Deoxyuridine and 5-Iodo-2′-Deoxyuridine Drugs Which Are Subject to Both Splanchnic and Extra-Splanchnic Elimination," *Sel. Cancer Ther.*, 5:193–203.

79. Wagner, J. G., M. C. Rogge, R. B. Natale, K. S. Albert and G. J. Szpunar. 1987. "Single Dose and Steady-State Pharmacokinetics of Adinazolam after Oral Administration to Man," *Biopharm. Drug Dispos.*, 8:405–425.

80. Wagner, J. G., T. L. Ling, E. J. Mroszczak, D. Freedman, A. Wu, B. Huang, I. J. Massey and R. R. Roe. 1987. "Single Intravenous Dose and Oral Steady-State Pharmacokinetics of Nicardipine in Healthy Subjects," *Biopharm. Drug Dispos.*, 8:133–148.

81. Wagner, J. G. 1988. "Modeling First-Pass Metabolism," *Pharmacokinetics*, A. Pecile and A. Rescigno, eds., *NATO ASI Series*, Plenum Publishing Corporation, pp. 129–149.

82. Wagner, J. G. 1989. "Relationships between First-Order and Michaelis-Menten Kinetics," *J. Pharm. Sci.*, 78:521–522.

83. Ferry, J. J. and J. G. Wagner. 1987. "The Nonlinear Pharmacokinetics of Prednisone

and Prednisolone. II. Plasma Protein Binding of Prednisone and Prednisolone in Rabbit and Human Plasma," *Biopharm. Drug Dispos.*, 8:261–272.

84. Ferry, J. J., Jr. and J. G. Wagner. 1988. "The Non-Linear Pharmacokinetics of Prednisone and Prednisolone. III. Experiments Using the Rabbit as an Animal Model," *Biopharm. Drug Dispos.*, 9:363–376.
85. Wagner, J. G., G. J. Szpunar and J. J. Ferry. 1985. "A Nonlinear Pharmacokinetic Model: I. Steady-State," *J. Pharmacokin. Biopharm*, 13:73–92.
86. Bergstrom, R. F. 1980. "The Pharmacokinetics of Penicillamine in a Female Mongrel Dog and Normal Volunteers," Ph.D. dissertation, The University of Michigan, p. 317.
87. van Ginneken, C. A. M. and F. G. M. Russel. 1989. "Saturable Pharmacokinetics in the Renal Excretion of Drugs," *Clin. Pharmacokin*, 16:38–54.
88. Wagner, J. G. 1983. "Modified Wagner-Nelson Absorption Equation for Multiple-Dose Regimens," *J. Pharm. Sci.*, 72:578–579.
89. Wagner, J. G. 1983. "The Wagner-Nelson Method Applied to a Multicompartment Model with Zero Order Input," *Biopharm. Drug Dispos.*, 4:359–373.
90. Wagner, J. G. 1974. "Application of the Wagner-Nelson Absorption Method to the Two Compartment Open Model," *J. Pharmacokin. Biopharm.*, 2:469–486.
91. Wagner, J. G. 1984. "Estimation of Drug Absorption Kinetics with Emphasis on Theophylline," *Sustained Release Theophylline in the Treatment of Chronic Reversible Airways Obstruction*, International Workshop, Mont. St. Marie, Canada. I. H. G. Jonkman, I. W. Jenne and F. E. Simons, eds., Amsterdam: Excerpta Medica, pp. 113–120.
92. Wagner, J. G. 1984. "Effect of Using an Incorrect Elimination Rate Constant in Application of the Wagner-Nelson Method to Theophylline Data in Cases of Zero Order Absorption," *Biopharm. Drug Dispos.*, 5:75–83.
93. Shumaker, R. C., H. Boxenbaum and G. A. Thompson. 1988. "ABSPLOTS: A LOTUS 123 Spreadsheet for Calculating and Plotting Drug Absorption Rates," *Pharm. Res.*, 5:247–248.
94. Schuirmann, D. J. 1987. "A Comparison of the Two One-Sided Tests Procedure and the Power Approach for Assessing the Equivalence of Average Bioavailability," *J. Pharmacokin. Biopharm.*, 15:657–680.
95. Yuh, L. 1990. "Robust Procedures in Comparative Bioavailability Studies," *Drug Inform. J.*, 24:747–751.
96. Sinko, P. J., G. D. Leesman and G. L. Amidon. 1991. "Predicting Fraction Dose Absorbed in Humans Using a Macroscopic Mass Balance Approach," *Pharm. Res.*, 8:979–988.
97. Dressman, J. B. and D. Fleisher. 1986. "Mixing-Tank Model for Predicting Dissolution Rate Control of Absorption," *J. Pharm. Sci.*, 75:109–116.
98. Hintz, R. J. and K. C. Johnson. 1989. "The Effect of Particle Size Distribution on Dissolution Rate and Oral Absorption," *Int. J. Pharm.*, 51:9–17.
99. Lu, A. T. K., M. E. Frisella and K. C. Johnson. "Dissolution Modelling: Factors Affecting the Dissolution Rates of Polydisperse Powders," personal communication, October, 1991.
100. Dedrick, R. L. and K. B. Bischoff. 1968. "Pharmacokinetics in Applications of Artificial Kidney," *Chem. Engr. Prog. Symp. Ser.*, 64:32–44.

101. Bischoff, K. B. and R. L. Dedrick. 1968. "Thiopental Pharmacokinetics," *J. Pharm. Sci.*, 57:1347–1357.

102. Bischoff, K. B., R. L. Dedrick and D. S. Zaharko. 1970. "Preliminary Model for Methotrexate Pharmacokinetics," *J. Pharm. Sci.*, 59:149–154 and *Cancer Chemother. Rep., Part 1,* 54:95.

103. Chen, H. S. G. and J. F. Gross. 1979. "Physiologically Based Pharmacokinetic Models for Anticancer Drugs," *Cancer Chemother. Pharmacol.*, 2:85–94.

104. Gerlowski, L. E. and R. K. Jain. 1983. "Physiologically Based Pharmacokinetic Modelling: Principles and Applications," *J. Pharm. Sci.*, 72:1103–1126.

105. Cowles, A. L., H. H. Borgstedt and A. J. Gillies. 1971. "Tissue Weights and Rates of Blood Flow in Man for the Prediction of Anesthetic Uptake and Distribution," *Anesthesiology,* 35:523–526.

106. Reeves, P. T., R. F. Minchin and K. F. Ilett. 1988. "Measurement of Organ Blood Flow in the Rabbit," *J. Pharmacol. Methods,* 20:187–196.

107. Altman, P. L. and D. S. Dittmer, eds. 1971. *Biological Handbook: Respiration and Circulation.* Bethesda, MD: Federation of American Societies for Experimental Biology.

108. Harris, P. A. and J. F. Gross. 1975. "Preliminary Pharmacokinetic Model for Adriamycin (NSC-123127)," *Cancer Chemother. Rep.*, 59:819–825.

109. Adolph, E. F. 1949. "Quantitative Relations in the Physiological Constituents of Mammals," *Science (Lond.)*, 109:579–585.

110. Travis, C. C., R. K. White and R. C. Ward. 1990. "Interspecies Extrapolation in Pharmacokinetics," *J. Theor. Biol.*, 142:285–304.

111. Wagner, J. G. 1988. "Modeling First-Pass Metabolism," *Pharmacokinetics,* A. Pecile and A. Rescigno, eds., *NATO ASI Series,* Plenum Publishing Corporation, pp. 129–149.

112. Gibaldi, M. and J. R. Koup. 1981. "Pharmacokinetic Concepts-Drug Binding, Apparent Volume of Distribution and Clearance," *Eur. J. Clin. Pharmacol.*, 20:299–305.

113. Chen, H. S. G. and J. F. Gross. 1979. "Physiologically-Based Pharmacokinetic Models for Anticancer Drugs," *Cancer Chemother. Pharmacol.*, 2:85–94.

114. Rowland, M., L. Z. Benet and G. G Graham. 1973. "Clearance Concepts in Pharmacokinetics," *J. Pharmacokin. Biopharm.*, 1:123–136.

115. Yacobi, A. and G. Levy. 1975. "Comparative Pharmacokinetics of Coumarin Anticoagulants: XIV. Relationship between Protein Binding, Distribution and Elimination Kinetics of Warfarin in Rats," *J. Pharm. Sci.*, 64:1660.

116. Levy, G. 1976. "Clinical Implications of Interindividual Differences in Plasma Protein Binding of Drugs and Endogenous Substances," Chapter 9 in *The Effect of Disease States on Drug Pharmacokinetics,* L. Z. Benet, ed., Washington, DC: American Pharmaceutical Association, pp. 137–151.

117. Benya, T. J. and J. G. Wagner. 1975. "Rapid Equilibration of Warfarin between Rat Tissue and Plasma," *J. Pharmacokin. Biopharm.*, 3:237–255.

118. Gerlowski, L. E. and R. K. Jain. 1983. "Physiologically-Based Pharmacokinetic Modelling: Principles and Applications," *J. Pharm. Sci.*, 72:1103–1126.

119. Albert, K. S., M. R. Hallmark, E. Sakmar, D. J. Weidler and J. G. Wagner. 1975. "Pharmacokinetics of Diphenhydramine in Man," *J. Pharmacokin. Biopharm.*, 3:159–170.

120. Norbury, H. M., R. A. Franklin and D. F. Graham. 1983. "Pharmacokinetics of the New Analgesic, Meptazinol, after Oral and Intravenous Administration to Volunteers," *Eur. J. Clin. Pharmacol.*, 25:77–80.

121. Wagner, J. G. 1976. "Rapid Method of Obtaining Area under Curve for Any Compartment of Any Linear Pharmacokinetic Model in Terms of Rate Constants," *J. Pharmacokin. Biopharm.*, 4:251–285.

122. Judd, R. L. and A. J. Pesce. 1982. "Free Drug Concentrations Are Constant in Serial Fractions of Ultrafiltrate," *Clin. Chem.*, 28:1726–1727.

123. Tozer, T. N., J. G. Gambertoglio, D. E. Furst, D. S. Avery and N. H. G. Holford. 1983. "Volume Shifts and Protein Binding Estimates Using Equilibrium Dialysis: Application to Prednisolone Binding in Humans," *J. Pharm. Sci.*, 72:1442–1446.

124. Behm, H. L. and J. G. Wagner. 1979. "Errors in Interpretation of Data from Equilibrium Dialysis Protein Binding Experiments," *Res. Comm. Chem. Path. Pharmacol.*, 26:145–160.

125. Wilkinson, G. R. 1983. "Plasma and Tissue Binding Considerations in Drug Disposition," *Drug. Metab. Rev.*, 14:427–465.

126. Wagner, J. G., G. K. Aghajanian and O. H. L. Bing. 1968. "Correlation of Performance Test Scores with Tissue Concentration of Lysergic Acid Diethylamide in Human Subjects," *Clin. Pharmacol. Ther.*, 9:635–638.

127. Metzler, C. M. 1969. "A Mathematical Model for the Pharmacokinetics of LSD Effect," *Clin. Pharmacol. Ther.*, 10:737–739.

128. Hill, A. V. 1910. "The Possible Effects of Aggregation of the Molecules of Hemoglobin on Its Dissociation Curves," *J. Physiol.*, 40:IV–VII.

129. Wagner, J. G. 1968. "Kinetics of Pharmacologic Response I. Proposed Relationships between Response and Drug Concentration in the Intact Animal and Man," *J. Theor. Biol.*, 20:171–201.

130. Sheiner, L. B., D. R. Stanski, S. Vogel, R. D. Miller and J. Ham. 1979. "Simultaneous Modeling of Pharmacokinetics and Pharmacodynamics," *Clin. Pharmacol. Ther.*, 25:358–371.

131. Paalzow, L. K., G. H. M. Paalzow and P. Tfelt-Hansen. 1972. "Kinetics of Drug Action," Chapter 24 in *Pharmacokinetics, A Modern View,* Plenum Press, pp. 327–343.

132. Greenblatt, D. J., B. T. Ehrenberg, J. Gunderman, A. Locniskar, J. M. Scavone, J. S. Harmatz and R. T. Shader. 1989. "Pharmacokinetic and Electroencephalographic Study of Intravenous Diazepam, Midazolam, and Placebo," *Clin. Pharmacol. Ther.*, 45:356–364.

133. Rowland, M. and T. N. Tozer. 1989. *Clinical Pharmacokinetics: Concepts and Applications, Second Edition.* Philadelphia, PA: Lea & Febiger, pp. 348, 358.

134. Wagner, J. G., M. C. Rogge, R. B. Natale, K. S. Albert and G. J. Szpunar. 1987. "Single Dose and Steady-State Pharmacokinetics of Adinazolam after Oral Administration to Man," *Biopharm. Drug Dispos.*, 8:405–425.

135. Eichelbaum, M. 1988. "Pharmacokinetics and Pharmacodynamic Consequences of Stereoselective Drug Metabolism in Man," *Biochem. Pharmacol.*, 37:93–96.

INDEX

9 781566 760324